Artificial Intelligence-Enhanced Software and Systems Engineering

Volume 7

The book series AI-SSE publishes new developments and advances on all aspects of Artificial Intelligence-enhanced Software and Systems Engineering—quickly and with a high quality. The series provides a concise coverage of the particular topics from both the vantage point of a newcomer and that of a highly specialized researcher in these scientific disciplines, which results in a significant cross-fertilization and research dissemination. To maximize dissemination of research results and knowledge in these disciplines, the series will publish edited books, monographs, handbooks, textbooks and conference proceedings. Of particular value to both the contributors and the readership are the short publication timeframe and the world-wide distribution, which enable both wide and rapid dissemination of research output.

Tirimula Rao Benala · Satchidananda Dehuri ·
Rajib Mall · Margarita N. Favorskaya
Editors

Boosting Software Development Using Machine Learning

Springer

Editors
Tirimula Rao Benala
Department of Information Technology
JNTU-GV College of Engineering
Vizianagaram
Jawaharlal Nehru Technological University
Gurajada Vizianagaram
Vizianagaram, Andhra Pradesh, India

Rajib Mall
Department of Computer Science
and Engineering
Indian Institute of Technology Kharagpur
Kharagpur, West Bengal, India

Satchidananda Dehuri
Department of Computer Science
Fakir Mohan University
Balasore, Odisha, India

Margarita N. Favorskaya
Department of Informatics
and Telecommunications
Siberian State Aerospace University
Krasnoyarsk, Russia

ISSN 2731-6025 ISSN 2731-6033 (electronic)
Artificial Intelligence-Enhanced Software and Systems Engineering
ISBN 978-3-031-88187-9 ISBN 978-3-031-88188-6 (eBook)
https://doi.org/10.1007/978-3-031-88188-6

This Springer imprint is published by the registered company Springer Nature Switzerland AG
The registered company address is: Gewerbestrasse 11, 6330 Cham, Switzerland

To my beloved wife, Smt. Srilakshmi Benala,
in loving memory of my late mother,
Smt. Krishna Kumari, and to my father,
Sri. Narasingham Benala.

Tirimula Rao Benala

To my wife, Dr. Lopamudra Pradhan, and daughters Rishna Dehuri and Khushyansei Dehuri.

Satchidananda Dehuri

Preface

The rapid evolution of software engineering has been profoundly influenced by the transformative role of machine learning (ML) and artificial intelligence (AI). Indeed, practitioners and researchers increasingly leverage data-driven methods to streamline every phase of the software development life cycle (SDLC)—from requirements elicitation and architectural design to coding, testing, and deployment. This volume, *Boosting Software Development Using Machine Learning*, brings together a collection of original research studies and insightful case analyses that underscore the profound impact of AI-ML in contemporary software engineering. This volume aims to serve as a valuable resource for academics, industry professionals, and students eager to deepen their understanding of AI-ML-enhanced software development practices by showcasing a broad range of theoretical foundations, practical applications, and empirical findings.

In Chap. 1, Benala and Dehuri trace the progression of software development from classical, structured frameworks—such as the Waterfall model—to the agile and DevOps paradigms that dominate modern practice. Their work culminates in a thought-provoking discussion on generative AI, which promises to automate many repetitive development tasks while fostering innovation and code quality. This emphasis on the potential of AI to automate tasks can inspire the audience and foster a sense of optimism about the future of software development.

Chapter 2, by Avvari et al., demonstrates how AI methods can optimize key SDLC stages, from automated requirements gathering to intelligent testing tools such as Mabl. Their findings highlight increased efficiency and improved reliability, thanks to AI-driven predictive analytics and anomaly detection. The authors' practical examples affirm that strategic integration of AI/ML can significantly enhance productivity and software quality across diverse organizational settings, providing the audience with a comprehensive understanding of the practical applications of AI in software engineering.

Moving toward systematic SDLC integration, Mohapatra et al. present a comparative analysis of traditional and ML-augmented SDLC methodologies in Chap. 3. By illustrating how real-time data feeds into iterative feedback loops, the authors demonstrate that machine learning streamlines development processes and enables rapid adaptability. Their work is particularly salient for organizations facing ever-evolving requirements in complex software systems.

In Chap. 4, Vineetha K V and Samuel discuss formalizing and clarifying requirements using ontologies. They argue that this structured representation improves consistency and efficiency, enabling automated translations to design models (e.g., UML) and expediting code generation. This chapter exemplifies how advanced AI approaches can facilitate bridging the gap between requirement documents and their implementation.

Chapter 5, by Mustyala et al., delves into the challenges of accurate software development effort estimation, especially as complex projects grow. Their survey of expert judgment, algorithmic models, and ML-based methods shows that ML techniques often yield superior accuracy, thus minimizing common pitfalls such as overestimating or underestimating project resources. Their case studies illustrate ML's potential to optimize budgeting, scheduling, and resource allocation.

Extending this focus on estimation, Kaushik et al. explore, in Chap. 6, a stacked-model framework incorporating support vector regression, ridge regression, K-nearest neighbors, and linear regression at Level-0, followed by stochastic gradient descent at Level-1. Tested on multiple datasets and evaluated using metrics such as MAE, RMSE, R^2, PRED, and MMRE, this novel framework demonstrates promising improvements in cost estimation accuracy.

In Chapter 7, Biswas et al. focus on design quality and provide an extensive literature review. They explore supervised, feature-based, clustering, graph-oriented, and ontology-based approaches for identifying design patterns in code. This comprehensive assessment emphasizes the importance of automated design pattern detection in ensuring code maintainability, clarity, and overall system robustness.

Turning to measurements and metrics, Goyal, in Chap. 8, examines how advanced ML methods—from dimensionality reduction to deep neural models—are revolutionizing the way software organizations approach fault prediction and effort estimation. The chapter surveys state-of-the-art ML-assisted measurements, guiding researchers and practitioners toward data-centric strategies for enhancing software reliability.

A broader look at optimization is offered in Chap. 9 by Panda et al. Their comparative study spans established (e.g., tabu search, genetic algorithms, particle swarm optimization) and more recent metaheuristic algorithms, addressing challenges like hybridization, parallelization, and scalability. The authors underscore metaheuristics' algorithmic and practical benefits by contextualizing these methods in software engineering tasks—such as debugging, test case creation, and prioritization.

In Chapter 10, Vashishtha et al., Shifting to Software Security, argue for the synergy between ML-driven anomaly detection and traditional security measures. They explore unsupervised learning, neural networks, and associated ethical considerations, ultimately offering guidance on how developers can integrate cutting-edge AI with existing security infrastructures to create resilient software products.

Exploring the realm of text analytics and social media, Moturi et al. propose in Chap. 11 a hybrid architecture combining convolutional neural networks (CNNs) with long short-term memory (LSTM) to capture local and sequential features.

Experimental evaluations on IMDB data verify the model's efficacy, signaling new opportunities for large-scale sentiment classification applications.

Chapter 12, by Panchumarthy and Benala, offers an extensive survey of optimization strategies—from hardware to hybrid approaches—that enhance AI workloads' cost efficiency, performance, and scalability. Their review is indispensable for decision-makers grappling with how best to deploy and manage resource-intensive AI applications.

Padwekar et al. then highlight the nexus between innovation and AI in Chap. 13, "Opportunity Discovery for Effective Innovation using Artificial Intelligence". By harnessing large language models (LLMs) for advanced text analysis, the authors demonstrate how underexplored market and technology challenges may be identified more rapidly. Their synthesis of innovation literature and AI-driven text analytics provides practical insights for corporate strategists and national policy planners.

Finally, Chap. 14, by Awasthi et al., explores the interplay between open innovation and ML-based techniques. Through topic modeling and an analysis of the extant literature, the authors illustrate how organizations can leverage ML to enhance knowledge exchange, streamline the inflow and outflow of new ideas, and, ultimately, promote sustained competitive advantage.

These chapters underscore the increasingly indispensable role of ML and AI in modern software engineering. By offering fresh insights, cutting-edge methodologies, and robust case studies, *Boosting Software Development Using Machine Learning* aspires to guide researchers and practitioners in harnessing the promise of data-driven automation. However, the road ahead is not without challenges. Ethical considerations, algorithmic bias, and data security demand robust governance frameworks. The need for interdisciplinary expertise in AI, software engineering, and domain knowledge also emphasizes the importance of training and collaboration. Ultimately, the future of software development with ML lies in striking a balance between leveraging advanced technologies and adhering to ethical, secure, and inclusive practices, ensuring sustainable innovation.

We thank all the contributing authors whose diligence and expertise have shaped this volume. We trust that readers will discover both inspiration and practical tools for advancing the science and practice of AI-augmented software development within these pages.

Vizianagaram, Andhra Pradesh, India — Tirimula Rao Benala
Balasore, Odisha, India — Satchidananda Dehuri
Kharagpur, West Bengal, India — Rajib Mall
Krasnoyarsk, Russia — Margarita N. Favorskaya

Contents

About the Editors

Tirimula Rao Benala is an Assistant Professor of Information Technology at JNTU-GV College of Engineering Vizianagaram (Autonomous), Jawaharlal Nehru Technological University Gurajada Vizianagaram, Vizianagaram-535003, India. He received his Ph.D. in 2020 from the Jawaharlal Nehru Technological University in Kakinada, India. He was at the Center for Theoretical Studies, Indian Institute of Technology Kharagpur, as a Visiting Scholar in 2017. He was a visiting fellow at Chicago State University in 2018 under the US-India 21st Century Knowledge Initiative grant. He is a senior member of IEEE and CSI. He is a Life member of I.E. (I) and IETE. He is a professional member of ACM. He has received several National Awards: Young Engineer of the Year Award in 2012 from the Institution of Engineers (I), National Award for Academic Excellence for Highest Publication in CSIC for 2012, National Award for Academic Excellence for International Paper Presentation for 2012, Young Engineer of the Year Award in 2011 from Government of Andhra Pradesh and Institution of Engineers (India). He has received the Best Teacher Award in 2009 from Anil Neerukonda Institute of Technology and Sciences, affiliated to Andhra University, Andhra Pradesh, and the Best Teacher Award in 2022 from Jawaharlal Nehru Technological University Kakinada, Andhra Pradesh.

He has published about 30 research papers in refereed journals and conferences. He has guided 75 postgraduate students and currently guides seven postgraduate students and four doctoral students. He has Twenty Years of teaching experience. His main research interests are Computational Intelligence, Empirical Software Engineering, Machine Learning, large language models, and Program Analysis.

Satchidananda Dehuri (Senior Member IEEE) has been a Professor in the Department of Computer Science (Erstwhile Department of Information and Communication Technology), Fakir Mohan University, Balasore, Odisha, India since 2013. Before this appointment, for a short stint (i.e., from October 2012 to May 2014), he was an Associate Professor in the Department of Systems Engineering at Ajou University, South Korea. He received his M.Tech. and Ph.D. in Computer Science from Utkal University, Vani Vihar, Odisha, in 2001 and 2006, respectively.

He visited as a BOYSCAST Fellow to the Soft Computing Laboratory, Yonsei University, Seoul, South Korea, under the BOYSCAST Fellowship Program of DST, Government of India in 2008. In 2010, he received the Young Scientist Award in Engineering and Technology for the year 2008 from Odisha Vigyan Academy, Department of Science and Technology, Government of Odisha. In 2021, he received a Teachers Associateship and Research Excellence (TARE) Fellowship from ANRF (Erstwhile SERB), DST, Government of India for three years to carry out intensive research on Higher Order Neural Networks for Big Data Analysis at host Institute, ISI Kolkata and Parent Institute, Fakir Mohan University, Balasore. His research interests include Multi-objective Optimization, Machine Learning, and Data Science. He has already published 275 research papers in reputed journals and conference proceedings. Under his direct supervision, 20 Ph.D. Scholars have been successfully awarded in Computer Science. He has completed three research projects from DST, UGC, and DRDO. His h-index as per Google Scholar is more than 31. As a part of Academic Collaboration, he has visited Ireland, New Zealand, Hong Kong, France, South Korea, and Nepal.

Rajib Mall obtained his professional degrees, Bachelor's, Master's, and Ph.D. from the Indian Institute of Science, Bangalore. He has been working as a faculty in the Department of Computer Science and Engineering at IIT, Kharagpur, for the last 30 years. He has published about 150 refereed journal and conference papers. His main research interests are Program Analysis and Testing.

Dr. Margarita N. Favorskaya is a Professor and Head of the Department of Informatics and Computer Techniques at Reshetnev Siberian State University of Science and Technology. Russian Federation. Professor Favorskaya has been a member of the KES organization since 2010, an IPC member, and the Chair of invited sessions of over 30 international conferences. She serves as a reviewer in international journals (*Neurocomputing, Knowledge Engineering and Soft Data Paradigms, Pattern Recognition Letters, Engineering Applications of Artificial Intelligence*), an associate editor of *Intelligent Decision Technologies Journal, International Journal of Knowledge-Based and Intelligent Engineering Systems, International Journal of Reasoning-Based Intelligent Systems*, an Honorary Editor of the *International Journal of Knowledge Engineering and Soft Data Paradigms*, the Reviewer, Guest Editor, and Book Editor (Springer). She is the author or the co-author of 200 publications and 20 educational manuals in computer science. She co-authored/co-edited around 20 books/conference proceedings for Springer in the last 10 years. She supervised nine Ph.D. candidates and is presently supervising four Ph.D. students. Her main research interests are digital image and video processing, remote sensing, pattern recognition, fractal image processing, artificial intelligence, and information technologies.

Chapter 1
Transforming Software Development: From Traditional Methods to Generative Artificial Intelligence

Tirimula Rao Benala and Satchidananda Dehuri

Abstract Over the past decades, software development processes have evolved rapidly, propelled by the increasing complexity of software systems, the demand for faster delivery, and the need to adapt to rapidly changing customer expectations. This chapter examines pivotal milestones, starting with structured approaches such as the Waterfall and V-Model, progressing through iterative methodologies such as Agile and DevOps, and culminating in the rise of microservices and cloud-native architectures. With the recent integration of generative artificial intelligence (AI) technologies, development processes have been revolutionized by automating tasks, enhancing collaboration, and enabling engineers to concentrate on strategic problem-solving. Furthermore, it explores how generative AI disrupts existing workflows, requiring software engineers to develop new skills and mindsets. Further, this chapter discusses the opportunities and challenges associated with AI-driven tools and processes as the industry continues to innovate and expand the horizons of software development. While comprehensively analyzing these transformations, the chapter accentuates the significance of balanced adoption to position technological advancements in synchronization with societal values and industry demands.

Keywords Software development · Generative artificial intelligence · Agile methodologies · DevOps · Code automation · AI-assisted tools

T. R. Benala (✉)
Department of Information Technology, JNTU-GV College of Engineering Vizianagaram, Jawaharlal Nehru Technological University Gurajada Vizianagaram, Dwarapudi, Vizianagaram, Andhra Pradesh, India
e-mail: btirimula.it@jntugvcev.edu.in

S. Dehuri
Department of Computer Science, Fakir Mohan University, Vyasa Vihar, Balasore, Odisha, India

T. R. Benala et al. (eds.), *Boosting Software Development Using Machine Learning*, Artificial Intelligence-Enhanced Software and Systems Engineering 7,
https://doi.org/10.1007/978-3-031-88188-6_1

1.1 Introduction

The dynamic software development industry has been constantly adapting to meet the ever-evolving demands of technology and society. The desire for improved efficiency, superior quality, and innovative solutions has propelled the transformations from the rigid frameworks of early programming paradigms to today's adaptable models. Traditional methodologies such as structured programming and linear models like the Waterfall and V-Model established the core principles of discipline and careful planning. However, these approaches could not effectively accommodate changing requirements and rapid technological progress.

The dawn of the twenty-first century marked a significant shift with the emergence of Agile methodologies, which prioritized collaboration, flexibility, and customer-focused practices. Frameworks like Scrum and Extreme Programming (XP) transformed software development by creating environments that could quickly adapt to change and deliver incremental value through iterative cycles. This progression continued with the adoption of integrating DevOps practices, bridging the gap between development and operations to support continuous delivery and deployment.

Today, the industry is entering another pivotal phase with the integration of generative artificial intelligence (AI) into software development. Generative AI technologies are driven by sophisticated machine learning models like OpenAI's GPT-4. They are reshaping the way software is designed, developed, and maintained. These tools can generate code, automate debugging, optimize performance, and even contribute to architectural design. The role of software engineers is shifting from coding and reviewing to managing and orchestrating AI-assisted development processes.

The present chapter studies the transformative journey of software development, from traditional structured methodologies like the Waterfall and V-Model to iterative frameworks such as Agile and DevOps. It examines the emergence of microservices architecture as a solution for scalability and flexibility, as well as the impact of cloud-native development, which leverages containers, orchestration, and serverless computing to promote agility and resource efficiency. Finally, it discusses the revolutionary impact of generative AI, with tools like GitHub Copilot enhancing automation and productivity, and redefining engineering roles. The chapter also addresses challenges such as ethical considerations and the need for skill adaptation, advocating for a balanced approach to innovation and integration. It emphasizes the importance of aligning technological advancements with societal values and industry demands, underscoring the dynamic interplay of technology, efficiency, and adaptability in contemporary software engineering.

1.2 The Conventional Era

In the early days of software engineering, development methodologies relied heavily on traditional engineering practices. The waterfall model, introduced by Winston W. Royce in 1970, exemplifies this era, with its structured linear and sequential phases: requirements analysis, system design, implementation, integration and testing, deployment, and maintenance [1]. While this approach provides a clear framework, it often struggled to adapt to changing needs, making it unsuitable for dynamic projects.

Building on the waterfall model, the V-Model integrated testing at each development stage, focusing on verification and validation [2]. This enhancement improved product quality by aligning development tasks with corresponding testing phases. However, like the waterfall model, it presented challenges in accommodating evolving project needs and was less effective in situations where requirements were unclear from the outset. These shortcomings highlighted the need for modern methodologies that prioritize flexibility, collaboration, and iterative progress.

1.3 The Shift to Iterative and Incremental Approaches

In response to the limitations of linear development models, the software industry adapted methodologies that emphasized increased adaptability and flexibility. A major advancement was the introduction of the Spiral Model by Barry W. Boehm in 1986 [3]. This risk-driven model combined iterative development with systematic risk management, allowing teams to refine requirements and designs through repeated cycles of planning, risk analysis, engineering, and evaluation. Continuous stakeholder feedback was central to this approach, enabling projects to adapt to changing needs and mitigate risks throughout the lifecycle.

Similarly, the Rational Unified Process (RUP) offered a customizable framework that also emphasized iterative development and risk reduction [4]. Divided into inception, elaboration, construction, and transition phases, RUP is a disciplined yet adaptable approach to software development.

These iterative methodologies recognized that developing software is not a one-size-fits-all process and that embracing change is crucial to meet user needs effectively.

1.4 The Agile Movement

The early 2000s marked a transformative shift with the introduction of the Agile Manifesto in 2001 [5]. Agile methodologies, including Scrum, Extreme Programming (XP), and Kanban, prioritized customer collaboration, adaptive planning, and early delivery of valuable software.

- Scrum introduced iterative cycles called sprints, promoting self-organizing teams and regular stakeholder involvement [6].
- XP focused on technical excellence through practices such as test-driven development, continuous integration, and pair programming [7].
- Kanban, inspired by Lean manufacturing, offered a visual system for workflow management, emphasizing continuous delivery and improvement [8].

Agile revolutionized software development by valuing individuals and interactions over processes and tools, allowing teams to respond quickly to changing requirements and feedback. However, Agile also posed challenges, necessitating a major cultural shift within organizations toward greater flexibility, collaboration, and rapid adaptation to change. Additionally, its emphasis on iterative development led to limited documentation occasionally, leading to potential ambiguities. The success of Agile depends heavily on the team's expertise and collaborative ability, meaning that achieving targets without a skilled team can be difficult.

1.5 The Era of DevOps and Continuous Delivery

As the pace of software delivery increased, the need for seamless collaboration between development and operations became apparent. DevOps emerged as a culture and practice designed to unify these traditionally siloed teams [9]. By promoting collaboration, automation, and continuous integration/continuous deployment (CI/CD), DevOps enabled organizations to rapidly and reliably deploy software.

Technologies such as Infrastructure as Code (IaC) and containerization further restructured deployment processes, reducing errors and improving scalability. This era underscored the importance of end-to-end responsibility for software, from its conception to its operation, ensuring efficiency and reliability in the development lifecycle.

1.6 Microservices Architecture: Breaking Down the Monolith

With the increasing complexity of software systems, the limitations of traditional monolithic architectures—where all components are tightly integrated—have become more evident. Monolithic frameworks often face major challenges in scalability, maintainability, and rapid deployment, which are crucial in today's fast-paced technological landscape [10]. To address these issues, microservices architecture emerged as a transformative approach that decomposes applications into smaller, independently deployable units.

A defining feature of microservices architecture is the decentralization of functionality. Each microservice encapsulates a specific business capability, functioning autonomously and enabling organizations to achieve greater agility [11]. These services interact through well-defined APIs, leveraging protocols such as REST or gRPC to ensure loose coupling and seamless interoperability.

One of the primary advantages of microservices is independent scalability. Each service can be scaled based on its individual resource requirements, optimizing resource utilization and enhancing performance under viable loads [12]. Furthermore, this architecture offers unparalleled flexibility, as development teams can select the most suitable programming languages or technologies for each service's unique demands.

Microservices also enhance system resilience through fault isolation. Since services operate independently, the failure of one does not necessarily impact the entire system, thereby improving overall reliability [13]. However, transitioning to microservices introduces the complexity of managing distributed systems. Organizations often rely on sophisticated monitoring and orchestration tools, such as Kubernetes, to address the increased operational overhead [14]. Readers may refer to [15] for more details on microservices.

Despite their numerous benefits, microservices are not without challenges. The architecture increases complexity in debugging and testing, and the overhead from inter-service communication can affect performance. Consequently, careful planning and strategic implementation are crucial for organizations considering a shift from monolithic systems to a microservices-based approach.

1.7 Cloud-Native Development: Building Scalable and Resilient Software Systems

The widespread adoption of cloud computing has reshaped the software development landscape, paving the way to cloud-native practices that enable the creation of scalable, adaptable, and resilient applications. These applications are specifically designed to leverage the dynamic features of cloud platforms, including elastic scalability, managed services, and containerization. By embracing cloud-native methodologies, organizations can streamline their workflows and respond swiftly to shifting market conditions and technological innovations [16, 17]. The main characteristics of cloud-native development are as follows:

- **Containers**: Applications are packaged within lightweight, portable containers (e.g., Docker) that provide consistent runtime environments. This ensures seamless deployment and operation across diverse cloud platforms.
- **Orchestration**: Automated tools such as Kubernetes automate essential tasks, including deployment, scaling, and operation of containerized workloads. This reduces manual intervention, increase efficiency, and optimizes resource utilization.
- **Serverless Computing**: Cloud-native applications often utilize serverless architectures, offloading infrastructure management to cloud providers. This shift allows development teams to concentrate on writing core business logic and delivering value without being encumbered by server maintenance tasks.
- **Scalability**: Cloud-native systems dynamically allocate resources based on workload demands, ensuring optimal performance and efficiency across varying conditions.
- **Cost-Effectiveness**: The pay-as-you-go model of cloud services helps reduce operational overhead, enabling organizations to align expenses with actual usage.
- **Agility**: Modular cloud-native architectures support rapid iteration and continuous delivery, accelerating the deployment of new features and improving responsiveness to user needs.

Cloud-native development has become a cornerstone for organizations aiming to achieve global scalability, high availability, and rapid innovation. Organizations can better align their software systems with contemporary market and technological imperatives by fully exploiting cloud technologies' capabilities.

1.8 Serverless Computing: Shifting from Infrastructure Management to Event-Driven Business Logic

Serverless computing revolutionizes cloud-based development by abstracting the complexities of infrastructure management, enabling developers to focus solely on crafting business logic. Instead of managing servers, developers create small, modular, event-triggered functions that execute in response to specific triggers such as HTTP requests, file uploads, or database updates. This paradigm, supported by platforms like AWS Lambda, Google Cloud Functions, and Azure Functions, automates resource provisioning and scaling, streamlining the development process [18].

Serverless computing is characterized by the following:

- **Event-Driven Execution**: Functions are triggered automatically by defined events, promoting a reactive and loosely coupled architectural design.
- **Automatic Scaling**: Computational resources scale up or down in real time to accommodate demand, eliminating the need for manual developer intervention.
- **No Server Management**: Developers are freed from managing hardware or server maintenance, as cloud providers oversee deployment and operational tasks. Developers focus specifically on writing and refining business logic.
- **Cost-Efficiency**: Serverless billing models are usage-based, charging only for the execution time and frequency of functions, offering significant savings over traditional always-on server models.
- **High Scalability**: Serverless platforms effortlessly handle substantial, unpredictable workloads, often supporting millions of parallel requests.

While serverless computing offers significant benefits, it is not without drawbacks. A notable issue is cold start latency, which can adversely affect performance when functions are inactive for extended periods and require time to initialize. This can be problematic for latency-sensitive applications. Additionally, managing application states in a stateless environment can be challenging, often requiring external services like databases for state persistence. Finally, reliance on proprietary platforms increases the risk of vendor lock-in, making it crucial for organizations to prioritize portability and interoperability to ensure long-term flexibility and sustainability.

1.9 Entering the Age of Generative AI

Advances in large language models (LLMs) are reshaping software engineering (SE) by automating the processes of code generation, debugging, and maintenance. A systematic review of 395 studies published between 2017 and 2024 exemplifies the role of LLMs in expanding the SE field [19]. Encoder-only models are designed to decipher codes; thus, they are effective in localizing bugs and detecting vulnerabilities. Decoder-only architectures, such as GPT derivatives, exhibit generative capabilities for tasks including code completion and synthesis. Nevertheless, encoder–decoder models balance comprehension and generation, and, are therefore, suitable for intricate tasks such as program repair.

A critical contribution of [19] lies in its emphasis on data quality. The study categorized datasets as open-source repositories, constructed datasets, and industrial datasets, emphasizing that the use of industrial data in academic research has been insufficient. Under-utilization of these data underscores the need for stronger collaboration between academia and industry to advance practical, scalable LLM-driven solutions.

Optimization and evaluation also feature prominently. Techniques such as parameter-efficient, fine-tuning and prompt engineering can aid in tailoring LLMs for SE-specific tasks. Nevertheless, ethical concerns arise in using LLMs, particular in sensitive security contexts. Navigating these challenges warrants thorough evaluation of emerging architectures and deployment of robust safeguards.

Generative AI, driven by LLMs, is poised to redefine conventional workflows. In addition to automating repetitive tasks and elevating code quality, generative tools foster collaboration and accelerate innovation across the software lifecycle. Researchers and practitioners can embrace these breakthroughs to harness LLMs' transformative potential, ultimately shaping a more efficient, adaptive, and collaborative era of software development.

1.10 Leveraging Generative AI as a Software Engineer

Generative AI offers transformative capabilities across various stages of software development, driving efficiency, creativity, and teamwork. Below are some of the critical areas where these technologies are making an impact, alongside examples of tools and their applications.

1.10.1 Automating Repetitive Coding Tasks

Tool Example: GitHub Copilot.

How It Works: GitHub Copilot, powered by OpenAI's Codex, leverages natural language input, code context, and comments to generate relevant code snippets [20]. It aids in creating boilerplate code, functions, and even algorithms.

Application: For instance, a backend engineer can quickly set up a new microservice by instructing Copilot to generate boilerplate code for an Express.js server. This significantly reduces the time required for routine setups.

By shifting the focus from writing code line-by-line to refining AI-generated output, developers can significantly enhance their productivity while maintaining control over the codebase.

1.10.2 Debugging and Error Detection

Tool Example: DeepCode (Now Part of Snyk).

How It Works: DeepCode uses machine learning models trained on extensive open-source datasets to identify bugs, security vulnerabilities, and performance bottlenecks. It provides actionable suggestions to detect and fix these issues [21].

Application: Full-stack developers can integrate DeepCode into their Continuous Integration/Continuous Deployment (CI/CD) pipelines to automatically scan for potential bugs and vulnerabilities during development cycles.

By automating code reviews for quality and security, engineers can focus more on building features and less on manually detecting flaws, enhancing both speed and reliability in software delivery.

1.10.3 Improving Code Quality and Optimization

Tool Example: SonarQube.

How It Works: SonarQube is a tool that performs static code analysis to help developers maintain high-quality code by identifying issues such as code smells, redundancies, and performance bottlenecks. It ensures adherence to coding standards and best practices [22].

Application: Developers can incorporate SonarQube into their continuous integration processes to ensure consistent code quality throughout the development lifecycle, preventing inefficient or problematic code from reaching production.

Consistent code quality ultimately leads to more maintainable and reliable robust software systems.

1.10.4 AI-Assisted Code Review and Collaboration

Tool Examples: Codacy, Code Climate.

How It Works: Codacy and Code Climate are AI-powered quality tools that automate the review of code changes, offering insights into maintainability, complexity, and adherence to coding standards. These tools integrate seamlessly into Git workflows, allowing for real-time reviews of pull requests [23, 24].

Application: In collaborative coding environments, engineers can use these tools to maintain consistent code quality and minimize the introduction of technical debt.

This fosters smoother collaboration and accelerates development.

1.10.5 Automating Data Cleaning and Preprocessing

Tool Example: DataRobot.

How It Works: DataRobot leverages AI to automate data preprocessing, cleaning, and transformation—tasks that are typically time-consuming in data analytics workflows. It manages missing data, normalizes datasets, and detects outliers without requiring manual input [25].

Application: These tools help save time of data analysts during the data preparation phase, enabling them to focus on extracting insights and building predictive models.

Automating these processes not only enhances efficiency but also improves the accuracy of data-driven decisions.

1.10.6 The Key to Unlocking Generative AI's Potential

The true power of generative AI lies in understanding how existing AI tools can augment workflows, rather than learning how to construct or program LLMs. Tools such as GitHub Copilot, DeepCode, and SonarQube incorporate advanced AI functions to simplify complex tasks, empowering engineers and analysts to focus on higher-level problem-solving and creativity.

1.11 The Future of Programming with Generative AI

With continuous advancements, generative AI is poised to transform the role of software engineers. Examples of potential shifts in the future of programming with AI are as follows:

1.11.1 Shift Toward High-Level Problem-Solving

AI will increasingly take on routine coding tasks, freeing engineers to concentrate on system architecture and design, and solving complex business problems. Leveraging AI tools, engineers can guide the development process, rather than writing every line of code.

1.11.2 AI-Assisted Collaboration

AI will become more integrated into collaborative coding environments. Real-time suggestions, error detection, and code optimization might equip engineers to work alongside AI for rapidly iterating and refining codebases.

1.11.3 Natural Language Programming

Tools such as GitHub Copilot are paving the way for a future where developers can write code using natural language [20]. This trend is expected to continue, allowing even non-engineers to easily access programming and easing the learning curve for new developers.

1.11.4 AI-Driven Software Maintenance

AI systems might proactively monitor live applications, identifying potential issues, security vulnerabilities, and performance bottlenecks. In some cases, autonomous systems may implement fixes, updates, or optimizations without human intervention [26].

1.11.5 Implications of Generative AI in Software Development

- Boosting Productivity: Developers can focus more on complex problem-solving and innovation by automating routine tasks.
- Enhancing Quality: AI-driven tools can identify bugs and suggest optimizations, resulting in more robust software.
- Democratizing Development: Low-code and no-code platforms powered by AI allow individuals with minimal programming experience to create applications, broadening the pool of innovators.

1.11.6 Challenges and Considerations

While generative AI presents substantial advantages, it also presents challenges to be addressed by the industry:

- Ethical Concerns: Ensuring that AI-generated code adheres to ethical standards and avoids reinforcing biases or unethical practices.
- Security Risks: Proper management of AI tools is essential to prevent introducing vulnerabilities or malicious code.
- Skill Evolution: Developers will need to adapt their skill sets to collaborate effectively with AI tools, focusing on oversight and strategic decision-making.

1.11.7 Looking Forward

The shift from traditional methods to the age of generative AI signifies both a technological evolution and a fundamental change in how we approach problem-solving and creativity in software development. As AI becomes more integrated into workflows, it is crucial to balance the advantages with thoughtful consideration of ethical, security, and professional implications.

1.11.8 Organizations Should

- Invest in Education and Training: Teams must be equipped with skills to responsibly and effectively use AI tools.
- Establish Ethical Guidelines: Policies governing the use of AI in software development to ensure compliance with legal and ethical standards should be established and enforced.
- Foster a Collaborative Culture: Organizations must foster a culture where human expertise and AI capabilities work together, with each enhancing the other.

1.12 Conclusions

Software development has been shaped by a dynamic interplay of technological advancements and changing societal demands. From the rigid frameworks of traditional methodologies to the flexibility of Agile and DevOps practices, each phase has built upon the last. The rise of microservices and cloud-native architectures reflects the industry's drive for scalability, resilience, and resource efficiency. Today, generative AI heralds a paradigm shift, transforming SE by automating repetitive tasks, improving collaboration, and enabling engineers to focus on higher-level problem-solving and creativity. While these innovations hold great potential for productivity and innovation, they also present challenges, such as ethical concerns, as well as the need for new skill sets and dependence on AI tools. Moving forward, the key to success lies in balancing the benefits of cutting-edge technologies with their broader societal and professional implications. By investing in education, promoting ethical practices, and encouraging collaborative approaches, the industry can continue to innovate responsibly, expanding the boundaries of what is possible while aligning with societal values. The future holds promise to redefine both the route of software development as well as the role of software engineers in shaping the digital landscape.

References

1. W.W. Royce, Managing the development of large software systems: concepts and techniques, in *Proceedings of the 9th International Conference on Software Engineering* (1987), pp. 328–338
2. J.J. Marciniak, *Encyclopedia of Software Engineering* (Wiley, 2002)

3. B.W. Boehm, A spiral model of software development and enhancement. IEEE Comput. **21**(5), 61–72 (1988)
4. P. Kruchten, *The Rational Unified Process: An Introduction* (Addison-Wesley Professional, 2004)
5. K. Beck, M. Beedle, A. van Bennekum, A. Cockburn, W. Cunningham, M. Fowler, J. Grenning, J. Highsmith, A. Hunt, R. Jeffries, J. Kern, B. Marick, R.C. Martin, S. Mellor, K. Schwaber, J. Sutherland, D. Thomas, Manifesto for Agile Software Development (2001). https://agilemanifesto.org/
6. K. Schwaber, J. Sutherland, *The Scrum Guide* (2020, November). Scrum.org. https://scrumguides.org/
7. K. Beck, *Extreme Programming Explained: Embrace Change*, 2nd edn (Addison-Wesley Professional, 2004)
8. D.J. Anderson, *Kanban: Successful Evolutionary Change for Your Technology Business* (Blue Hole Press, 2010)
9. G. Kim, J. Humble, P. Debois, J. Willis, N. Forsgren, The DevOps handbook: How to create world-class agility, reliability, & security in technology organizations. IT Revolution (2021)
10. S. Newman, *Building Microservices: Designing Fine-Grained Systems* (O'Reilly Media, 2015)
11. N. Dragoni, S. Giallorenzo, A.L. Lafuente, M. Mazzara, F. Montesi, R. Mustafin, L. Safina, Microservices: yesterday, today, and tomorrow, in *Present and Ulterior Software Engineering* (Springer, 2017), pp. 195–216
12. I. Nadareishvili, R. Mitra, M. McLarty, M. Amundsen, *Microservice Architecture: Aligning Principles, Practices, and Culture* (O'Reilly Media, 2016)
13. J. Thönes, Microservices. IEEE Softw. **32**(1), 116–116 (2015)
14. B. Burns, B. Grant, D. Oppenheimer, E. Brewer, J. Wilkes, Borg, Omega, and Kubernetes. Commun. ACM **59**(5), 50–57 (2016)
15. M. Fowler, J. Lewis, *Microservices*. Martinfowler.com (2020). Retrieved December 26, 2024, from https://martinfowler.com/articles/microservices.html
16. G. Toffetti, S. Brunner, M. Blöchlinger, J. Spillner, T.M. Bohnert, Self-managing cloud-native applications: design, implementation, and experience. Futur. Gener. Comput. Syst. **72**, 165–179 (2017)
17. N. Kratzke, P.C. Quint, Understanding cloud-native applications after 10 years of cloud computing-a systematic mapping study. J. Syst. Softw. **126**, 1–16 (2017)
18. Z. Li, L. Guo, J. Cheng, Q. Chen, B. He, M. Guo, The serverless computing survey: a technical primer for design architecture. ACM Comput. Surv. (CSUR) **54**(10s), 1–34 (2022)
19. X. Hou, Y. Zhao, Y. Liu, Z. Yang, K. Wang, L. Li, X. Luo, D. Lo, J. Grundy, H. Wang, Large language models for software engineering: a systematic literature review. ACM Trans. Softw. Eng. Methodol. **33**(8), 1–79 (2024)
20. GitHub, Introducing GitHub Copilot: Your AI pair programmer. *GitHub Blog* (2021, June 29). https://github.blog/2021-06-29-introducing-github-copilot-ai-pair-programmer/
21. Snyk. Snyk Code: Snyk's developer-first SAST powered by DeepCode AI. Retrieved November 10, 2024, from https://snyk.io/product/code/
22. SonarSource. SonarQube: continuous code quality. Retrieved November 10, 2024, from https://www.sonarqube.org/
23. Codacy. Automated Code Reviews & Code Quality. [Online]. Available: https://www.codacy.com/
24. Code Climate. Velocity and quality: Engineering intelligence tools. Retrieved November 10, 2024, from https://codeclimate.com/
25. DataRobot, DataRobot AI Platform. Retrieved November 10, 2024 (n.d.). From https://www.datarobot.com/platform/
26. Gartner. Predicts 2022: Artificial Intelligence. Retrieved November 10, 2024, from https://www.gartner.com/en/topics/artificial-intelligence

Chapter 2
Case Study: Transforming Operational and Organizational Efficiency Using Artificial Intelligence and Machine Learning

Vindhya Avvari, Abhik Choudhury, Arjun Karat, and Tirimula Rao Benala

Abstract Machine learning (ML) and artificial intelligence (AI) are revolutionizing the world, creating a new era in the realm of technology. They are transforming the way we approach problem-solving and decision-making in the digital age. AI involves the development of systems with human-like intelligence, while ML enables systems to learn and improve using data, making predictions and decisions without explicit programming. Virtual assistants, robots, speech recognition, and autonomous vehicles are some of the areas where AI can be applied. ML has applications in image recognition, natural language processing, recommendation systems, and predictive analytics. It has the potential to enhance organizational efficiency and decision-making. This article presents how the successful implementation of AI and ML in the software development lifecycle can foster improvements in the software industry. It also summarizes real-world case studies of companies that leveraged these tools to automate, optimize, enhance, and improve operational efficiency.

Keywords Machine learning · Artificial intelligence · Virtual assistants · Predictive analytics · Software development lifecycle

---

V. Avvari (✉)
University of Southern California, Los Angeles, CA, USA
e-mail: avvari@usc.edu

A. Choudhury
IBM, King of Prussia, PA, USA
e-mail: abhikcho@in.ibm.com

A. Karat
Salesforce Inc, Palo Alto, CA, USA
e-mail: akarat@salesforce.com

T. R. Benala
Department of Information Technology, JNTU-GV College of Engineering Vizianagaram, Jawaharlal Nehru Technological University Gurajada Vizianagaram, Dwarapudi, Vizianagaram, Andhra Pradesh, India
e-mail: btirimula.it@jntugvcev.edu.in

T. R. Benala et al. (eds.), *Boosting Software Development Using Machine Learning*, Artificial Intelligence-Enhanced Software and Systems Engineering 7,
https://doi.org/10.1007/978-3-031-88188-6_2

2.1 Introduction

2.1.1 Background

In the software industry, operational and organizational efficiency refers to the streamlining of software development teams' processes, optimizing resource utilization, and delivering high-quality products and services on time. For this, software development teams have adopted agile methodologies, DevOps practices, lean principles, efficient resource utilization, quality assurance and testing, metrics, and performance monitoring. Software development projects often involve complex requirements, technologies, and integration points. Planning, coordination, and expertise are also crucial for successful deliverables. Some of the common challenges in this context are: (a) Development teams may face an overwhelming amount of work, especially when working on tight deadlines or multiple projects simultaneously; (b) in dynamic environments, requirements may change frequently because of evolving business needs, stakeholder feedback, or market demands; (c) developers may encounter repetitive tasks, such as manual testing, debugging, or code refactoring, which can consume time and resources. To meet the target Go Live dates without compromising the quality of the product, teams must reduce time spent on non-value-added activities and allocate more time and effort for problem-solving, innovation, and producing high-quality deliverables. In the modern world, artificial intelligence (AI) and machine learning (ML) are transforming the software development life cycle (SDLC) by introducing automation and optimization into various stages of the development process, leading to faster delivery, better quality software, and improved efficiency across development teams. This study discusses the different SDLC stages and how AI and ML tools are redefining each step of SDLC by optimizing production processes, reducing costs, enhancing decision-making, and improving overall employee productivity. We investigated specific instances where AI and ML have been successfully implemented for presenting concrete evidence of their benefits. We also explored how different organizations have leveraged these technologies to optimize their software development processes, resulting in enhanced productivity and efficiency. This was elaborated by breaking down various phases of the SDLC and providing detailed contributions of AI and ML in each case. This will provide insights for understanding how these technologies help enhance planning and requirements gathering, improve design processes, streamline implementation and coding, automate testing, and facilitate efficient deployment of software systems.

2.2 Literature Review

2.2.1 The Importance of AI and ML in Software Development

The integration of AI in software development is reshaping various aspects of the development process, leading to increased efficiency and productivity. AI-Powered tools are revolutionizing code generation and autocompletion by assisting developers in generating code snippets, autocompleting lines, and suggesting relevant functions. This speeds up the coding process and reduces the likelihood of errors, as exemplified by the use of AI-driven IDEs like Microsoft's Visual Studio IntelliCode. Code suggestion tools improve the speed and accuracy of creating and maintaining software by helping programmers prevent code smells, make the code modular, migrate to new API versions, and so on [1–3]. This study discusses handy tools that a developer can benefit from during the implementation phase, such as TabNince Code Documentation and OpenAI's Codex. AI is also playing a pivotal role in automated testing, enabling more comprehensive test coverage and early detection of bugs. AI algorithms for regression testing, performance testing, and security testing, such as those utilized by Applitools and Mabl, exemplify this advancement.

AI-Based tools are transforming bug detection and resolution by analyzing code to identify potential bugs or security vulnerabilities. Platforms like DeepCode utilize ML to learn from historical data, predicting and preventing future software issues. Natural language processing (NLP) technologies further enhance developer productivity by enabling code interaction using spoken or written language, thereby making programming more accessible. For instance, GitHub's Copilot leverages NLP to assist developers in writing code using natural language.

Furthermore, AI systems are employed in code review and quality assurance processes, helping analyze code for adherence to best practices and style guidelines. Tools such as DeepCode Code review assistant automatically review code, ensuring better quality and consistency across projects. Predictive analytics powered by AI algorithms help in project planning and management by analyzing historical project data to predict potential delays and estimate completion times. Automated documentation tools, such as Swagger and Sphinx, leverage AI to generate and update documentation automatically based on code changes, reducing manual effort.

AI-Driven virtual assistants also provide developers with contextual information and assist in problem-solving, streamlining the development process and enhancing productivity. Moreover, AI is utilized to automate the deployment process, ensuring smooth integration and delivery. Predictive analysis tools identify potential issues before deployment, minimizing disruptions. AI can also personalize development environments based on individual preferences, providing developers with a more efficient workspace.

Lastly, AI is integrated into applications for enhancing user experience (UX) through intelligent features, such as personalized recommendations, natural language interfaces, and predictive user behavior analysis. For example, users of Netflix often see personalized content recommendations, which has been possible through AI algorithms, enabling enhanced overall user satisfaction and engagement of Netflix users. In essence, the integration of AI in software development is revolutionizing the way developers work, driving efficiency, and innovation across the entire development lifecycle.

2.3 Stages of SDLC

According to Nayan B Ruparelia, "SDLC has been defined as a conceptual framework that evaluates the structure of the steps involved in the development of an application, from planning until deployment" [1]. SDLC is applicable to various hardware and software applications [6]. Several models describe various approaches to the SDLC process. A SDLC model is generally used to describe the steps that are followed within the life-cycle framework [1]. The initial comprehensive depiction of a SDLC model was introduced by Herbert Benington in 1956 [4]. Subsequently, SDLC models have considerably advanced. In the modern day, the concept of SDLC is encapsulated in these main steps: gathering requirements, design, implementation, testing, and delivery. In this study, we focus on how each step of the SDLC is revolutionized and optimized with the help of AI and ML by providing examples from the real world (Fig. 2.1).

The importance of using AI in software engineering lies in its ability to generate significant cost savings and foster innovation in a competitive market. This is achieved

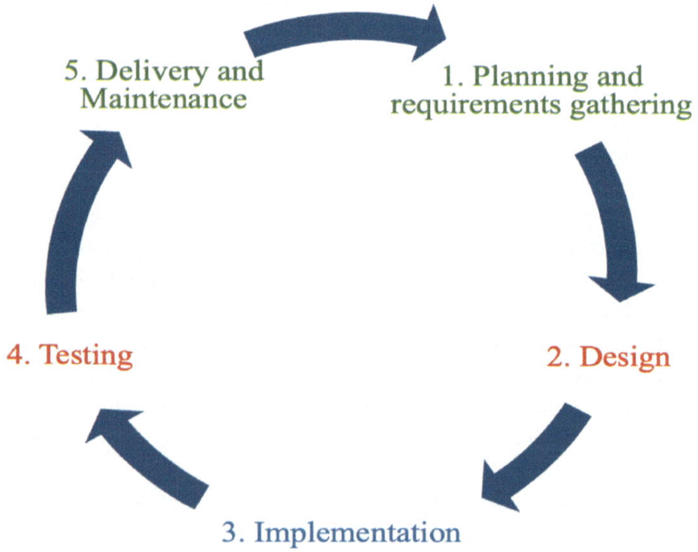

Fig. 2.1 Stages of software development life cycle

by improving efficiency, minimizing the occurrence of bugs, and expediting the development process. Several considerations, together with the recent substantial investments in this field, highlight the potential and significance of AI for transforming the future of software development [5].

2.3.1 Planning and Gathering of Requirements

In a software engineering project, business requirements are gathered from stakeholders; effective techniques used for gathering requirements can empower individuals to identify and prioritize requirements, incorporate modifications, and ensure that the product meets the set expectations [7]. Business requirements represent the needs, gaps, objectives, goals, or pain points that the team is trying to address. These requirements drive the development of a software solution. Typically, functional as well as non-functional requirements are gathered either through direct interactions such as interviews, workshops, surveys, or from an existing well-documented business requirements document. This planning phase involves activities such as gathering, analyzing, documenting, and validating requirements. Using AI-powered tools that leverage NLP, this process can be automated, reducing the man hours that the developers spend performing the above-mentioned activities. AI Tools can transcribe the requirements from speech and extract valuable insights from stakeholder meetings and meeting transcripts. Popular AI-powered tools that help in planning and requirements gathering are discussed below:

IBM Engineering Requirements Management DOORS Next

This tool provides AI capabilities for analyzing and managing requirements. It includes features such as natural language understanding, semantic analysis, and requirement tracing, helping teams streamline the process of gathering requirements and the documentation process. Developed and administered by the Jet Propulsion Laboratory (JPL), the Mars2020 project effectively executed the landing of the Perseverance rover and its flying partner Ingenuity on the martian surface on February 18, 2021 [8]. The NASA/JPL Mars 2020 rover utilized IBM's DOORS Next product as the requirements management tool for storing dictionary artifacts.

Specright

Specright includes AI features for analyzing textual requirements and specifications, and it is primarily focused on product specification management. It offers functionalities such as text parsing, entity recognition, and requirement validation to help organizations effectively manage complex product requirements. A case study was conducted on the American fast food chain Jack in the Box, which has a huge supply chain to manage, but because of outdated systems and limited accessibility, spec compliance reporting was almost impossible. The process was manual and time-consuming, making quality assurance difficult and supplier relationships strained.

With Specright, quality managers saved 50%–60% of their time in managing product quality, resulting in improved supplier relationships and customer satisfaction [9].

reQtest

reQtest is a software testing and requirements management platform with AI-powered features for analysis of requirements. It offers functionalities such as NLP for automatically extracting and categorizing requirements from textual documents, along with intelligent recommendation systems for identifying potential gaps or inconsistencies in requirements.

Amazon Comprehend

This cloud-based NLP service provides APIs for tasks such as entity recognition, key phrase extraction, sentiment analysis, and language detection, which can be used to analyze textual documents with requirements and extract valuable insights.

Google Cloud Natural Language

It offers APIs for various text analysis tasks, such as entity recognition, syntax analysis, and sentiment analysis. It can be integrated into software applications or platforms for performing advanced NLP-based requirements gathering and documentation.

2.3.1.1 Speech to Text

Some advanced AI tools can greatly improve the efficiency and accuracy of requirements documentation by capturing spoken requirements in written form. By leveraging speech recognition technology, teams can streamline the documentation process, improve collaboration, and ensure that all requirements are accurately captured and documented for posterity. Examples of such tools are discussed below:

Google Cloud Speech-to-Text

This cloud-based service converts audio input into text using ML models. It supports multiple languages and can transcribe real-time streaming or pre-recorded audio and can be integrated into applications or workflows. According to HG Insights, more than 700 companies are utilizing this tool as of 2024.

IBM Watson Speech to Text

The IBM Developer site explains this tool's capabilities as, "text recognition, audio preprocessing, noise removal, background noise separation, and semantic sentence conversation." It converts speech into text using AI-powered speech recognition and transcription. It enables transcription of spoken requirements and can integrate the text into documentation processes. It is being used by several customers including CVS to capture medical data, and TEDx to extract meaningful excerpts from large volumes of videos.

Amazon Transcribe

This speech recognition service offered by Amazon Web Services (AWS) can transcribe audio recordings into text with high accuracy and supports multiple languages and dialects. It can be used to document spoken requirements and extract valuable insights from recorded conversations.

Microsoft Azure Speech Service

It provides speech recognition capabilities for converting spoken language into text. It can accurately translate spoken words into text in over 100 languages and dialects. By enabling search or analytics on transcribed text, a user can get additional value out of spoken audio. Apart from documentation, teams can also utilize this tool to refer to transcribed data and find the relevant information from audio files more quickly.

2.3.2 Architectural, System, Database, and User Interface Design

AI and ML are making significant strides in various business areas, offering noteworthy advantages beyond automation. Herein, we delve deeper into how AI/ML can empower software and data engineers alike in each design domain with specific use cases and examples.

2.3.2.1 Architectural Design

Early-Stage Feasibility Analysis

In the initial system/architecture design stages, AI can be extremely beneficial in various aspects of the overall development [17]. Imagine feeding an AI tool information about project requirements, performance goals, and existing technology constraints. The AI could subsequently create multiple architectural design versions depending on the requirements, exploring different trade-offs between scalability, security, and maintenance. This allows engineers to quickly assess the feasibility of various approaches, help early decision-making in terms of potential design constraints, and identify potential problems early on.

Automated Performance Simulations

Performance is an extremely important aspect of software architecture [18]. AI can be used to simulate the behavior of a proposed architecture under different situations and data and load conditions. This allows engineers and architects to identify potential blockages and performance issues by simulating various scenarios before any code is written. ML models can be trained on historical performance data to predict the behavior of specific architecture patterns in conjunction with business/data scenarios under real-world scenarios, enabling informed design choices and modifications.

Microservices Dependency Analysis

As an architectural style, microservices break down large applications into smaller, independent components, with each component having its own specific responsibilities [19]. To fulfill a single user request, a microservices-based application may be needed to coordinate multiple internal microservices. AI can analyze the dependencies between these services, identifying potential conflicts or areas where better collaboration may be required. This enables engineers and architects to optimize the microservices architecture for better performance and maintenance. AI-Aided tools can analyze communication patterns between microservices and identify potential problems such as service bottlenecks and circular dependencies; they can even allow engineers and architects to modify the architecture for improved communication and isolation of individual services for improving overall robustness and reliability.

Microservices Deployment and Scaling Recommendations

Despite its numerous advantages, deploying and scaling microservices can be a complex task. Monitoring and tracking requests through multiple microservices can become extremely difficult when there are more microservices. To ensure that resources are allocated efficiently and that the system can effectively handle peak loads, AI can be used to analyze historical usage data and recommend optimal deployment strategies for individual services. Specific ML models can predict resource allocation and usage patterns for each microservice based on historical behaviors, allowing engineers to automate deployment configurations and scale decisions for optimal resource utilization.

AI-Assisted Refactoring

Codebase refactoring involves re-evaluating and restructuring how code is organized into packages and how those packages interact. It helps split a package into more manageable pieces for users, improves naming, and helps lighten dependencies. It is often a complex and time-consuming task. AI can help analyze code dependencies and suggest refactoring strategies to improve modularity, reduce technical debt, and ensure better maintenance. ML models can be trained on historical refactoring projects to identify code that is not maintainable, clean, and has potential bugs and unusual patterns; they can recommend appropriate refactoring techniques based on established architectural guidelines and best practices.

Dynamic Resource Allocation

Dynamic resource allocation in the cloud involves automatically adjusting resource usage to meet fluctuating user demands. AI can be leveraged to access system performance in real-time and allocate resources dynamically to various components based on their current load. This ensures that the utilization of various resources is optimized and prevents performance bottlenecks during peak usage periods. Software engineers can use AI-powered resource management tools to automatically scale resources such as memory usage, disk space, type of workload, or computing based on predicted load patterns, improving system responsiveness and scalability.

Some AI/ML-based tools that are used in a multitude of industry-wide architecture designs are discussed below:

Autodesk Archimate

This tool leverages AI and ML to assist with architecture definition and analysis based on the Archimate modeling language. It can automatically generate architecture diagrams from code and specifications, identify potential gaps and inconsistencies in the design, and recommend best practices for architecture implementation.

CloudMile SkyWalking

This AI-powered application performance management tool analyzes application behavior and infrastructure performance. It helps software architects identify bottlenecks and inefficiencies in the microservices architecture, suggesting optimization strategies for improved scalability and resource utilization.

Medusa by GitLab

This AI-powered security testing tool integrates with GitLab's CI/CD pipeline. It scans code for identifying potential vulnerabilities and security risks early in the development process, helping architects identify design flaws that might introduce security weaknesses later.

2.3.2.2 System Design

Some of the areas where AI/ML methodologies and tools have found their way into scalable and system design are discussed below:

Predictive Load Forecasting

Load forecasting analyzes historical load data to predict future energy consumption [20]. These predictions can be used to adjust physical resources, reducing task wait times and optimizing energy efficiency. Accurately predicting system load is a crucial aspect of the overall system design, choice of components, performance, and system maintenance. AI can help analyze historical usage data and identify usage behaviors. This can allow engineers to enable proactive resource allocation and prevent system overload to develop models by predicting future load spikes. ML models can be trained on historical traffic patterns and external factors (e.g., seasonality, marketing campaigns) to forecast future demands, allowing engineers to scale resources, such as servers and databases, ahead of time to maintain optimal performance.

Dynamic Autoscaling

AI can be used to automate the process of scaling system resources based on real-time load [21]. This ensures that the system can handle peak traffic without compromising performance or incurring unnecessary costs during low-traffic periods. AI-Powered autoscaling tools can monitor system metrics (CPU utilization, memory usage) and automatically adjust resource allocation based on predefined thresholds, optimizing resource utilization and cost efficiency.

Anomaly Detection in System Behavior

AI can be used to analyze system logs and identify anomalies that might lead to potential failures in future. This allows engineers to detect and address issues before they become critical outages. ML models can be trained on historical log data to identify patterns that deviate from normal system behavior, enabling early detection of potential failures and proactive troubleshooting.

Self-healing Systems

AI can be used to design systems that can automatically recover from failures [22]. This involves building redundancy and implementing logic that allows the system to identify and re-route around failing components. AI-Powered self-healing systems can leverage ML to diagnose failures and automatically trigger corrective actions (e.g., restarting services and rerouting traffic) to minimize downtime and ensure service continuity.

Some AI/ML-based tools that can aid in system design are discussed below:

HPE Machine Learning Anomaly Detection

This tool analyzes system logs and network traffic to identify anomalies that might indicate security threats or performance issues.

IBM Cloud Watson AIOps

This suite of AI-powered tools helps automate IT operations tasks, including anomaly detection, event correlation, and root cause analysis, helping system designers build more resilient and self-healing systems.

Microsoft Azure Sentinel

This cloud-based security information and event management tool leverages AI and ML to analyze security data from various sources, helping system designers identify and respond to security threats more effectively.

2.3.2.3 Database Design

By leveraging AI/ML in database design, software engineers can streamline the design process, improve data quality, and ensure optimal performance for data-driven applications.

Data Type and Cardinality Recommendations

AI can analyze data samples and recommend optimal data types (e.g., integer, string) and cardinalities (e.g., one-to-one and one-to-many) for database columns [23]. This ensures efficient storage utilization and simplifies query execution. ML models can be trained on historical data and existing database schemas to identify patterns and suggest data types with minimal redundancy and optimal storage efficiency based on the expected data distribution.

Entity Relationship Diagram Generation

AI can analyze existing data or user requirements to automatically generate initial entity relationship (ER) diagrams, a foundational step in database design [24]. This can save engineers time and effort, especially for complex schemas with many entities and relationships. AI tools can process data descriptions and constraints to suggest relationships between entities, providing a starting point for engineers to refine and complete the ER diagram.

Data Cleansing and Anomaly Detection

AI can be used to identify and address data inconsistencies and anomalies within a database. This can help ensure data integrity and improves the accuracy of queries and reports. ML models can be trained on clean data samples to identify patterns and deviations, flagging potential data errors and inconsistencies such as missing values or outliers.

Data Profiling and Lineage Tracking

AI can be used to automatically generate comprehensive data profiles that summarize the characteristics of data stored in a database [25]. This includes information on data types, value ranges, and null value percentages. Additionally, AI can assist in tracking data lineage, helping engineers understand the origin and transformations applied to data as it flows through the system. AI-Powered data profiling tools can analyze data statistics and relationships, providing insights into data distribution, potential biases, and data quality issues.

Below we discuss some real-world examples of AI/ML tools used in database design.

Amazon Redshift Spectrum

This managed data warehouse service from AWS leverages ML to optimize query execution plans based on real-time data statistics and workload patterns.

Stream DB

This AI-powered database platform uses ML to automatically optimize data structures and query execution for complex analytical workloads.

Denodo Platform

This data visualization tool utilizes ML to discover and understand data across heterogeneous sources, simplifying database design by providing a unified view of data for application development.

2.3.2.4 User Interface Design

By embracing AI/ML, software engineers can create user interfaces (UIs) that are not only visually appealing but also personalized, user-friendly, and accessible to a wider audience. AI serves as a powerful tool to streamline workflows, optimize UX, and ultimately empower engineers to design intuitive and engaging UIs.

Predictive UI Customization

Imagine a UI that adapts to individual user preferences in real time [26]. AI can analyze user behavior data and predict which UI elements a user is most likely to interact with. This allows engineers to design UIs that dynamically adjust layouts, prioritize content, and even suggest relevant actions based on user context. ML models can be trained on user interaction data to predict user intent and preferences, enabling personalized UI layouts and content recommendations that cater to individual user needs.

A/B Testing with Advanced User Behavior Analysis

A/B testing is an essential component of UI design [27]. AI can revolutionize A/B testing by analyzing not just conversion rates but also user interactions and emotional responses to different UI variations. This provides richer data to engineers for understanding user behavior and make informed design decisions. AI-Powered A/B testing tools can analyze user recordings, such as eye tracking and mouse movements, and facial expressions to gauge user engagement and emotional responses to different UI variations, providing deeper insights compared with traditional conversion rate metrics.

Automated UI Component Generation

AI can be used to generate basic UI components based on pre-defined design principles and UI libraries [28]. This allows engineers to focus on the more complex aspects of UI design, such as layout and interaction design. ML models can be trained on existing UI component libraries to generate variations of buttons, forms, and other UI elements based on specified functionalities and design styles, saving development time.

AI-Powered Accessibility Testing

Accessibility is crucial for inclusive design. AI can be used to analyze UI designs and identify potential accessibility issues for users with disabilities. This enables engineers to proactively address accessibility concerns and ensure a more user-friendly experience. AI-Powered accessibility testing tools can analyze color contrast, screen reader compatibility, and keyboard navigation functionality, flagging potential issues that might hinder accessibility for users with visual impairments or motor skill limitations.

Some real-world examples of AI/ML tools used in UI design are discussed below:

Adobe XD's Auto-Animate Feature

This feature leverages AI to automatically generate animation transitions between different UI states based on design principles, streamlining the animation creation process for engineers.

Figma's Smart Layout tool.

This tool utilizes ML to automatically arrange UI elements based on pre-defined design patterns and spacing rules, assisting engineers in creating consistent and visually appealing UIs.

UXPin Merge

This AI-powered design collaboration tool uses ML to identify inconsistencies and potential usability issues across different design versions, helping engineers maintain design consistency and user-friendliness.

2.3.3 Implementation (Coding)-Executable Code, Unit Test Cases, and Documentation

2.3.3.1 Code Documentation Assistant (LLM)

AI/ML-Powered platforms such as TabNine Code Documentation Assistant [11] can significantly improve operational efficiency across various stages of the SDLC. Some of them are listed below:

Automated Code Documentation

AI/ML-Powered tools can automatically generate documentation for code snippets, functions, and modules. By analyzing code context, these tools can suggest comprehensive descriptions, parameter explanations, and usage examples, reducing the manual effort required for documentation tasks.

Real-Time Code Assistance

AI/ML algorithms can provide intelligent suggestions and corrections as developers write code, helping prevent errors, improve code quality, and adhere to coding standards. This real-time assistance streamlines the coding process and enhances developer productivity [12].

Code Optimization

AI/ML can analyze code patterns and identify opportunities for optimization, such as redundant or inefficient code segments. By providing suggestions for code refactoring and performance improvements, these tools contribute to the development of more efficient and scalable software solutions.

Bug Detection and Resolution

AI/ML algorithms can analyze code repositories and identify potential bugs, vulnerabilities, or code smells. By detecting issues early in the development process, these tools enable developers to address them proactively, reducing the likelihood of bugs reaching production environments.

Predictive Maintenance

AI/ML can analyze historical data from software projects to predict potential maintenance issues or areas of code that may require attention. By identifying patterns and trends, these tools help teams prioritize tasks, allocate resources efficiently, and prevent downtime or system failures.

Automated Testing

AI/ML-powered testing tools can automate test case generation, execution, and result analysis. By intelligently identifying test scenarios, generating test data, and adapting test suites over time, these tools improve test coverage, reduce manual effort, and accelerate the testing process.

Continuous Integration and Deployment (CI/CD)

AI/ML can optimize CI/CD pipelines by analyzing code changes, performance metrics, and user feedback to automate release processes and deployment strategies. These tools help teams achieve faster delivery cycles, smoother deployments, and more reliable software releases.

NLP for Requirements Analysis

AI/ML techniques, such as NLP, can parse and analyze natural language requirements documents, user stories, or feedback to extract actionable insights and prioritize development tasks. By understanding and interpreting textual data, these tools facilitate better communication and collaboration between stakeholders and development teams.

OpenAI's Codex (Using ChatGPT)

OpenAI's Codex is a descendant of OpenAI's generative pre-trained transformer (GPT) models, similar to ChatGPT [13]. Codex is specifically trained on a vast amount of code from various programming languages, including Python, JavaScript, and Java, which allows it to understand and generate code based on natural language prompts. While ChatGPT is designed for general conversational tasks, Codex is tailored specifically for coding-related tasks, making it a powerful tool for software development.

Platforms similar to OpenAI's Codex can significantly impact the SDLC by leveraging AI/ML technologies to enhance operational efficiency in various ways:

Code Generation

Codex can generate code snippets or even entire functions based on natural language descriptions or specifications provided by developers. This accelerates the development process by automating repetitive coding tasks and reducing the need for manual coding from scratch.

Code Reviews and Suggestions

Codex can analyze code submissions and provide intelligent suggestions for improvement, optimization, or addressing potential issues. This helps developers identify and address coding errors, maintain code quality, and adhere to best practices, leading to fewer bugs and faster development cycles.

Automated Testing

Codex can assist in test case generation, execution, and analysis by automatically generating test scenarios, identifying edge cases, and providing insights into test coverage. This streamlines the testing process, improves code reliability, and accelerates the feedback loop between the development and testing phases.

Documentation and Comments

Codex can generate documentation and comments for code snippets, functions, or modules based on natural language descriptions or code context. This automates the documentation process, ensures consistency in documentation standards, and facilitates knowledge sharing among team members.

Code Refactoring and Optimization

Codex can analyze codebases and suggest refactoring or optimization to improve code readability, performance, or maintainability. This helps developers identify areas for improvement, implement coding best practices, and enhance overall code quality.

Natural Language Interfaces

Codex can serve as a natural language interface for interacting with development tools, repositories, or project management systems. Developers can use plain language commands or queries to perform tasks such as code searches, version control operations, or project status updates, enhancing productivity and collaboration.

Code Completion and Contextual Suggestions

Codex can provide intelligent code completion suggestions based on the current context, including variable names, function signatures, or library usage. This accelerates coding workflows, reduces cognitive load, and helps developers write code more efficiently.

Knowledge Sharing and Collaboration

Codex can facilitate knowledge sharing and collaboration by providing access to a vast repository of code examples, documentation, and best practices. Developers can leverage Codex to learn new concepts, explore alternative solutions, and collaborate with peers across different projects or domains.

2.3.3.2 Code Review Assistant

AI/ML-Powered platforms such as DeepCode, a code review assistant, can enhance the SDLC and boost operational efficiency in various ways [14].

Automated Code Analysis

DeepCode uses AI/ML algorithms to analyze code repositories and identify potential issues, bugs, vulnerabilities, or code smells. By scanning code changes in real time, DeepCode provides actionable insights to developers, enabling them to address issues early in the development process and ensure code quality.

Intelligent Code Suggestions

DeepCode offers intelligent suggestions for code improvements, optimizations, or refactoring based on its analysis of code patterns and best practices. Developers can leverage these suggestions to write cleaner, more efficient code, reducing technical debt and enhancing overall code quality.

Contextual Recommendations

DeepCode provides contextual recommendations tailored to specific programming languages, frameworks, or project requirements. By understanding the context of the codebase, DeepCode offers relevant insights and suggestions that align with the project's goals and coding standards.

Code Review Automation

DeepCode automates parts of the code review process by flagging potential issues and providing explanations or remediation suggestions. This streamlines the code review process, reduces manual effort, and accelerates the feedback loop between developers and reviewers.

CI/CD Integration

DeepCode integrates with CI/CD pipelines to perform automated code analysis as part of the build and deployment process. By automatically scanning code changes before they are merged into the main branch or deployed to production, DeepCode helps prevent regressions and ensures code stability.

Customizable Rules and Policies

DeepCode allows teams to define custom rules, policies, or coding standards tailored to their specific project requirements. Developers can configure DeepCode to enforce coding best practices, security guidelines, or performance optimizations, ensuring consistency and adherence to standards across the codebase.

Learning and Adaptation

DeepCode continuously learns from code patterns, developer interactions, and feedback to improve its analysis capabilities over time. By adapting to evolving project requirements and coding practices, DeepCode becomes more effective in identifying issues and providing relevant recommendations as the project progresses.

Developer Education and Training

DeepCode serves as a valuable educational tool by providing explanations, examples, and insights into coding principles and best practices. Developers can learn from DeepCode's suggestions and improve their coding skills, leading to a more proficient and productive development team.

Overall, platforms such as DeepCode leverage AI/ML technologies to enhance operational efficiency in the SDLC by automating code analysis, providing intelligent recommendations, streamlining code review processes, and promoting coding best practices. By integrating DeepCode into their development workflows, teams can improve code quality, reduce time-to-resolution for issues, and deliver more reliable and secure software products.

2.3.3.3 Testing-Test Plans, Test Cases, and Defect Reports

Mabl's AI-Powered Testing Platform (LLM)

Mabl's AI-powered testing platform leverages AI/ML technologies to streamline the SDLC and increase operational efficiency through a variety of improvements [15].

Test Automation

Mabl's platform leverages AI/ML algorithms to automate the creation, execution, and maintenance of test scripts. By analyzing application behavior and user interactions, Mabl can autonomously generate and adapt test cases, reducing the manual effort required for test script development and maintenance.

Self-healing Tests

Mabl uses AI/ML to detect and address test failures automatically. When a test fails, Mabl's platform analyzes the underlying cause and adjusts the test script accordingly to accommodate changes in the application's behavior or UI. This self-healing capability reduces the need for manual intervention and ensures test stability over time.

Intelligent Test Prioritization

Mabl prioritizes test execution based on factors such as code changes, application usage patterns, and business impact. By intelligently scheduling tests, Mabl focuses resources on critical areas of the application, accelerating feedback cycles and enabling faster release cycles.

Dynamic Test Data Generation

Mabl's platform generates realistic test data dynamically, using AI/ML techniques to mimic real-world scenarios and edge cases. This ensures comprehensive test coverage and reduces the risk of overlooking potential issues related to data dependencies or boundary conditions.

Continuous Regression Testing

Mabl integrates seamlessly with CI/CD pipelines to perform automated regression testing. By automatically executing regression tests after each code change or deployment, Mabl helps detect and prevent regressions early in the development process, minimizing the risk of introducing defects into production environments.

Predictive Analytics

Mabl analyzes historical test data and performance metrics to provide predictive insights into future test outcomes and application behavior. By forecasting potential issues or bottlenecks, Mabl helps teams proactively address areas of concern and optimize test strategies for maximum efficiency.

Root Cause Analysis

Mabl's platform employs AI/ML algorithms to identify the root causes of test failures or performance issues. By analyzing test results, logs, and system metrics, Mabl helps pinpoint underlying issues, facilitating faster resolution and improving overall application reliability.

Collaborative Testing

Mabl enables collaboration among development, testing, and operations teams by providing real-time visibility into test results, trends, and anomalies. By fostering collaboration and communication, Mabl helps teams identify and address issues more effectively, leading to faster resolution and improved software quality.

Overall, Mabl's AI-powered testing platform plays a critical role in optimizing the SDLC by automating testing processes, improving test coverage, enhancing test reliability, and enabling faster delivery of high-quality software products. By leveraging AI/ML technologies, Mabl helps organizations achieve greater operational efficiency and agility in software development and delivery.

Applitools

Applitools primarily utilizes computer vision and image processing techniques for its visual testing capabilities, rather than specifically leveraging LLMs like GPT

[16]. While LLMs are adept at NLP tasks, such as understanding text and generating human-like responses, computer vision techniques are better suited for analyzing and comparing visual elements within graphical UIs.

Here is how Applitools is playing a significant role in the SDLC:

Automated Visual Testing

Applitools enables automated visual testing of web and mobile applications across different browsers, devices, and screen resolutions. By automatically detecting visual differences between expected and actual application screenshots, Applitools helps identify UI defects, layout issues, and styling inconsistencies early in the development process.

Accelerated Testing Cycles

Applitools streamlines the testing process by automating visual validation tasks, reducing the time and effort required for manual testing. By providing rapid feedback on UI changes and regressions, Applitools accelerates testing cycles, enabling faster release cycles and time-to-market for software products.

Improved Test Coverage

Applitools enhances test coverage by focusing on visual aspects of applications, complementing traditional functional testing approaches. By verifying the visual correctness and consistency of UI components, Applitools helps ensure comprehensive test coverage and a better UX for end-users.

Cross-Browser and Cross-Device Compatibility

Applitools facilitates testing across various browsers, devices, and screen sizes, ensuring consistent UXs across different platforms. By automatically capturing and comparing screenshots across multiple environments, Applitools helps identify compatibility issues and ensures application functionality across diverse user configurations.

Regression Testing Optimization

Applitools automates regression testing by comparing baseline and current versions of application UIs, detecting visual changes and regressions introduced by code changes. By prioritizing visual discrepancies and highlighting areas of concern, Applitools helps teams focus their efforts on critical issues, reducing time spent on manual regression testing.

Integration with CI/CD Pipelines

Applitools integrates seamlessly with CI/CD pipelines, enabling automated visual validation as part of the build and deployment process. By automatically triggering visual tests and providing immediate feedback on code changes, Applitools helps teams maintain code quality and stability throughout the development lifecycle.

Collaborative Testing and Reporting

Applitools facilitates collaboration among development, testing, and design teams by providing centralized test management and reporting capabilities. By enabling stakeholders to view test results, track issues, and collaborate on resolution efforts, Applitools fosters communication and transparency, leading to more efficient bug resolution and improved software quality.

By ensuring visual correctness and consistency across applications, Applitools helps deliver high-quality software products that meet user expectations and drive business success. Other tools worth mentioning are AccelQ, Functionalize, Aqua ALdM, Sauce Labs, and TestComplete.

2.4 Deployment—Deployed Software, User Manuals, and Training Materials

AI/ML can significantly improve efficiency and effectiveness throughout the SDLC. By leveraging AI/ML, software engineers can streamline the deployment process, ensure the smooth operation of deployed software, and deliver exceptional UXs through personalized user manuals and training materials. Below, we present a breakdown of its contributions from a software engineering perspective:

Deployed Software

Predictive Anomaly Detection and Proactive Rollbacks

AI can analyze application logs and system metrics in real time to identify potential performance issues or bugs that might arise after deployment. This allows engineers to proactively detect anomalies and initiate rollbacks if necessary, minimizing downtime and impact on users. ML models can be trained on historical deployment data to identify patterns that indicate performance regressions or errors, enabling engineers to predict potential issues and trigger automated rollbacks before they significantly impact users.

AI-Powered A/B Testing and Feature Rollouts

A/B testing for new features after deployment is crucial. AI can automate the A/B testing process and analyze results in real time. This allows engineers to identify the optimal feature variations and roll them out to the entire user base with greater confidence. AI-Powered A/B testing tools can analyze user interactions and engagement metrics within the A/B groups, providing insights beyond conversion rates and enabling engineers to identify feature variations that offer the most value to users.

AI-Powered User Manual Generation

Generating user manuals can be time-consuming. AI can analyze application functionality and user interactions to automatically generate draft user manuals. This provides a starting point for technical writers and saves development time. ML models can be trained on existing user manuals and documentation to learn the structure and language used to explain functionalities. Using this knowledge, they can generate documentation outlines and draft content for new features or functionalities.

Context-Aware and Personalized Training Materials

AI can personalize training materials based on user roles and skill levels. This ensures that users receive the information they need most effectively. ML models can analyze user behavior data and interaction patterns to identify knowledge gaps and tailor training materials accordingly. This allows engineers to create targeted training modules or suggest relevant documentation sections based on specific user needs and actions within the application.

Below we discuss some real-world examples of AI/ML tools used for software deployment:

Rollbar

This platform leverages AI to analyze application logs and identify potential issues after deployment. It provides real-time insights and helps engineers pinpoint the root cause of errors, enabling faster troubleshooting and resolution.

Pendo

This user onboarding platform utilizes AI to personalize the UX by guiding users through relevant features based on their actions and needs. This helps users learn the software quickly and efficiently.

2.5 Survey and Results

A survey was conducted among software engineering teams and IT leaders regarding the usage of generative AI and the challenges associated with it [10]. The following data were curated from the Gartner Peer Community, highlighting the common use cases and challenges associated with generative AI. Of all the software departments, 60% utilize generative AI for AI-assisted pair programming, 52% use it for code generation, 45% for document generation, 38% for productivity (generating meeting summaries, etc.), 30% for test data generation, and 30% for technical document generation.

The survey on challenges revealed that 66%, 43%, and 38% of the respondents cited unfavorable outcomes, sufficient corporate governance policies, and resistance from the leadership, respectively; 35%, 32%, and 31% identified security concerns, ethical concerns, and the absence of practical use cases, respectively [10].

In general, respondents thought AI will have a good effect on software engineering. Out of the 110 respondents, 23% were very positive, while 1% felt somewhat negative. However, a large majority, 70%, felt somewhat positive, while 6% were neutral about it.

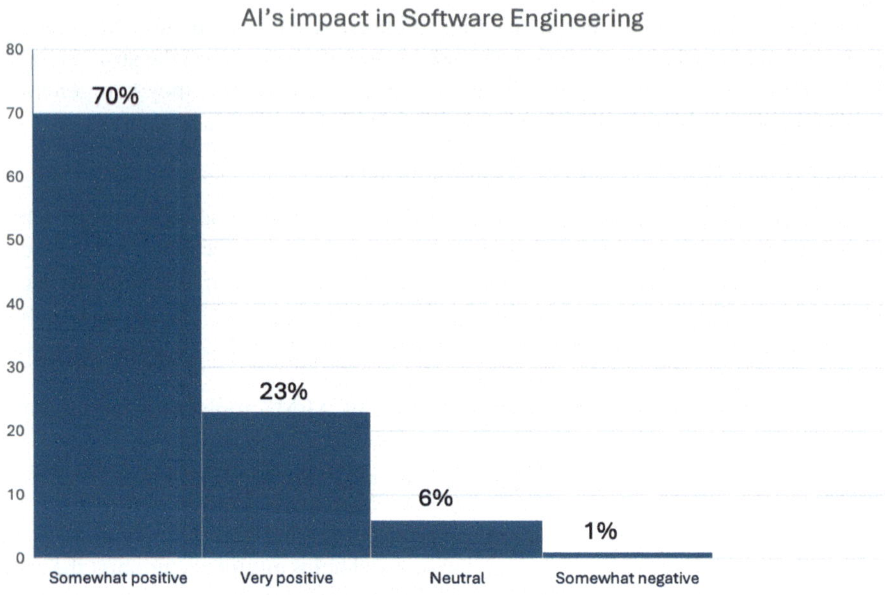

Fig. 2.2 Impact of AI in software engineering

The survey respondents comprised a mix of C-suite employees (10%), VPs (41%), Directors (31%), and Managers (18%), of which 63% are from North America (Fig. 2.2).

2.5.1 Lessons Learned

Techniques used in generative AI represent artifacts from data; this information is then used to create entirely new and distinct artifacts that mimic yet differ from the original data. While many software engineering leaders are using it for code generation, pair programming, code document generation, and more, others are concerned with undesirable and biased results. Leaders in software engineering may learn how generative AI can benefit the field and provide some challenges [10].

2.6 Threats to Validity

When the validity of this study is considered, several potential threats may be seen as affecting the robustness and reliability of the findings such as internal validity, external validity, construct validity, and conclusion validity.

2.6.1 Internal Validity

Selection Bias: While our research was predominantly focused on the success stories, there was a risk of bias in underreporting failures or negative outcomes.

Confounding Variables: Organizational culture, team composition, and resource accessibility are a few factors that may affect how well AI and ML tools work in the SDLC. These variables may confound results, making it hard to attribute improvements solely to AI and ML integration.

2.6.2 External Validity

Generalizability: The case studies under examination were taken from particular organizational and industry contexts, which may restrict the applicability of study findings to other industries, project sizes, and team configurations.

Technological Evolution: Rapid advancements in AI and ML may render the tools and methodologies evaluated in this study outdated quickly, affecting the relevance and longevity of the conclusions.

2.6.3 Construct Validity

Measurement of Efficiency and Productivity: The metrics used might not fully capture the complexities of processes in the SDLC. Subjective measures such as perceived productivity gains or user satisfaction could introduce variability and bias.

Tool Selection: The specific AI and ML tools chosen for evaluation may not represent the full spectrum of available technologies. This selective focus could result in an incomplete understanding of the potential benefits and limitations of AI and ML in the SDLC.

2.6.4 Conclusion Validity

Causal Inferences: Establishing causal relationships between AI and ML integration and improvements in software development is challenging. Correlational findings do not necessarily imply causation, and the study may overlook intervening variables.

Data Reliability: Relying on secondary data sources and publicly available information for case studies introduces potential threats to data reliability, as inaccuracies, omissions, or biases in the original sources could impact the validity of the findings.

Addressing these threats to validity is critical for enhancing the credibility of future research.

2.7 Conclusion

This paper highlights the transformative potential of AI and ML within the SDLC. Through automation and optimization, AI and ML technologies have significantly enhanced various stages of SDLC, leading to faster delivery times, improved software quality, and increased operational efficiency. From planning and requirements gathering to design, implementation, testing, and deployment, AI tools such as TabNine, OpenAI's Codex, DeepCode, Mabl, and Applitools have demonstrated substantial benefits by reducing manual effort and enabling smarter decision-making.

The integration of AI in planning phases through tools such as IBM Engineering Requirements Management DOORS Next has streamlined requirement analysis by leveraging NLP. In the design stages, AI has facilitated architectural assessments and system simulations to efficiently predict performance outcomes. Implementation phases have been revolutionized by intelligent code generation and real-time assistance from platforms such as TabNine and OpenAI's Codex. Automated testing tools such as Mabl have enhanced test coverage while ensuring rapid feedback loops.

Moreover, real-world case studies underscore the practical applications of these technologies for improving resource utilization and optimizing workflows across development teams. Companies that have effectively employed AI solutions report notable gains in productivity, cost-efficiency, and overall software performance.

However, some challenges still remain. Security risks associated with AI usage need mitigation through robust governance policies. Ethical concerns regarding data privacy necessitate vigilant oversight to prevent unintended consequences. Future advancements should focus on enhancing contextual awareness in AI systems for better adaptive learning capabilities.

Addressing these challenges will ensure that organizations can fully harness the potential of AI/ML in software development while maintaining ethical standards and security protocols. Ultimately, the strategic adoption of AI/ML within SDLC signifies a significant leap toward more innovative, efficient, and resilient software engineering practices.

By continually evolving these technologies to meet emerging needs and incorporating comprehensive safeguards, the software industry can look forward to a future defined by enhanced capabilities and unprecedented growth.

2.8 Future Recommendations

As we look to the future of software development, the integration of AI and ML offers unprecedented opportunities for enhancing efficiency, productivity, and innovation. By addressing key areas such as expanding AI training datasets, integrating AI into project management, and understanding industry-specific tradeoffs, we can maximize the benefits of these technologies while mitigating potential challenges.

Here are some strategic recommendations to guide this integration and ensure the most effective use of AI and ML in software development.

Expand AI Training Datasets: Expanding AI training datasets is essential for improving model accuracy and applicability. However, this must be coupled with stringent guardrails and monitoring mechanisms to ensure data quality, mitigate biases, and maintain data security.

AI in Project Management: Integrating AI into project management represents significantly more advanced applications beyond traditional SDLC processes. AI will enable project managers to make more informed decisions, anticipate challenges, and ensure smoother project execution, leading to more efficient management of software development projects.

Tradeoffs between Different Industries: Certain industries will benefit with the advancement of AI but some may not. Industry-specific analyses are required to help develop tailored strategies to maximize the benefits of AI. Industry collaborations should be encouraged to further drive the adoption of AI best practices. This will ensure increased productivity across industries and benefit everyone involved.

References

1. N.B. Ruparelia, Software development lifecycle models. ACM SIGSOFT Softw. Eng. Notes **35**(3), 8–13 (2010). https://doi.org/10.1145/1764810.1764814
2. E. Murphy-Hill, A.P. Black, Refactoring tools: fitness for purpose. IEEE Softw. **25**(5), 38–44 (2008). https://doi.org/10.1109/ms.2008.123
3. M. Fowler, K. Beck, J. Brant, W. Opdyke, D. Roberts, *Refactoring: Improving the Design of Existing Code* (Addison-Wesley Professional, n.d.)
4. R. Kneuper, Sixty years of software development life cycle models. IEEE Ann. Hist. Comput. **39**(3), 41–54 (2017). https://doi.org/10.1109/mahc.2017.3481346
5. F. Ribeiro, J.N.C. De Macedo, K. Tsushima, R. Abreu, J. Saraiva, GPT-3-powered type error debugging: investigating the use of large language models for code repair, in *Proceedings of the 16th ACM SIGPLAN International Conference on Software Language Engineering* (2023b), pp. 111–124. https://doi.org/10.1145/3623476.3623522
6. M.K. Sharma, A study of SDLC to develop well engineered software. Int. J. Adv. Res. Comput. Sci. **8**(3), 520 (n.d.)
7. A. Bashir, N. Latif, A.M. Khan, M. Sohail, Smart cities paradigm with AI-enabled effective requirements engineering. Int. Conf. Emerg. Technol. **14**, 176–182 (2023). https://doi.org/10.1109/icet59753.2023.10375008
8. M. Muszynski, E. Fosse, A. Plave, G. Pyrzak, Flight software dictionary development for the Mars2020 rover, in *2022 IEEE Aerospace Conference (AERO)*, vol. 239 (2022), pp. 1–14. https://doi.org/10.1109/aero53065.2022.9843621
9. *Jack in the Box Case Study | Specright*. (online). https://specright.com/case-studies/jack-in-the-box-case-study
10. *Generative AI for Software Engineering Teams*. (2023, August 3). Gartner. Retrieved July 5, 2024, from https://www.gatner.com/docment/4599599?ref=solrAll&refval=406177457
11. *Tabnine Docs*. (2024). https://docs.tabnine.com/main
12. *Tabnine: AI Chat & Autocomplete for JavaScript, Python, Typescript, Java, PHP, Go, and more - Visual Studio Marketplace*. (2024). https://marketplace.visualstudio.com/items?itemName=TabNine.tabnine-vscode

13. *OpenAI Codex: Advancing Code Generation with AI*. (n.d.). OpenAI. Retrieved July 30, 2024, from https://openai.com
14. S *DeepCode AI: AI-powered code analysis for secure development*. (n.d.). Snyk. Retrieved July 26, 2024, from https://snyk.io
15. *MABL: AI-powered test automation for continuous Delivery*. (n.d.). Mabl. Retrieved July 26, 2024, from https://www.mabl.com
16. *AppliTools: Revolutionizing visual testing and monitoring with AI*. (n.d.). Applitools. Retrieved August 1, 2024, from https://applitools.com
17. Y. Zhang, S. Huang, W. Wang, Q. He, L. Zhang, AI-driven microservice dependency analysis for improved system performance and maintainability. IEEE Trans. Software Eng. **46**(10), 1101–1112 (2020)
18. X. Luo, Y. Wu, L. Zhang, X. Ma, AI-driven load forecasting and dynamic resource allocation in cloud environments. Futur. Gener. Comput. Syst. **114**, 463–473 (2021)
19. M. Shahin, M.A. Babar, L. Zhu, Continuous architecture in the age of DevOps and AI: designing scalable and adaptive software systems. IEEE Softw. **34**(2), 28–35 (2017)
20. C. Riley, D. Zowghi, J. Grundy, Machine learning-based automated refactoring for enhanced software maintainability. ACM Trans. Softw. Eng. Meth-Odology (TOSEM) **27**(3), 1–38 (2018)
21. Dynamic scaling of cloud applications using machine learning in AWS. J. Cloud Comput. Mach. Learn. **12**(3), 45–60 (n.d.). https://www.researchgate.net
22. S. Caron, M. Marie, Anomaly detection in system log data using machine learning techniques. J. Mach. Learn. Syst. Monit. **15**(3), 223–239 (n.d.)
23. S. Johnston, R. Hilderman, E. Milios, A study on advanced AI techniques for automated code Generation. IEEE Trans. Softw. Eng. **43**(10), 42–53 (n.d.). https://doi.org/10.1109/TSE.2017.7997723
24. A. Unuriode, O. Durojaiye, B. Yusuf, L. Okunade, The integration of artificial I intelligence Into D database systems (Ai—Db integration review). SSRN Electron. J. (2024). https://doi.org/10.2139/ssrn.4744549
25. R. Riggs, *Predictive User Interface Design: Machine Learning Techniques in Web Development*. (2023). Retrieved June 1, 2024, from https://codeconspirators.com/predictive-user-interface-design-ml-techniques-in-web-development
26. J. Zhou, L. Zhang, Research on the optimization of a/b testing system based on dynamic strategy. Distribution **12**(3), 345–360 (2021). https://doi.org/10.3390/info12080345
27. R. Kohavi, A. Deng, B. Frasca, T. Walker, Y. Xu, N. Pohlmann, Online controlled experiments at large scale, in *KDD 2013. The 19th ACM SIGKDD International Conference on Knowledge Discovery and Data Mining*. 2013, IL, Chicago, United States of America (2013). https://bit.ly/ExPScale
28. C. Wang, J. Thompson, B. Lee, Data formulator: AI-powered concept-driven visualization authoring. IEEE Trans. Visual Comput. Graphics **30**(1), 1128–1138 (2024). https://doi.org/10.1109/tvcg.2023.3326585

Chapter 3
Revolutionizing Software Development: The Transformative Influence of Machine Learning Integrated SDLC Model

Hitesh Mohapatra, Subhadip Pramanik, and Soumya Ranjan Mishra

Abstract This research paper investigates the transformative influence of machine learning (ML) on software development life cycle (SDLC) models, illustrating a profound shift from traditional practices. Conventional SDLC models, characterized by their sequential and phase-driven nature, typically involve distinct stages such as requirements analysis, design, implementation, testing, and maintenance. The integration of ML into these models introduces a more dynamic and iterative approach, where machine learning algorithms continuously refine and optimize the development process based on real-time data and feedback. This integration enhances predictive accuracy, automates decision-making processes, and fosters adaptive system performance, addressing the complexities of modern software requirements. The paper concludes that the incorporation of ML into SDLC models not only revolutionizes software development practices but also significantly improves efficiency, responsiveness, and innovation, marking a critical evolution in meeting contemporary technological demands.

Keywords SDLC · Machine learning · Software development · Ethics · Accuracy · Efficiency · ML-SDLC

3.1 Introduction

The advent of Artificial Intelligence (AI) has brought about a paradigm shift in various sectors, and software development is no exception. The traditional Software Development Life Cycle (SDLC) models, which have been the backbone of software engineering for decades, are now being transformed by AI's generative capabilities. This transformation is not just a minor adjustment, but a fundamental change in how

H. Mohapatra · S. Pramanik · S. Ranjan Mishra (✉)
School of Computer Engineering, KIIT Deemed to be University, Bhubaneswar Odisha, India
e-mail: soumyaranjanmishra.in@gmail.com

S. Pramanik
e-mail: subhadip.pramanikfcs@kiit.ac.in

T. R. Benala et al. (eds.), *Boosting Software Development Using Machine Learning*, Artificial Intelligence-Enhanced Software and Systems Engineering 7,
https://doi.org/10.1007/978-3-031-88188-6_3

software is developed [1]. AI's generative capabilities are enabling developers to automate tasks that were previously manual and time-consuming. For instance, AI can generate code based on high-level descriptions, reducing the need for detailed specifications and manual coding. This not only speeds up the development process but also reduces the likelihood of human error. Moreover, AI can learn from past projects and use this knowledge to improve future ones. It can analyze historical data to identify patterns and trends, predict potential issues, and suggest optimal solutions. This ability to learn and adapt makes AI a powerful tool for improving the quality and efficiency of software development. Furthermore, AI's impact extends beyond the technical aspects of software development. It is also changing the way teams work and collaborate [2]. With AI handling more of the technical work, developers can focus more on creative and strategic tasks. This shift is leading to more innovative solutions and a more fulfilling work experience for developers. AI is transforming traditional SDLC models, AI is making software development faster, more efficient, and more innovative.

The transformative influence of AI on SDLC models is indeed multi-faceted. It not only automates mundane tasks but also aids in complex decision-making processes. For instance, AI can generate code snippets, suggest optimal algorithms, detect bugs, and even predict future issues based on historical data. AI's ability to automate mundane tasks is a significant boon for software developers [3]. Tasks such as code generation, which were once time-consuming and prone to human error, can now be automated with AI. This automation not only speeds up the development process but also improves the quality of the software by reducing the likelihood of human error. In addition to automating tasks, AI also aids in complex decision-making processes [4]. By analyzing large amounts of data, AI can provide insights that would be difficult, if not impossible, for humans to discern. These insights can guide developers in choosing the most effective algorithms, optimizing code, and making other critical decisions [5]. One of the most impressive aspects of AI's transformative influence on SDLC models is its ability to generate code snippets. Given a high-level description of a function, AI can generate a corresponding code snippet, effectively turning a human-readable description into executable code. This capability can significantly speed up the coding process and allow developers to focus on more complex and creative aspects of software development [6]. AI's ability to detect bugs is another significant advantage. By analyzing code, AI can identify potential bugs that might be overlooked by human developers. This early detection can save considerable time and effort in the debugging process. Perhaps one of the most promising aspects of AI's influence on SDLC models is its ability to predict future issues based on historical data. By analyzing past projects and their associated issues, AI can identify patterns and trends that can help predict potential problems in future projects. This predictive capability can enable proactive problem-solving and improve the overall quality of the software [7].

Despite the promising prospects, the integration of AI into SDLC models also presents challenges. Concerns regarding the transparency of AI decisions, the need for human oversight, and the potential for job displacement are some of the issues that need to be addressed. However, the opportunities for innovation and growth far outweigh these challenges. The integration of AI into SDLC models is not without its hurdles [8]. One of the primary concerns is the transparency of AI decisions. AI algorithms, particularly those based on machine learning, can be complex and difficult to understand. This lack of transparency, often referred to as the "black box" problem, can make it challenging to understand why an AI system made a particular decision [9]. This can be particularly problematic in situations where the AI's decision has significant implications, such as identifying bugs or suggesting changes in code [10]. Another challenge is the need for human oversight. While AI can automate many aspects of the SDLC, it is not infallible. There is still a need for human oversight to ensure that the AI is functioning correctly and making appropriate decisions. This need for oversight can add complexity to the SDLC and require additional resources [11].

The potential for job displacement is also a concern. As AI becomes more capable, there is a fear that it could replace human developers. However, it is important to note that while AI can automate certain tasks, it cannot replace the creativity, critical thinking, and problem-solving abilities of human developers [12]. Instead, AI is likely to change the nature of software development jobs, with developers spending less time on mundane tasks and more time on higher-level, creative tasks. Despite these challenges, the integration of AI into SDLC models offers significant opportunities for innovation and growth. AI can automate and optimize many aspects of the SDLC, leading to increased efficiency and productivity [13]. It can also provide valuable insights and predictions, enabling proactive problem-solving and continuous improvement. Furthermore, by freeing developers from mundane tasks, AI allows them to focus on more creative and strategic aspects of software development, leading to more innovative solutions [14].

3.2 Literature Review

The Software Development Lifecycle (SDLC) has been a guiding framework for project delivery in the tech industry, evolving over time in response to changes in project scale and complexity. Agile and other methodologies have reduced the time spent on each stage by focusing on iterative development and quick delivery of incremental software [15]. The rise of an AI ecosystem, particularly generative AI technologies like ChatGPT, has raised questions about how these advancements will transform the software industry and the SDLC itself, and what impact this will have on the work of tech professionals worldwide. This paper [1] proposes a new SDLC model, anticipating that the stages of the SDLC will undergo a fundamental shift due to recent advances in generative AI [16]. Challenges exist, particularly due to minimal interaction between construction and design teams, leading to issues like production delays and rework. The paper explores how Artificial Intelligence (AI) techniques can automate processes to overcome these challenges, and identifies areas still open for further research [17]. This paper presents a deep learning-based approach to automate the selection of Software Development Life Cycle (SDLC) models, aiming to streamline the software development process by recommending the most appropriate model based on user requirements [18]. The approach utilizes an Artificial Neural Network (ANN) to predict the relevance of four SDLC models-Waterfall, Incremental, Evolutionary, and Hybrid-to a given project based on various parameters. The proposed system employs an ANN with four layers: input, hidden, dropout, and output. The input layer receives eight parameters related to the project, which are then processed through the hidden and dropout layers before the output layer predicts the relevance of each SDLC model [19].

The system was trained and validated using a dataset generated based on guidelines from previous research, with 6000 rows of data distributed across the four models [20]. This paper [21] presents a unified framework, MLASDLC (Machine Learning Application Software Development Life Cycle), aimed at facilitating the planning, development, and deployment of machine learning applications through parallel processes for software and machine learning engineering. The framework integrates concepts from standard software development life cycle methodologies (SDLC), development operations (DevOps), and machine learning operations (MLOps) to address challenges in machine learning production, such as data collection and preparation, model entanglement, and technical debt [22]. The MLASDLC framework is designed to transition from coding-centric development cycles to data-centric engineering, facilitating the continuous agile development, training, and deployment of machine learning application components in parallel. The framework incorporates agile methodologies, DevOps practices, and MLOps principles to manage the iterative and non-linear nature of machine learning model development and to automate key steps in the development process [23].

SDLC models have evolved over decades to improve software development efficiency and effectiveness. The traditional Waterfall model introduced a linear and sequential approach, which was later enhanced by iterative models like Spiral and V-Model. The Agile methodology brought flexibility and customer-centric development, while DevOps integrated development and operations for continuous delivery and improvement [24]. AI has been increasingly integrated into software development processes to automate repetitive tasks, enhance decision-making, and improve code quality. Early applications of AI focused on automated code generation, bug detection, and project management tools [25]. Recent advancements have extended AI's role into requirement analysis, design optimization, and predictive maintenance [26]. Generative AI models, such as Generative Adversarial Networks (GANs) and Generative Pre-trained Transformers (GPT), have demonstrated capabilities in creating new, coherent, and contextually relevant content. These models have been applied in various domains, including natural language processing, image generation, and software code synthesis [27].

The Software Development Lifecycle (SDLC) has undergone significant transformations over the years to address the growing demands of technology projects. The shift from traditional, linear approaches like the Waterfall model to more iterative and flexible methodologies such as Agile and DevOps reflects the industry's need for faster and more adaptive development processes [28]. Agile methodologies, in particular, emphasize iterative development and continuous feedback, which have proven to be effective in managing complex projects and delivering incremental value to stakeholders. The rise of Artificial Intelligence (AI) has introduced new dimensions to the SDLC, influencing both its structure and execution. Generative AI technologies, such as Generative Adversarial Networks (GANs) and Generative Pretrained Transformers (GPT), are at the forefront of this transformation. These models excel in creating new, contextually relevant content, which extends their applicability beyond traditional tasks to more complex functions like code generation and design optimization. As AI continues to evolve, its role in the SDLC is becoming more integral, driving changes in how software development processes are approached and managed [29].

One significant area of impact is the automation of repetitive and mundane tasks. AI-driven tools are now capable of automating code generation, bug detection, and even project management tasks. This automation not only speeds up the development process but also reduces the likelihood of human error, leading to higher quality software. Furthermore, AI's capability to analyze and process large amounts of data allows for more informed decision-making, enhancing the overall efficiency of the SDLC [30]. However, integrating AI into the SDLC is not without its challenges. Ensuring data security and managing potential biases in AI algorithms are critical concerns that must be addressed. The reliance on AI tools also necessitates a shift in skill sets among developers, who must now become adept at using these advanced technologies. Additionally, maintaining transparency in automated processes is essential to build trust and ensure that AI-driven decisions align with project goals and ethical standards [31].

The introduction of intelligent agents within the SDLC represents a notable advancement. These agents can perform tasks autonomously, such as continuous testing and code review, providing real-time feedback that enhances development speed and quality. The potential for intelligent agents to drive the development process further underscores the need for ongoing research into their capabilities and limitations. Recent studies suggest that AI's role in the SDLC is likely to expand, with future research focusing on optimizing AI integration and addressing associated challenges. Exploring new methodologies and frameworks that leverage AI can lead to more effective and efficient development processes. Additionally, understanding the implications of AI on team dynamics, project management, and overall productivity will be crucial as the technology continues to advance [32].

The evolution of the SDLC, influenced by advancements in AI, reflects a broader trend towards more adaptive and efficient software development practices. As AI technologies become more sophisticated, their integration into the SDLC will likely continue to reshape the field, offering both opportunities and challenges for tech professionals. Future research and development will play a critical role in harnessing the full potential of AI while addressing the complexities and concerns that arise from its use.

3.3 Proposed Study

3.3.1 ML Integrated Planning Phase

Integrating machine learning into the planning phase of the Software Development Life Cycle (SDLC) can greatly enhance project outcomes. Machine learning algorithms can leverage historical data to predict project timelines, costs, and potential risks, leading to more accurate planning [28]. They optimize resource allocation by analyzing past projects to ensure efficient use of resources. Additionally, ML can identify and forecast potential risks, enabling proactive risk management. It can also assist in analyzing user requirements by processing large volumes of feedback, improving the clarity of project goals. For cost estimation, machine learning models can identify patterns from previous projects to create more reliable budget forecasts [29]. Lastly, ML can optimize project schedules by analyzing task dependencies and resource availability, resulting in more achievable timelines. Overall, incorporating machine learning into the planning phase helps in making data-driven decisions and reducing uncertainty, ultimately improving project planning and execution [30].

Planning with traditional SDLC and planning with an ML-integrated SDLC differ significantly in their approach and effectiveness. Traditional SDLC relies on historical data and expert judgment, which can be less precise and prone to inefficiencies. In contrast, an ML-integrated SDLC utilizes machine learning algorithms to analyze extensive data sets, offering more accurate predictions and insights [31]. For risk management, ML can detect complex patterns and forecast potential issues more effectively than manual assessments. Resource allocation and cost estimation benefit from ML's ability to identify usage patterns and trends, leading to more efficient management and reliable forecasts. Additionally, ML models optimize project schedules by analyzing task dependencies and historical performance, resulting in more realistic timelines [32]. Requirement analysis also improves as ML processes large volumes of data to identify key requirements and gaps. Overall, ML integration enhances planning accuracy and efficiency by leveraging data-driven insights and advanced analytics. Table 3.1 highlights the key differences between conventional and ML-based planning approaches, showing how machine learning enhances various aspects of project planning [33].

A quantitative analysis comparing conventional planning and ML-based planning in the SDLC highlights notable differences in efficiency, accuracy, and effectiveness. Conventional planning often has accuracy varying by 20–30% of actual outcomes due to manual estimates and historical data, whereas ML-based planning improves accuracy to within 5–10% by leveraging advanced data analysis and pattern recognition [34]. Planning duration with conventional methods can span several weeks to months, but ML tools can reduce this time by 30–50%, automating data analysis and providing quicker insights. Risk identification in conventional planning may miss 15–20% of potential risks, while ML models can detect 25–40% more risks through complex data analysis, enhancing risk mitigation strategies [35]. Resource

Table 3.1 Qualitative analysis between conventional planning phase with SDLC and ML integrated SDLC

Metric	Conventional planning	ML-based planning
Accuracy of estimates	±(20–30)% of actual outcomes	±(5–10)% of actual outcomes
Time to complete planning	Several weeks to months	Reduced by (30–50)% (faster completion)
Risk identification	Misses (15–20)% of potential risks	Identifies (25–40)% more risks
Resource allocation efficiency	˜(70–80)% efficiency	(85–95)% efficiency
Cost estimation accuracy	±(15–20)% of actual costs	±(5–10)% of actual costs
Schedule adherence	Overruns of (10–20)%	Overruns reduced to (5–10)%

allocation efficiency with conventional planning is around 70–80%, whereas ML-based approaches improve this to 85–95% by optimizing based on detailed data. Cost estimates in conventional planning might be accurate within 15–20% of actual costs, but ML models refine accuracy to within 5–10% by analyzing financial data and trends [36]. Finally, conventional planning may result in schedule overruns of 10–20%, whereas ML-based planning reduces overruns to 5–10% by providing better timelines and predicting delays. Overall, ML-based planning offers significant improvements in accuracy, efficiency, and effectiveness, leading to more successful project outcomes compared to conventional methods [37].

3.3.1.1 ML Integrated Requirements Analysis Phase

In conventional requirement analysis within the Software Development Life Cycle (SDLC), the process involves systematically gathering, documenting, and analyzing user needs and system requirements. This phase typically begins with identifying stakeholders and conducting interviews, surveys, or workshops to collect their input [38]. Requirements are then documented in detail, often using techniques such as use cases, user stories, and functional specifications. Analysts review and refine these requirements through iterative discussions and validations with stakeholders to ensure clarity and completeness [39]. This manual approach, while thorough, can be time-consuming and prone to errors or omissions due to its reliance on human judgment and the potential for miscommunication. The focus is on understanding user needs and translating them into a structured format that guides subsequent design and development phases. Table 3.2 presents the ML-integrated requirement analysis offers significant advantages over traditional methods by enhancing accuracy, efficiency, and adaptability [40].

In an ML-integrated requirement analysis phase of the Software Development Life Cycle (SDLC), machine learning algorithms enhance the process by analyzing

Table 3.2 Qualitative analysis between conventional requirement analysis and ML integrated analysis

Aspect	Traditional requirement analysis	ML-integrated requirement analysis
Data handling	Manual collection and documentation of requirements through interviews, surveys, and workshops	Automated processing of large volumes of data using ML algorithms and Natural Language Processing (NLP)
Accuracy	Accuracy can vary due to human errors and subjective interpretations	Increased accuracy with ML models that analyze data patterns and trends, reducing human error
Efficiency	Time-consuming, often taking weeks to months to gather and document requirements	Faster and more efficient with ML tools that automate data extraction and analysis, reducing the time needed
Requirement extraction	Manual extraction from documents, meetings, and stakeholder inputs, which can be prone to missing details	Automated extraction using NLP and ML, leading to more comprehensive and complete requirement gathering
Risk identification	Risks are identified through manual assessment and historical knowledge, potentially missing subtle issues	ML models identify and predict risks by analyzing complex patterns in historical and current data, leading to better risk management
Stakeholder communication	Direct interaction with stakeholders is required, which can be inconsistent and time-consuming	ML can process stakeholder feedback more efficiently, ensuring that diverse inputs are considered and analyzed
Adaptability to changes	Changes in requirements can be challenging to incorporate once documented, requiring manual updates	ML models can adapt more dynamically to changes in requirements by continuously learning from new data and feedback
Data insights	Limited to the insights gathered through manual analysis and subjective interpretation	Provides deeper insights by analyzing large data sets and identifying patterns that may not be apparent manually
Cost	Higher cost due to labor-intensive processes and potential for errors leading to rework	Reduced costs through automation and improved accuracy, leading to fewer revisions and more efficient processes

large volumes of data to extract and refine requirements [41]. ML models can process historical project data, user feedback, and industry standards to identify patterns and trends, providing insights that may not be immediately apparent through manual analysis. Natural Language Processing (NLP) techniques can be used to automate the extraction of requirements from unstructured text, such as emails or documents, improving the efficiency and accuracy of capturing user needs [42]. Additionally, ML algorithms can continuously learn and adapt from ongoing inputs, enabling more dynamic and responsive requirement gathering. This approach not only accelerates the requirement analysis process but also enhances the quality of requirements by reducing human error and bias, leading to a more accurate and comprehensive understanding of project needs [43]. In the context of the Software Development Life Cycle (SDLC), ML-integrated requirement analysis presents several advantages over traditional methods. Traditional requirement analysis relies heavily on manual processes, involving direct interactions with stakeholders through interviews and surveys, and documenting requirements manually [44]. This approach can be time-consuming, often taking weeks or months, and may suffer from inaccuracies due to human error and subjective interpretation.

In contrast, ML-integrated requirement analysis automates the handling and processing of large volumes of data using advanced algorithms and Natural Language Processing (NLP). This automation enhances accuracy by minimizing human errors and speeds up the process significantly, reducing the time required for gathering and documenting requirements [45]. ML tools also excel at extracting requirements from diverse sources more comprehensively and dynamically adapting to changes, which can be challenging with traditional methods. Additionally, ML-based analysis provides deeper insights by recognizing patterns in data that might not be apparent through manual analysis, improving risk identification and management. Overall, ML-integrated requirement analysis reduces costs by streamlining processes, minimizing revisions, and enhancing overall efficiency compared to traditional approaches[46].

3.3.2 ML Integrated Design Phase

The design phase in the Software Development Life Cycle (SDLC) is a critical stage where the system's architecture and components are conceptualized and detailed. During this phase, the requirements gathered in the previous stages are translated into a structured blueprint that guides the development process[47]. This involves creating detailed design documents that outline the system's architecture, including hardware and software components, data flow, and user interfaces. Design considerations include ensuring scalability, performance, security, and integration with other systems. The phase typically results in design specifications such as wire-frames, data models, and interface designs, which serve as a road-map for developers [48].

Effective design is crucial for setting a solid foundation for the subsequent development and implementation stages, helping to ensure that the final product meets the desired requirements and functions as intended.

In the ML-integrated design phase of the Software Development Life Cycle (SDLC), machine learning enhances the design process by leveraging advanced data analysis and predictive capabilities. During this phase, ML algorithms can analyze historical project data, user feedback, and design patterns to inform and optimize design decisions [49]. Machine learning models help in identifying potential design issues and suggesting improvements by recognizing patterns and trends from previous projects. For example, ML can assist in creating more efficient system architectures by predicting performance bottlenecks and optimizing resource allocation [50]. Additionally, ML tools can automate the generation of design documentation, such as wire-frames and data models, based on historical design successes and user requirements. This approach not only accelerates the design process but also enhances the quality and accuracy of the design by incorporating data-driven insights and reducing the potential for human error. Overall, ML integration in the design phase ensures that the system's design is more robust, scalable, and aligned with user needs [51].

Table 3.3 highlights the differences between conventional and ML-integrated design phases in the Software Development Life Cycle (SDLC). Conventional design relies on manual techniques and human expertise, which can lead to variability in creativity and innovation, limited data handling, and potential inaccuracies due to human error and subjective judgment [52]. Risk management is typically based on manual assessments, which can miss certain issues. Design changes can be challenging and time-consuming, with documentation often being labor-intensive and prone to inconsistencies [53]. In contrast, ML-integrated design leverages machine learning algorithms and extensive data analytics to enhance accuracy and creativity. ML tools provide data-driven insights and innovative solutions, improving design accuracy and risk management by predicting potential issues. The flexibility of ML-based approaches allows for rapid adaptation to changes, and documentation is automated, ensuring consistency and accuracy. Overall, ML integration offers significant improvements in efficiency, accuracy, and flexibility compared to traditional design methods [54]. The numerical time differences between conventional design and ML-integrated design phases in the Software Development Life Cycle (SDLC) can vary depending on the complexity of the project and the specific tools used (Ref: Table 3.4).

3.3.3 ML Integrated Implementation (or Coding) Phase

The Implementation phase in the Software Development Life Cycle (SDLC) is where the actual coding of the software occurs. During this phase, developers write the source code based on the design specifications outlined in the previous phase. This involves translating design documents into a functional program, adhering to coding standards, and integrating various components and modules. The implementation

Table 3.3 Qualitative analysis between conventional design phase and ML integrated design phase

Aspect	Conventional design phase	ML-integrated design phase
Approach	Manual design techniques based on human expertise and standard practices	Uses machine learning algorithms and data analytics to guide design decisions
Creativity and innovation	Relies on designer's skills and creativity; involves brainstorming and iterative reviews	Enhances creativity with data-driven insights and innovative solutions suggested by ML tools
Data handling	Limited use of data analytics; decisions based on historical data and anecdotal evidence	Extensive use of data analytics; ML models process large volumes of data to inform design choices
Design accuracy	Potential for human error and oversight; decisions influenced by subjective judgment	Improved accuracy with data-driven insights; designs are more aligned with best practices and user needs
Risk management	Risks identified through manual assessment and experience; potential for missed issues	Enhanced risk management through ML predictions; identifies and mitigates risks based on data analysis
Flexibility	Design changes can be time-consuming and challenging; manual updates required	Increased flexibility with rapid iteration and adaptation; ML tools quickly adjust designs based on new data
Documentation	Manual creation and updating of design documentation; can be time-consuming and inconsistent	Automated generation and updating of design documentation; consistent and accurate records maintained

Table 3.4 Quantitative analysis

Aspect	Conventional design phase	ML-integrated design phase
Design time	4–12 weeks (depending on project size)	2–6 weeks (reduced by 30–50% with ML)
Time for design changes	1–3 weeks per change cycle	1–2 weeks per change cycle (faster with ML)
Documentation time	2–4 weeks (manual creation and updates)	1–2 weeks (automated generation)

phase is critical as it transforms design into a working system, and it often includes unit testing to ensure each component functions correctly. Collaboration among team members is essential to address any issues and to ensure that the code meets the project requirements and quality standards [27].

In the ML-integrated Implementation phase of the SDLC, the focus is on integrating machine learning models into the software system. This phase involves coding not only the application logic but also embedding the machine learning algorithms into the system's architecture. Developers work on implementing data pipelines for training and inference, ensuring that the model interfaces correctly with the application. Additionally, they must optimize the model for performance and scalability, handle model versioning, and address issues related to data preprocessing and feature engineering [33]. Effective testing is crucial, including both traditional software testing and evaluation of the model's accuracy and effectiveness. This phase requires collaboration between data scientists and software engineers to ensure that the machine learning components are seamlessly integrated and function as intended within the broader system. When comparing the traditional Implementation phase in the SDLC to an ML-integrated approach, several key differences emerge. In the traditional approach, the focus is primarily on coding the application logic based on predefined specifications, where the main challenge lies in ensuring that all software components work together seamlessly. In contrast, the ML-integrated approach involves additional complexities, such as embedding machine learning models, managing data pipelines, and ensuring that the models perform well in real-world conditions. Traditional coding emphasizes functional correctness and efficiency, whereas the ML-integrated approach also requires attention to data handling, model accuracy, and adaptability. Additionally, the ML approach necessitates close collaboration between data scientists and software engineers, while the traditional method mainly involves developers [36]. Thus, while both approaches share the goal of creating a functional and reliable system, the ML-integrated phase requires addressing the unique challenges of combining predictive models with software development. Table 3.5 highlights the distinct considerations and challenges inherent in each approach, underscoring the additional layers of complexity in ML-integrated implementation.

Table 3.5 compares traditional and ML-integrated Implementation phases in the SDLC, highlighting the key differences. Traditional implementation focuses on coding application logic based on predefined designs, with an emphasis on functional correctness and efficiency. In contrast, ML-integrated implementation involves embedding machine learning models, managing data pipelines, and optimizing performance, requiring collaboration between developers and data scientists. The ML approach adds complexity through the need for model evaluation, adaptability to evolving data, and balancing software performance with model accuracy, making it a more intricate process compared to traditional software development [29].

Table 3.5 Qualitative analysis between conventional implementation phase and ML integrated implementation phase

Aspect	Traditional implementation	ML-integrated implementation
Focus	Coding based on predefined design and specifications	Integrating machine learning models into the software system
Key activities	Writing application logic, ensuring functional correctness	Coding ML algorithms, managing data pipelines, and optimizing models
Complexity	Generally, less complex, with a focus on software functionality	Higher complexity due to data handling, model integration, and performance optimization
Collaboration	Primarily involves developers	Requires close collaboration between developers and data scientists
Testing	Emphasizes unit and integration testing for software components	Involves traditional testing plus model evaluation (e.g., accuracy, precision)
Adaptability	Changes are typically driven by feature updates or bug fixes	Must adapt to evolving data, requiring model retraining and updates
Performance	Concerns focuses on software efficiency and resource management	Balances software performance with model accuracy and inference speed
End-goal	Deliver a functional, reliable software product	Deliver a system that not only functions but also makes accurate predictions or decisions

3.3.4 ML Integrated Testing Phase

The conventional testing phase in the SDLC is akin to a quality check for the software. After the code has been developed, testers step in to ensure that everything functions as intended. This phase is comparable to building a piece of furniture and then testing it to confirm that it stands firm, all components fit together, and nothing is missing or malfunctioning. Testers meticulously search for bugs, glitches, or any issues that could cause the software to fail or behave unexpectedly. They employ various methods, such as running the software under different conditions, to assess its performance. If problems are identified, the software is sent back to the developers for corrections. This phase is critical in ensuring that the final product is reliable, secure, and meets its intended purpose before being released to users [26]. The ML-integrated testing phase in the SDLC goes beyond traditional software testing by focusing on both the software and the machine learning model embedded within it. In this phase, testers not only check that the software functions correctly, but also ensure that the machine learning model performs accurately and consistently. This is like making sure a self-driving car not only runs smoothly but also makes the right decisions on the road. Testers evaluate the model's predictions under various scenarios, checking for issues like bias or poor generalization to new data. They also monitor how the model interacts with the software, ensuring that any changes in data or code don't negatively impact performance. If any problems arise, the model might need retraining or adjustments, making this phase critical in delivering a system that is both reliable and intelligent [16]. Comparing the conventional testing phase with the ML-integrated testing phase reveals distinct differences in scope and complexity. Traditional testing focuses primarily on validating that the software functions as intended, identifying bugs, and ensuring stability under various conditions. In contrast, the ML-integrated testing phase adds an additional layer of complexity by requiring validation of both the software and the embedded machine learning model. This approach not only checks the software's functionality but also evaluates the model's accuracy, consistency, and decision-making capabilities. Testers must consider issues like data bias, model generalization, and how changes in data or code may impact the model's performance. As a result, while conventional testing aims to ensure that the software is reliable and functional, ML-integrated testing must also confirm that the system is intelligent and adaptive, requiring a broader and more nuanced approach to quality assurance. Table 3.6 compares the conventional testing phase with the ML-integrated testing phase in the SDLC, highlighting key differences in focus, complexity, and scope. Conventional testing primarily aims to ensure that the software functions correctly and is free of bugs, with moderate complexity focused on verifying software behavior under various conditions. Key activities include bug detection, performance testing, and user acceptance testing, using traditional testing tools like unit, integration, and system tests [54].

In contrast, ML-integrated testing adds significant complexity by not only testing the software but also evaluating the performance of the embedded machine learning model. This phase involves a broader scope, including data validation, model accuracy checks, and assessing how the model interacts with the software. It requires specialized methods for model evaluation, such as cross-validation and A/B testing, in addition to traditional testing approaches. Adaptability is also a critical concern, as the model may need retraining or adjustments based on new data or performance issues. The challenges in ML-integrated testing include addressing model bias, overfitting, and ensuring the model generalizes well to unseen data. The outcome of this phase is a system that is not only reliable and bug-free but also intelligent and capable of making accurate and fair predictions, highlighting the need for a more comprehensive testing approach compared to conventional methods. Table 3.6 presents the comparison of testing phase performance with and without ML in SDLC.

Table 3.6 Qualitative analysis between conventional testing phase and ML integrated testing phase

Aspect	Conventional testing	ML-based testing
Focus	Ensuring the software functions correctly and is free of bugs	Validating both software functionality and ML model performance
Complexity	Moderate; focuses on software behavior under various conditions	High; involves software testing plus evaluation of model accuracy, bias, and generalization
Key activities	Bug detection, performance testing, and user acceptance testing	Includes all conventional activities plus model evaluation, data validation, and retraining as needed
Testing scope	Primarily centered on software code and user interfaces	Expands to include data handling, model predictions, and interactions between software and the ML model
Adaptability	Software updates or fixes are based on identified bugs	Involves potential retraining of the model, adjustments to algorithms, and updates based on new data
Tools and methods	Traditional testing tools like unit, integration, and system tests	Combination of traditional tools with specialized methods for model testing, such as cross-validation and A/B testing
Challenges	Identifying hidden bugs, ensuring coverage across scenarios	Addressing issues like model bias, overfitting, and ensuring that the model adapts well to unseen data
Outcome	A reliable, bug-free software product	A reliable, intelligent system that not only functions correctly but also make accurate and fair predictions

3.3.5 ML Integrated Maintenance Phase

The conventional maintenance phase of the SDLC is where the software is looked after once it's been deployed and is in use. During this phase, the focus shifts to ensuring that the software continues to operate smoothly and meets users' needs over time. This involves fixing any bugs that weren't caught during testing, making necessary updates, and implementing new features as user requirements evolve. It's like regular upkeep for a car-addressing issues as they arise, performing routine checks, and making improvements to keep it running efficiently. The maintenance phase is crucial because it extends the software's life, ensures its reliability, and keeps it relevant as technology and user expectations change. The ML-integrated maintenance phase in the SDLC involves not just the upkeep of the software but also the ongoing care of the machine learning model embedded within it [32]. After the system is deployed, continuous monitoring is essential to ensure the model remains accurate and effective as new data comes in. This phase includes updating the software, retraining the model with fresh data, and adjusting algorithms to improve performance or address issues like model drift, where predictions may degrade over time. It's similar to maintaining a smart device-keeping the software up-to-date while also ensuring that the intelligent features, like voice recognition or recommendations, continue to work well. This phase is critical because it ensures that both the software and the ML model stay reliable, relevant, and capable of making accurate decisions as the environment and data evolve.

Comparing the conventional maintenance phase with the ML-integrated maintenance phase reveals key differences in scope and complexity. In conventional maintenance, the focus is primarily on fixing bugs, updating software, and adding new features to keep the system running smoothly and meeting user needs. It involves routine upkeep similar to maintaining any standard software product. On the other hand, the ML-integrated maintenance phase introduces additional challenges, as it requires not only software updates but also the ongoing management of the machine learning model [38]. This includes monitoring the model's performance, retraining it with new data, and making adjustments to prevent issues like model drift, where the model's accuracy might decline over time. While conventional maintenance ensures the software remains functional and relevant, ML-integrated maintenance also focuses on keeping the model accurate and adaptive, making it a more complex and dynamic process. Table 3.7 provides a comparative analysis of the conventional maintenance phase versus the ML-integrated maintenance phase in the SDLC, emphasizing their differences in focus, complexity, and approach. In conventional maintenance, the primary goal is to ensure the software remains functional and up-to-date through activities like bug fixing, updates, and feature enhancements. The complexity is moderate, as it mainly involves routine software upkeep and user feedback implementation. Monitoring is generally limited to software performance, and the tools used are standard maintenance tools. In contrast, the ML-integrated maintenance phase introduces additional complexity due to the dual focus on both software and the machine learning model. This phase involves not only standard maintenance tasks but also model retraining, data management, and continuous performance monitoring to address issues like model drift. The tools and methods used in this phase include specialized ML monitoring and retraining processes, which add to the overall complexity. The challenges are more intricate, as maintaining an accurate and adaptive ML model requires ongoing attention to data quality and model performance [42]. Ultimately, while conventional maintenance ensures a stable software product, ML-integrated maintenance aims to maintain a stable software system with an accurate and adaptive ML model, reflecting the increased demands of integrating machine learning into software systems. Table 3.7 presents the qualitative analysis between conventional maintenance phase and ML integrated maintenance phase.

Table 3.7 Qualitative analysis between conventional maintenance phase and ML integrated maintenance phase

Aspect	Conventional maintenance	ML-integrated Maintenance
Focus	Software updates, bug fixes, and feature enhancements	Software updates, model retraining, and performance monitoring
Complexity	Moderate; primarily involves software upkeep	High; involves both software maintenance and model management
Key activities	Bug fixing, routine updates, user feedback implementation	Bug fixing, model retraining, data management, and algorithm adjustments
Monitoring needs	Regular software performance checks	Continuous monitoring of both software and model accuracy
Adaptability	Updating based on user needs and software requirements	Adapting to new data, model drift, and evolving prediction needs
Tools and methods	Standard maintenance tools and software management practices	Combination of software tools and specialized ML monitoring and retraining processes
Challenges	Keeping the software functional and up-to-date	Preventing model degradation, managing data quality, and ensuring ongoing model relevance
Outcome	A stable, up-to-date software product	A stable software system with an accurate, adaptive ML model

3.4 Analysis of the Proposed Study

The planning stage of a conventional Waterfall Software Development Life Cycle (SDLC) includes Project Initiation, Project Scope Definition, Resource Allocation, Timeline and Milestones, Budgeting and Cost Estimation, Risk Management Planning, Quality Assurance Planning, Project Schedule Development, and Project Plan Finalization (Ref: Fig. 3.1).

Whereas, (ML)-based Software Development Life Cycle (SDLC) involves Project Initiation, Project Scope Definition, Resource Allocation, Timeline and Milestones, Budgeting and Cost Estimation, Risk Management, Quality Assurance Planning, Data Management Planning, Model Development Planning, Evaluation Metrics Definition, and Project Plan Finalization in its planning phase. Fig. 3.1 shows the time is reduced by transferring the knowledge of some components from similar ML-based SDLC in the same domain/task (Ref: Fig. 3.2).

The requirement analysis stage in conventional SDLC includes Initial Requirements Gathering, Requirements Elicitation, Requirements Analysis, Requirements

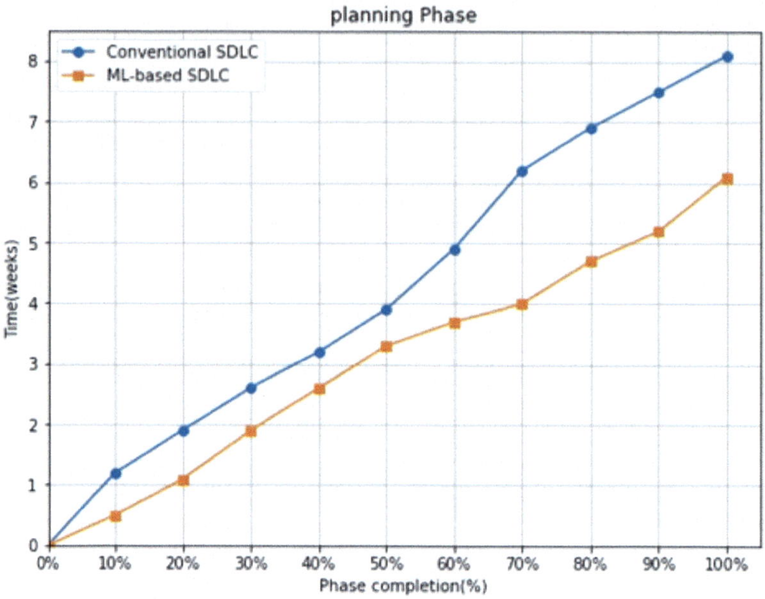

Fig. 3.1 Comparative analysis of planning phase performance between SDLC and ML-SDLC

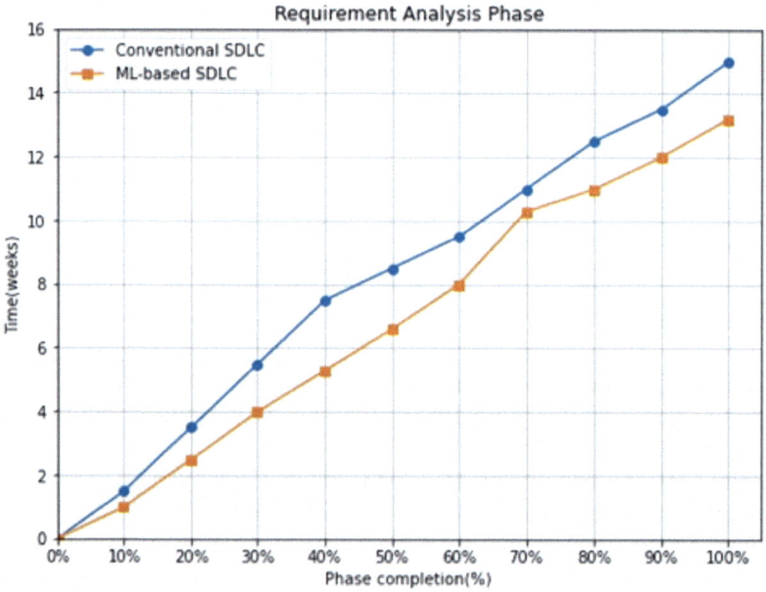

Fig. 3.2 Comparative analysis of requirement analysis phase performance between SDLC and ML-SDLC

Specification, Requirements Validation, Requirements Management Plan, and Finalize Requirements Document. Whereas requirement analysis stage in ML-based SDLC involves Problem Definition, Data Requirements Gathering, Data Quality Assessment, Data Preparation Requirements, Model Requirements Gathering, Model Evaluation Criteria, and Requirements Specification. Again, most of these components are inherited from similar tasks solved by another ML-based SDLC (Ref: Fig. 3.3).

The design stage in conventional waterfall SDLC consists of system design, architectural design, component design, interface design, data design, security design, design review and feedback, and finalized design documents. Whereas, design stage in ML-based waterfall SDLC consists of System Design, Architectural Design, Component Design, Model Design, Data Pipeline Design, Feature Engineering Design, Hyperparameter Tuning Plan, Design Review and Feedback, and Finalize Design Documents. Instead of building models from scratch, utilizing pre-trained models and leveraging well-known ML frameworks (like TensorFlow, PyTorch, or Scikit-learn) to avoid reinventing the wheel can significantly reduce the time (Ref: Fig. 3.4).

The implementation stage in conventional waterfall SDLC involves Coding, Unit Testing, Integration Testing, System Testing, Debugging and Fixing, Code Review and Refactoring, and Finalize Implementation. Whereas, the implementation stage in ML-based SDLC requires Data Preparation, Model Development, Model Training and Tuning, Model Testing and Evaluation, Deployment and Integration, Debugging and Fixing, and Finalize Implementation. Instead of building models from scratch,

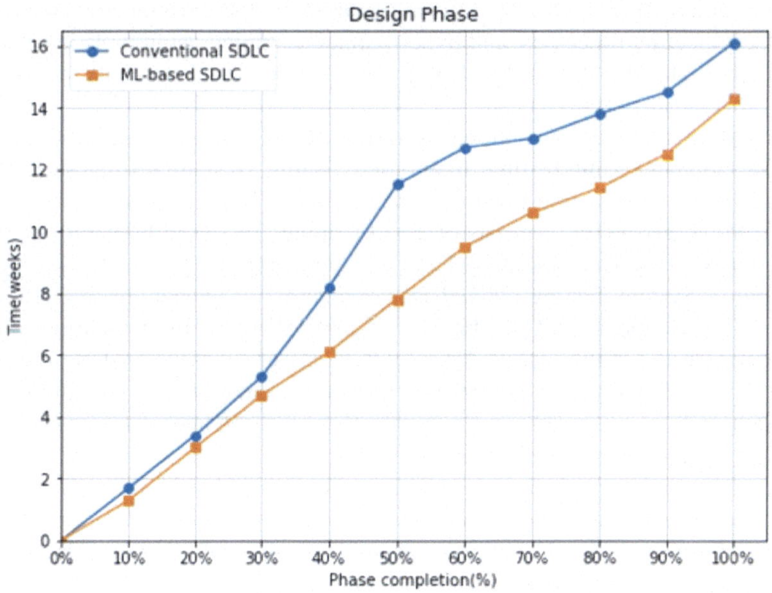

Fig. 3.3 Comparative analysis of design phase performance between SDLC and ML-SDLC

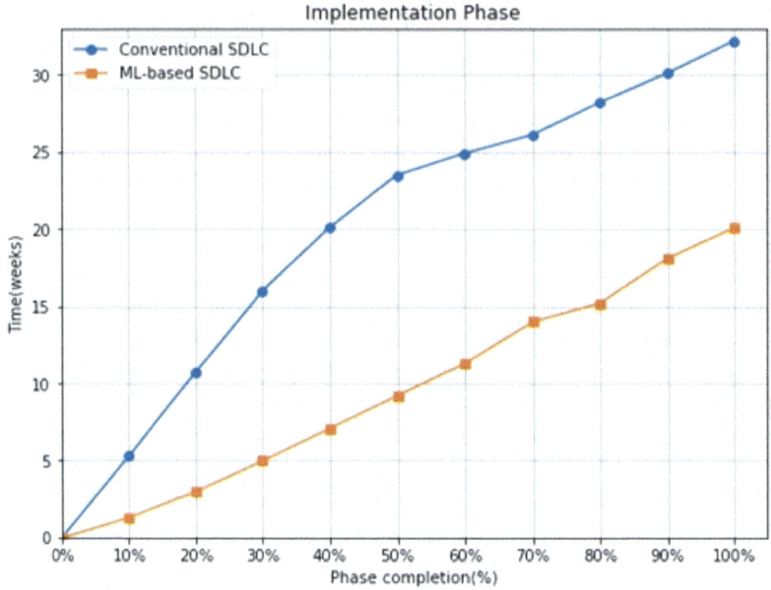

Fig. 3.4 Comparative analysis of implementation phase performance between SDLC and ML-SDLC

leveraging pre-trained models and applying transfer learning to adapt them to our specific problem can reduce time. Also, automating the data cleaning, transformation, feature engineering processes, and hyper-parameter tuning using tools like Pandas, Dask, or specialized platforms like DataRobot, AutoML can potentially speed up the whole process (Ref: Fig. 3.5).

The testing stage in conventional waterfall SDLC involves Test Planning, Test Case Development, Unit Testing, Integration Testing, System Testing, Acceptance Testing, Test Reporting and Feedback, and Finalize Testing. Whereas, the testing stage in ML-based SDLC consists of Test Planning, Test Data Preparation, Model Evaluation Metrics, Unit Testing (Model Components), Integration Testing (Model Integration), System Testing (Model Deployment), Acceptance Testing (Business Metrics), Test Reporting and Feedback, and Finalize Testing. Implementing continuous integration and continuous deployment (CI/CD) pipelines that include automated testing (like AutoML tools) ensures that tests and validation are run automatically with each code change, reducing manual effort and catching issues early makes optimizing required time (Ref: Fig. 3.6).

The maintenance stage in conventional waterfall SDLC includes Deployment and Rollout, Initial Bug Fixing, Stabilization and Optimization, Ongoing Maintenance and Support, Minor Enhancements and Updates, Major Version Updates, and Continuous Monitoring and Improvement. Whereas, the maintenance stage in ML-based SDLC consists of Model Deployment and Monitoring, Initial Model Tuning and Optimization, Ongoing Model Maintenance and Updates, Data Drift Detection

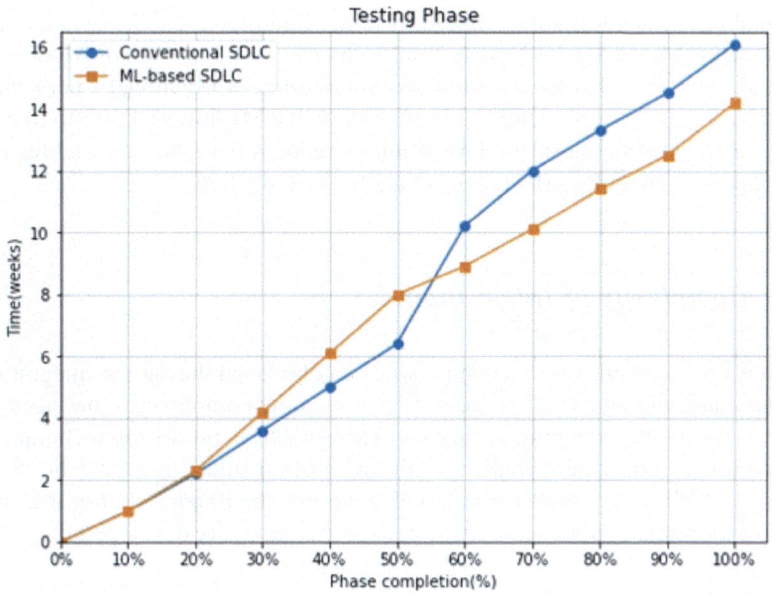

Fig. 3.5 Comparative analysis of testing phase performance between SDLC and ML-SDLC

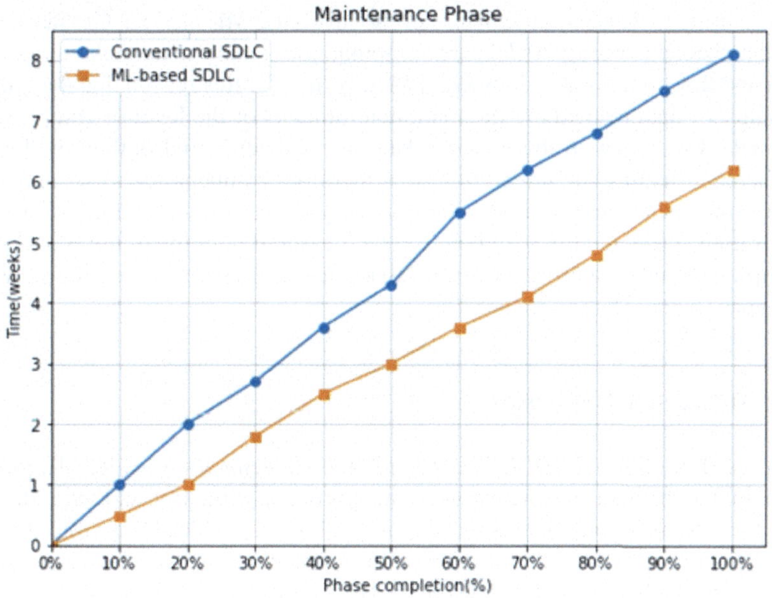

Fig. 3.6 Comparative analysis of maintenance phase performance between SDLC and ML-SDLC

and Handling, Model Retraining and Refreshing, Continuous Model Evaluation and Improvement, Minor Model Updates and Enhancements, and Major Model Overhauls and Upgrades. Setting up automated monitoring of the model's performance (e.g., accuracy, precision, recall) and using tools like MLflow or DVC to track and manage different versions of models definitely reduces both the time and resources required to keep models performing optimally in production.

3.4.1 Complexity of Integration

There are several challenges or complexities can be faced during the integration of ML with conventional SDLC process. Here, are some complexities that need to be addressed during this integration process. Though ML is promising techniques but that also invites many new challenges during implementation stage. The primary advantage of ML integration is the initial setup may be tough but after that all the project of the same domain (*same requirement* can be easily developed there after.

Algorithmic Complexity

It is more intricate based on the nature of the involving ML models. Greater models for instance deep learning models are computational intensive and take longer time to train and this increase development difficulty. It is for this reason that the quantity and quality of data needed are the main determinants of the level of complexity in the models. Dealing with the missing data, normalization and outliers will be the other factors affecting the integration. It is therefore cumbersome to set and adjust hyperparameters for the ML models because they all create another dimension on the configuration space. Complexity here can be measured in terms of the Big-O of the model training and data processing time/space and the hyperparameter optimization search space.

System Architecture Complexity

As discussed, traditional SDLC is centered on code generation, while ML pipeline calls for managing data data ingestion, training process, validation and retraining iterations. Include such components to raise archeological profiles. Therefore the complexity level of architectures is quite high. Traditionally in SDLC, data management is not as central whereas with ML projects data ingestion/feature engineering/model deployment means creation of new dependencies which must be well managed. This complexity can be evaluated by architectural complexity metrics such as Cyclomatic Complexity or Depth of Inheritance Tree to check the relatedness of some parts of the architecture.

Process Complexity

When adopting ML into an application, versions need to be kept not just for the code the program is written in, but also the dataset and models. Overhead comprises multiple sets of data and models as well as retrained models. Compared to classic SDLC, which is geared towards deploying software, ML integration also implies deploying models and monitoring their performance overtime (model drift). The latter adds another level of dynamic monitoring which complicates the issue.

Dependency Complexity

Implementation of ML usually comes with new tools like TensorFlow, PyTorch or Scikit-learn and these new tools may need to interface with conventional software architectures which might be complicated. Sometimes, many different ML models are needed and depend on cloud services (AWS, Google Cloud AI), which may cause problems with or combine use, security or cross-platform. To measure, dependency complexity, Dependency Graphs and analyzing the number of external libraries, frameworks as well as the nature of services necessary for interfacing can be measured.

Team Skill Set Complexity

In a team, there must be the proper knowledge of software development and machine learning solutions. The integration becomes a challenge though important for the success of the integration of the company's data scientists, the company's ML engineers and the company's software developers. ML projects demand more practice and talent development of team members to appreciate data and models specifics, and how to apply them. This learning curve makes the work even more challenging. This can be assessed by using competency frameworks of measuring the distance between the present human capital skills and the required ones. As the gap size increases, therefore, the integration becomes progressively more stringent.

Testing and Validation Complexity

ML models cannot be tested in the same way like any other software. In this case, we will have to check whether the models work well on other data sets; check accuracy, precision, recall, and more importantly confirm that the model provides a performance that is in line with the business requirements. In contrast with more conventional methodologies, such as linear regression or decision trees, the behavior of ML models is intrinsically more random because of random inclusion of instances in one model and this randomness can cause more unpredicted and complex behavior during the validation and debugging phases. Testing complexity can be evaluated by the Number of Testing Cycles, the type of data used for testing, and the Number of tracked performance indicators.

Scalability Complexity

One of the main limitations in dealing with data is the difficulty of scaling up the ML models. Handling large scale data and making sure that a system doesn't slow down during the inference phase adds more complications. When trained, models have to be updated with these new pieces of data, and managing how this can be done at scale can be difficult as well. Measures of scalability complexity can be expressed by the Throughput (requests/second) and Latency of the model inference and Time Complexity of the model retraining.

Integration with Legacy Systems

Adapting the existing systems as a part of the larger scope of ML systems can add some level of extra complications, especially in case of original application design that may not consider data processing and feeding elements in its base. Having to ensure clear transitions between the two systems makes it also challenging. When incorporating the ML into a traditional SDLC, the fit to the projects would not be same across the board. It is true that different projects have varied needs, limitations, and environments in which an ML-supported SDLC can be implemented. This variability calls for development of a qualitative approach as opposed to a quantitative model to solve the problem.

- **Assessing Project Suitability for ML Integration**: The first way to handle this problem is to think twice before deciding whether it is reasonable to integrate the ML as a solution into a particular task. It is also important to note that not all software projects should incorporate ML in some cases doing so will just bring extra overhead with little to show for it. To determine if ML is a good fit, the following factors should be considered: problem complexity, data availability, project scope and timeline, performance requirements, and team expertise.
- **Modular SDLC Frameworks**: Second way is, how this challenge can be addressed would be to adapt a modular SDLC framework that incorporates select ML models as required. This means, in practice, regarding the ML components (data pipelines, model training and validation and so on) as pluggable components to SDLC and not integrated components.
- **Tailored SDLC Models Based on Project Type**: Rather than using an ML-developed color change in one broad SDLC model, alternative specific types of SDLC models may be developed based on project type. The adaptability of the SDLC to specific software projects can be ensured by categorizing projects into types and modifying the SDLC model accordingly. data-driven projects, non-data-intensive projects, hybrid projects

- **Customizable ML Workflows:** In the cases where the projects need ML integration, the easily adaptability of the ML workflows should be maintained suiting the project requirements. Such must be broad enough to accommodate project's complexity and constraints, thereby providing the ability to manage ML development in a much more granualr manner.
- **Agile and Iterative Development**Another strategy is the Interaction of the ML with the SDLC using an agile and iterative development life cycle process. This makes it possible to adopt a step-by-step enhancement of ML into an organization's systems through the use of sprints or other development cycles. This means that processes are implemented in stages so that the team can work according to the change after some time and evaluation is made.
- **Use of ML Operations (Ops) for Integration Management**To cover that, managing the complexity of bringing ML into SDLC can be solved through using ML Ops frameworks. ML Ops is a process of integration and overall management of the building and deploying of the models and it is more or less related to the establishment of DevOps in the SDLC.

3.5 Conclusion

In conclusion, the comparative analysis between conventional SDLC stages and the ML-integrated SDLC model reveals significant distinctions and advancements in software development practices. While traditional SDLC methodologies emphasize a linear, structured approach with defined phases such as requirements gathering, design, implementation, testing, and maintenance, the ML-integrated SDLC model introduces a more dynamic and iterative process. This model incorporates machine learning techniques throughout the development lifecycle, enabling continuous learning and adaptation based on real-time data and feedback. The integration of ML into the SDLC framework enhances the ability to handle complex, data-driven applications and promotes greater flexibility and efficiency. Ultimately, the shift towards an ML-integrated approach signifies a transformative evolution in software development, highlighting the growing importance of adaptability and intelligence in meeting the demands of modern technology landscapes.

References

1. A.S. Pothukuchi, L.V. Kota, L.V., V. Mallikarjunaradhya, Impact of generative AI on the software development life cycle (SDLC) (2023)
2. S. Sharma, S. Pandey, Integrating AI techniques in SDLC: design phase perspective. Int. Symp. Women Comput. Inform
3. J. Dhami, N. Dave, O. Bagwe, A. Joshi, P. Tawde, Deep learning approach to predict software development life cycle model, in *2021 International Conference on Advances in Computing,*

Communication, and Control (ICAC3) (Mumbai, India, 2021), pp. 1–7. https://doi.org/10.1109/ICAC353642.2021.9697271
4. R. Ranawana, A.S. Karunananda, An agile software development life cycle model for machine learning application development, in *2021 5th SLAAI International Conference on Artificial Intelligence (SLAAI-ICAI)* (Colombo, Sri Lanka, 2021), pp. 1–6. https://doi.org/10.1109/SLAAI-ICAI54477.2021.9664736
5. Barry W. Boehm, A spiral model of software development and enhancement. Computer **21**(5), 61–72 (1988)
6. A. Manifesto, Manifesto for agile software development (2001)
7. Tim Menzies, Charles Pecheur, Verification and validation and artificial intelligence. Adv. Comput. **65**, 153–201 (2005)
8. M. Harman, S.A. Mansouri, Y. Zhang, Search-based software engineering: trends, techniques and applications. ACM Comput. Surv. (CSUR) **45**(1), 1–61 (2012)
9. I. Goodfellow, J. Pouget-Abadie, M. Mirza, B. Xu, D. Warde-Farley, S. Ozair, Y. Bengio, Generative adversarial nets. Adv. Neural Inf. Process. Syst. **27** (2014)
10. A. Radford, J. Wu, R. Child, D. Luan, D. Amodei, I. Sutskever, Language models are unsupervised multitask learners. OpenAI Blog **1**(8), 9 (2019)
11. M. Harman, The role of Artificial Intelligence in Software Engineering, in *2012 First International Workshop on Realizing AI Synergies in Software Engineering (RAISE)* (Zurich, Switzerland, 2012), pp. 1–6. https://doi.org/10.1109/RAISE.2012.6227961.keywords: Software;Software engineering;Learning systems;Machine learning;Optimization;Probabilistic logic
12. H. Sultanov, J.H. Hayes, Application of reinforcement learning to requirements engineering: requirements tracing, in *21st IEEE International Requirements Engineering Conference (RE)* (Rio de Janeiro, Brazil, 2013), pp. 52–61. https://doi.org/10.1109/RE.2013.6636705
13. J.J. Cuadrado-Gallego, P. Rodríguez-Soria, B. Martín-Herrera, Analogies and differences between machine learning and expert based software project effort estimation, in *2010 11th ACIS International Conference on Software Engineering, Artificial Intelligence, Networking and Parallel/Distributed Computing* (London, UK, 2010), pp. 269–276. https://doi.org/10.1109/SNPD.2010.47
14. D.G.E. Silva, M. Jino, B.T.D. Abreu, Machine learning methods and asymmetric cost function to estimate execution effort of software testing, in 2010 *Third International Conference on Software Testing, Verification and Validation* (Paris, France, 2010), pp. 275–284. https://doi.org/10.1109/ICST.2010.46
15. H.D. Tran, L.T.M. Hanh, N.T. Binh, Combining feature selection, feature learning and ensemble learning for software fault prediction, in *2019 11th International Conference on Knowledge and Systems Engineering (KSE)* (Da Nang, Vietnam, 2019), pp. 1–8. https://doi.org/10.1109/KSE.2019.8919292
16. V.H.S. Durelli et al., Machine learning applied to software testing: a systematic mapping study. IEEE Trans. Reliab. **68**(3), 1189–1212 (2019). https://doi.org/10.1109/TR.2019.2892517
17. Z. Wan, X. Xia, D. Lo, G.C. Murphy, How does machine learning change software development practices? IEEE Trans. Softw. Eng. **47**(9), 1857–1871 (2021). https://doi.org/10.1109/TSE.2019.2937083
18. F. Falcini, G. Lami, A.M. Costanza, Deep learning in automotive software. IEEE Software **34**(3), 56–63 (2017). https://doi.org/10.1109/MS.2017.79
19. A. Perini, A. Susi, P. Avesani, A machine learning approach to software requirements prioritization. IEEE Trans. Softw. Eng. **39**(4), 445–461 (2013). https://doi.org/10.1109/TSE.2012.52
20. R. Navarro-Almanza, R. Juarez-Ramirez, G. Licea, Towards supporting software engineering using deep learning: a case of software requirements classification, in *5th International Conference in Software Engineering Research and Innovation (CONISOFT)* (Merida, Mexico, 2017), pp. 116–120. https://doi.org/10.1109/CONISOFT.2017.00021

21. J. Guo, J. Cheng, J. Cleland-Huang, Semantically enhanced software traceability using deep learning techniques, In *2017 IEEE/ACM 39th International Conference on Software Engineering (ICSE)* (Buenos Aires, Argentina, 2017), pp. 3–14. https://doi.org/10.1109/ICSE.2017.9
22. M. White, C. Vendome, M. Linares-Vasquez, D. Poshyvanyk, Toward deep learning software repositories, in *2015 IEEE/ACM 12th Working Conference on Mining Software Repositories* (Florence, Italy, 2015), pp. 334–345. https://doi.org/10.1109/MSR.2015.38
23. V.-S. Ionescu, An approach to software development effort estimation using machine learning, in *2017 13th IEEE International Conference on Intelligent Computer Communication and Processing (ICCP)* (Cluj-Napoca, Romania, 2017), pp. 197–203. https://doi.org/10.1109/ICCP.2017.8117004
24. N. Maneerat, P. Muenchaisri, Bad-smell prediction from software design model using machine learning techniques, in *2011 Eighth International Joint Conference on Computer Science and Software Engineering (JCSSE)* (Nakhonpathom, Thailand, 2011), pp. 331–336. https://doi.org/10.1109/JCSSE.2011.5930143
25. H. Lal, G. Pahwa, Code review analysis of software system using machine learning techniques, in *2017 11th International Conference on Intelligent Systems and Control (ISCO)* (Coimbatore, India, 2017), pp. 8–13.https://doi.org/10.1109/ISCO.2017.7855962
26. J. Abbineni, O. Thalluri, Software defect detection using machine learning techniques, in *2018 2nd International Conference on Trends in Electronics and Informatics (ICOEI)* (Tirunelveli, India, 2018), pp. 471–475. https://doi.org/10.1109/ICOEI.2018.8553830
27. Y.A. Alshehri, K. Goseva-Popstojanova, D.G. Dzielski, T. Devine, Applying machine learning to predict software fault proneness using change metrics, static code metrics, and a combination of them, in *SoutheastCon 2018* (St. Petersburg, FL, USA, 2018), pp. 1–7. https://doi.org/10.1109/SECON.2018.8478911
28. G.P. Bhandari, R. Gupta, Machine learning based software fault prediction utilizing source code metrics, in *2018 IEEE 3rd International Conference on Computing, Communication and Security (ICCCS)* (Kathmandu, Nepal, 2018), pp. 40–45. https://doi.org/10.1109/CCCS.2018.8586805
29. G.P. Bhandari, R. Gupta, Measuring the fault predictability of software using deep learning techniques with software metrics, in *2018 5th IEEE Uttar Pradesh Section International Conference on Electrical, Electronics and Computer Engineering (UPCON)* (Gorakhpur, India, 2018), pp. 1–6. https://doi.org/10.1109/UPCON.2018.8597154
30. G. Huang, Y. Li, Q. Wang, J. Ren, Y. Cheng, X. Zhao, Automatic classification method for software vulnerability based on deep neural network. IEEE Access **7**, 28291–28298 (2019). https://doi.org/10.1109/ACCESS.2019.2900462
31. C. Fu, D. Qian, Z. Luan, Estimating software energy consumption with machine learning approach by software performance feature, in *2018 IEEE International Conference on Internet of Things (iThings) and IEEE Green Computing and Communications (GreenCom) and IEEE Cyber, Physical and Social Computing (CPSCom) and IEEE Smart Data (SmartData)* (Halifax, NS, Canada, 2018), pp. 490–496. https://doi.org/10.1109/Cybermatics_2018.2018.00106
32. Y. Tamura, M. Matsumoto, S. Yamada, Software reliability model selection based on deep learning, *2016 International Conference on Industrial Engineering, Management Science and Application (ICIMSA)* (Jeju, Korea (South), 2016), pp. 1–5. https://doi.org/10.1109/ICIMSA.2016.7504034
33. S. Jha et al., Deep learning approach for software maintainability metrics prediction. IEEE Access **7**, 61840–61855 (2019). https://doi.org/10.1109/ACCESS.2019.2913349
34. A. Andrzejak, L. Silva, Using machine learning for non-intrusive modeling and prediction of software aging, in *NOMS 2008-2008 IEEE Network Operations and Management Symposium* (Salvador, Brazil, 2008), pp. 25–32. https://doi.org/10.1109/NOMS.2008.4575113
35. T. Xie, The synergy of human and artificial intelligence in software engineering, in *2nd International Workshop on Realizing Artificial Intelligence Synergies in Software Engineering (RAISE)* (San Francisco, CA, USA, 2013), pp. 4–6. https://doi.org/10.1109/RAISE.2013.6615197

36. N. Nascimento, P. Alencar, C. Lucena, D. Cowan, Toward human-in-the-loop collaboration between software engineers and machine learning algorithms, in *2018 IEEE International Conference on Big Data (Big Data)* (Seattle, WA, USA, 2018), pp. 3534–3540. https://doi.org/10.1109/BigData.2018.8622107
37. R. Pandita, X. Xiao, H. Zhong, T. Xie, S. Oney, A. Paradkar, Inferring method specifications from natural language API descriptions, in *2012 34th International Conference on Software Engineering (ICSE)* (Zurich, Switzerland, 2012), pp. 815–825. https://doi.org/10.1109/ICSE.2012.6227137
38. M. Jahan, Z. Shakeri Hossein Abad, B. Far, Detecting emergent behaviors and implied scenarios in scenario-based specifications: a machine learning approach, in *IEEE/ACM 11th International Workshop on Modelling in Software Engineering (MiSE)* (Montreal, QC, Canada, 2019), pp. 8–14. https://doi.org/10.1109/MiSE.2019.00009
39. A.K. Dwivedi, A. Tirkey, R.B. Ray, S.K. Rath, Software design pattern recognition using machine learning techniques, in *IEEE Region 10 Conference (TENCON)* (Singapore, 2016), pp. 222–227. https://doi.org/10.1109/TENCON.2016.7847994
40. D. Zhang, Machine learning in value-based software test data generation, in *2006 18th IEEE International Conference on Tools with Artificial Intelligence (ICTAI'06)* (Arlington, VA, USA, 2006), pp. 732–736. https://doi.org/10.1109/ICTAI.2006.77
41. W.T. Tsai, K.G. Heisler, D. Volovik, I.A. Zualkernan, A critical look at the relationship between AI and software engineering, in *[Proceedings], IEEE Workshop on Languages for Automation@m_Symbiotic and Intelligent Robotics* (College Park, MD, USA, 1988), pp. 2–18. https://doi.org/10.1109/LFA.1988.24945
42. B. Lin, F. Zampetti, G. Bavota, M. Di Penta, M. Lanza, R. Oliveto, Sentiment analysis for software engineering: how far can we go?, in *2018 IEEE/ACM 40th International Conference on Software Engineering (ICSE)* (Gothenburg, Sweden, 2018), pp. 94–104. https://doi.org/10.1145/3180155.3180195
43. R. Robbes, A. Janes, Leveraging small software engineering data sets with pre-trained neural networks, in *2019 IEEE/ACM 41st International Conference on Software Engineering: New Ideas and Emerging Results (ICSE-NIER)* (Montreal, QC, Canada, 2019), pp. 29–32. https://doi.org/10.1109/ICSE-NIER.2019.00016
44. B. Schreck et al., Augmenting software project managers with predictions from machine learning, in *2018 IEEE International Conference on Big Data (Big Data)* (Seattle, WA, USA, 2018), pp. 2004–2011. https://doi.org/10.1109/BigData.2018.8622586
45. J. Shen, O. Baysal, M.O. Shafiq, Evaluating the performance of machine learning sentiment analysis algorithms in software engineering, in *IEEE International Conference on Dependable, Autonomic and Secure Computing, International Conference on Pervasive Intelligence and Computing, International Conference on Cloud and Big Data Computing, International Conference on Cyber Science and Technology Congress (DASC/PiCom/CBDCom/CyberSciTech)* (Fukuoka, Japan 2019), pp. 1023–1030. https://doi.org/10.1109/DASC/PiCom/CBDCom/CyberSciTech.2019.00185
46. M. White, Deep representations for software engineering, in *2015 IEEE/ACM 37th IEEE International Conference on Software Engineering* (Florence, Italy, 2015), pp. 781–783. https://doi.org/10.1109/ICSE.2015.248
47. H. Mohapatra, A.K. Rath, *Fundamentals of Software Engineering: Designed to Provide an Insight into the Software Engineering Concepts* (BPB Publications, 2020)
48. S.N. Ahsan, J. Ferzund, F. Wotawa, Automatic classification of software change request using multi-label machine learning methods, in *33rd Annual IEEE Software Engineering Workshop* (Skovde, Sweden, 2009), pp. 79–86. https://doi.org/10.1109/SEW.2009.15
49. L. van Rooijen, F.S.Bäumer, M.C. Platenius, M. Geierhos, H. Hamann, G. Engels, From user demand to software service: using machine learning to automate the requirements specification process, in *IEEE 25th International Requirements Engineering Conference Workshops (REW)* (Lisbon, Portugal 2017), pp. 379–385. https://doi.org/10.1109/REW.2017.26

50. C. Mills, S. Haiduc, A machine learning approach for determining the validity of traceability links, in *2017 IEEE/ACM 39th International Conference on Software Engineering Companion (ICSE-C)* (Buenos Aires, Argentina, 2017), pp. 121–123. https://doi.org/10.1109/ICSE-C.2017.86
51. K.W. Nafi, B. Roy, C.K. Roy, K.A. Schneider, [Research Paper] CroLSim: cross language software similarity detector using API documentation, in *2018 IEEE 18th International Working Conference on Source Code Analysis and Manipulation (SCAM)* (Madrid, Spain, 2018), pp. 139–148. https://doi.org/10.1109/SCAM.2018.00023
52. M. Wieloch, S. Amornborvornwong, J. Cleland-Huang, Trace-by-classification: a machine learning approach to generate trace links for frequently occurring software artifacts, in *7th International Workshop on Traceability in Emerging Forms of Software Engineering (TEFSE)* (San Francisco, CA, USA, 2013), pp. 110–114. https://doi.org/10.1109/TEFSE.2013.6620165
53. A. Sankaran, R. Aralikatte, S. Mani, S. Khare, N. Panwar, N. Gantayat, DARVIZ: deep abstract representation, visualization, and verification of deep learning models, in *2017 IEEE/ACM 39th International Conference on Software Engineering: New Ideas and Emerging Technologies Results Track (ICSE-NIER)* (Buenos Aires, Argentina, 2017), pp. 47–50. https://doi.org/10.1109/ICSE-NIER.2017.13
54. B.K. Sidhu, K. Singh, N. Sharma, A catalogue of model smells and refactoring operations for object-oriented software, in *2018 Second International Conference on Inventive Communication and Computational Technologies (ICICCT)* (Coimbatore, India, 2018), pp. 313–319). https://doi.org/10.1109/ICICCT.2018.8473027

Chapter 4
Generative Coding: Unlocking Ontological AI

Vineetha K V and Philip Samuel

Abstract The ultimate success of a software system is determined by its potential to fulfill the intended objectives. In the dynamic landscape of software development, the speed of application development and the adaptability to change are defining factors. The persistent demand for superior, swiftly deployable, and highly user—friendly software systems propel the evolution of Requirement Engineering (RE) to meet diverse development scenarios. Within the realm of RE, ontology emerges as a powerful tool, providing a structured and formal representation of a software system's requirements. This study delves into the pivotal role of formalizing requirement documents in the wider picture of software development. The emphasis is placed on how this formalization process becomes a catalyst for applying intelligence to develop code. By embracing a structured methodology in requirement documentation, this literature study highlights the potential to amplify efficiency and consistency in the translation of requirements into design models like Unified Modeling Language (UML) and code. In essence, this chapter emphasizes the pivotal connection between formalizing requirements through ontology and the subsequent automatic generation of code, showcasing the integral role of Artificial Intelligence (AI) in this process. This integrated approach not only aligns with the perpetual demand for better software but also provides a systematic and efficient pathway to bridge the conceptualization of requirements with the generation of code.

Keywords Generative AI · Requirement engineering · Ontology

Vineetha K V (✉) · P. Samuel
Department of Computer Science, Cochin University of Science and Technology, Kochi, India
e-mail: vineetharohith92@cusat.ac.in

P. Samuel
e-mail: philips@cusat.ac.in

T. R. Benala et al. (eds.), *Boosting Software Development Using Machine Learning*,
Artificial Intelligence-Enhanced Software and Systems Engineering 7,
https://doi.org/10.1007/978-3-031-88188-6_4

4.1 Introduction to Generative AI

Artificial Intelligence (AI) encompasses a diverse array of computational algorithms designed to undertake tasks traditionally requiring human intelligence. Within this broad domain, Machine Learning (ML) emerges as a significant subfield, focusing on the creation of algorithms that can independently resolve tasks by learning from data, rather than through explicit programming. Advancing further, Deep Learning (DL) represents a more sophisticated branch of ML, utilizing artificial neural networks to intricately model complicated data representations and autonomously unearth patterns and correlations within vast datasets. At the forefront of DL innovations, deep generative models [1] have been developed. These advanced models are capable of generating new, authentic content by learning from existing data, thereby opening up an expansive array of novel possibilities for AI applications. Figure 4.1 depicts the hierarchical progression from AI to Generative Models, illustrating the evolution towards more advanced autonomous problem-solving and creative capabilities in machines.

Generative AI is a subset of ML technology that uses patterns found in existing data to generate new content, such as text, images, videos or code [2–4]. These ML techniques are trained on enormous volume of data to find out the underlying patterns in it. Generative AI has made significant strides across various domains, with notable advancements in text, image, movie, and code generation [5]. In the text domain, models like ChatGPT, Google's Bard, DeepMind's Chinchilla, and Meta's LLaMA demonstrate the ability to produce contextually relevant text for

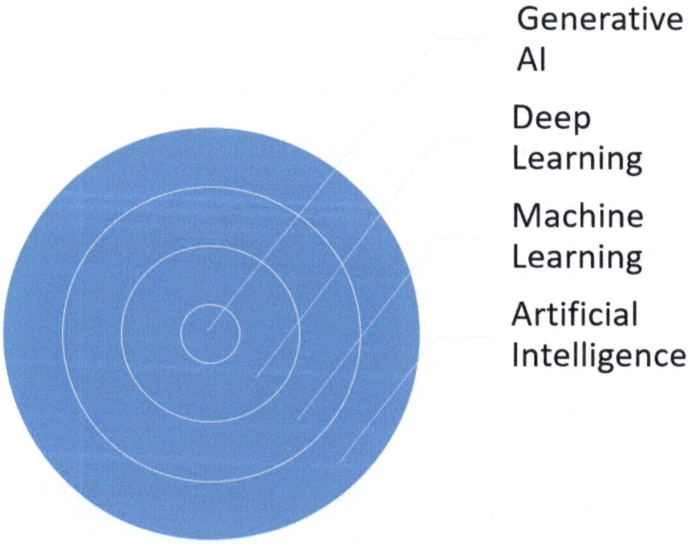

Fig. 4.1 AI and its major subfields

applications such as chatbots and content generation. In image generation, technologies such as Generative Adversarial Networks (GANs), exemplified by OpenAI's DALL-E, Midjourney, and Stable Diffusion, showcase the capacity to create diverse and realistic images based on textual prompts. AI's capabilities extend to multimedia content generation, including movies, where music and voice synthesis, alongside multi-modal approaches, facilitate the creation of comprehensive media experiences. Moreover, in coding, GitHub Copilot offers groundbreaking assistance by leveraging machine learning to provide developers with contextually relevant code suggestions and completions based on natural language prompts. These advancements collectively underscore the growing potential of generative AI in producing diverse and high-quality content across various media types.

4.2 Generative AI for Software Development

Generative AI presents promising avenues for enhancing software productivity through automation and optimization across several phases of the software development process. It streamlines processes by automating repetitive tasks like testing and requirements traceability, improving efficiency [6]. Furthermore, it elevates software quality by generating test suites directly from requirements and seamlessly automating workflows to route work products through production pipelines. However, the adoption of generative AI introduces new risks due to its non-deterministic and unexplainable nature. Despite these challenges, several generative software platforms have emerged, transforming simple instructions into computer code and facilitating tasks such as code generation, test case generation, traceability establishment, code explanation, refactoring of legacy code, software maintenance, and code improvement. These platforms offer promising solutions to optimize software development processes and augment the capabilities of software engineers.

Generative AI is revolutionizing software development by streamlining the process, improving application quality, and customizing it to meet the demands of each user. Beyond merely advancing technology, AI enhances human creativity and judgment, ensuring that advancements in this field continue to be ethically and technically sound.

Considering ongoing societal, academic, and industry discussions, Sauvola explores the potential impact of generative AI and large language models (LLMs) on software development [34]. The study presents four primary scenarios:

1. Classical software development is where all roles are manually controlled with the help of automation tools.
2. AI-in-the-loop, where humans lead, but AI begins to manage more significant tasks.
3. AI assumes specific roles, such as overseeing design, testing, and maintenance.
4. Human-in-the-loop, where AI handles most development processes, with humans playing a supervisory role.

These scenarios are assessed within the context of various software development organizations (SDOs).

4.3 Generative Coding

Generative coding refers to the process of using AI techniques, particularly generative models, to create or assist in the creation of code automatically. Unlike traditional coding where humans manually write instructions, generative coding entails machines independently creating code snippets, functions, or even complete programs based on provided requirement documents or examples.

Over the past decade, progress in ML, especially in the development of generative models like Generative Adversarial Networks (GANs), Variational Autoencoders (VAEs), and Transformers, has facilitated the emergence of generative coding. These models have demonstrated remarkable capabilities in generating diverse and realistic outputs across various domains, including natural language, images, and now, code.

Generative coding holds immense promise across multiple domains of software development. It can automate repetitive coding tasks, assist developers in prototyping and exploration, generate code snippets based on high-level specifications, and even optimize existing codebases. By augmenting human capabilities, generative coding has the potential to accelerate innovation, reduce development time, and enhance the overall quality of software.

While generative coding offers significant benefits, it also presents challenges and considerations. These include ensuring the generated code is syntactically and semantically correct, addressing ethical concerns surrounding code generation, and maintaining control over the output to align with project requirements and standards. Additionally, there is a need for interpretability and explainability in generative models to foster trust and facilitate collaboration between human developers and AI systems.

The novelty of our problem lies in combining ontological knowledge structures with generative AI, leading to more accurate, intelligent, structured and context-aware code generation that overcomes the current limitations in AI-driven programming. Our approach bridges the gap between abstract concepts and executable code, enhancing the AI's understanding of programming contexts.

The field of generative coding is rapidly evolving, with ongoing research focused on improving the capabilities, robustness, and usability of generative models for code generation. Future directions include exploring novel architectures tailored specifically for code generation, integrating generative coding tools into existing development environments, and advancing techniques for evaluating and validating generated code. Ultimately, the goal is to empower developers with powerful AI-assisted tools that enhance productivity and creativity in software development.

4.4 Significance of Formalization

The ultimate measure of success for a software system is its ability to meet the objectives it was designed for. Requirement Engineering (RE) [7] is the process of identifying the intended function and goals of a software system by recognizing the stakeholders involved, understanding their needs, and documenting them in a format suitable for analysis, facilitating effective communication among the stakeholders, and future reference [8]. In this phase, system requirements are identified and documented. This is a multidisciplinary and user-centric process, serving as the initial stages of developing software. Effective software systems adapt and develop as the operational environment undergoes changes. Therefore, it emphasizes the importance of efficiently handling changing requirements. The majority of requirement engineering work is concentrated in the initial stages of software development, as dealing with misunderstood or omitted requirements in later stages can incur substantial costs.

The novelty of many applications stems from the rapid development speed required and the extent to which they are likely to change. The demand for faster, better and more user-friendly software systems will persist, driving the evolution of RE to adapt to various development scenarios. According to Sommerville [9], the RE phase consists of the processes: elicitation, analysis and negotiation, specification, validation, and management. Bashar [7] outlines the fundamental activities of RE as requirements elicitation, modeling and analysis, communication, agreement, and, evolution.

In RE, If the requirements that have been captured are not fully understood, clients will be dissatisfied, leading to the necessity for changes in later phases. The RE phase has a very prominent role in software development. According to a study conducted by the Standish group in 1995 [10], even if the development process was good, the software projects may tend to fail. This study identifies that the majority of the reasons for the software project failure are related to RE. This study signifies the importance of the RE phase and the necessity to improve the RE phase. Significant research is going on in this area and since the requirement document is drafted in Natural Language (NL), most of the research makes employs Natural Language Processing (NLP) techniques in the RE phase.

The needs and expectations of a software client are captured in NL and are documented as Software Requirement Specification (SRS). Software clients convey their needs and expectations using NL, resulting in Software Requirement Specification (SRS) documents. However, NL-written SRS documents tend to be ambiguous, inconsistent, incomplete, vague, and subject to conflicts. They should ideally be accessible to stakeholders with limited domain knowledge but often pose challenges when converting them into UML models, especially for large systems. Identifying inconsistencies, duplicates, and missing specifications is also problematic in these documents. Furthermore, NL-written requirements follow an informal specification style and employ domain-specific vocabulary, diverging from common English. This unique ness makes them unsuitable for analysis using NLP tools trained on general

English documents. As a result, dealing with NL-written SRS documents necessitates careful handling and translation into more formal representations to facilitate effective software development.

Since the NL written requirement document is unstructured, automating the architecture generation from software requirements is quite challenging. The solutions that can be proposed are:

1. Document SRS formally. But this is not possible because formally written SRS will not be easily comprehensible to stakeholders (basically non–technical persons) with limited domain knowledge.
2. Initially document SRS in NL. Transform NL written requirement document to a formal one.
3. Make use of CNL (Constrained Natural Language) while documenting requirements.

A high-quality architecture can be produced if the requirement document is of sufficient quality. Here, we impose formalism for constructing a high-quality requirement document. Formalizing makes it easy to store and fetch information from the ontology and easy to infer new knowledge from it.

4.5 Fundamentals of Ontology

A formal, clear representation of a common conceptualization is known as an ontology [11]. A conceptualization refers to an abstract and concise representation of the world that we intend to model for a particular purpose. In this context, concepts within the domain are called classes, the properties that characterize different aspects of these concepts are known as slots, and the constraints on these slots are termed role restrictions (facets). When an ontology is combined with a collection of individual instances corresponding to these classes, it forms a knowledge base.

This encompasses the creation of machine-readable definitions for fundamental concepts within a domain and their interrelationships. The goal of developing an ontology is multi-fold: it facilitates a shared understanding of information structure among individuals or software entities, promotes the reuse of domain-specific knowledge, makes underlying domain assumptions explicit, separates domain and operational knowledge, and analyzes the domain knowledge.

An ontology is composed of classes, relations, slots, facets, formal axioms, and functions, and when combined with a set of specific instances, it forms a comprehensive knowledge base. Classes represent concepts in a domain and relations usually represent the binary associations between classes. Functions represent a particular case of relations. Slots, also known as properties or attributes, describe the characteristics of the defined classes. Formal axioms represent statements that are invariably true. Instances represent the individual elements within an ontology.

The process of creating an ontology involves identifying and defining the classes within the ontology, organizing these classes into a structured hierarchy, defining relationships between the classes, specifying slots along with permissible values for those slots, creating instances, and assigning specific values to the slots for each instance. The methodology [12] for generating an ontology includes 7 steps:

1. Identifying the domain and its scope: Define the domain that the ontology will cover, the intended use of the ontology, the kinds of queries should the information within the ontology be able to address, and, who are the intended users and maintainers of the ontology.
2. Consider the incorporation of existing ontologies: It is usually beneficial to examine others' work to see if we can refine and expand upon existing resources to suit our specific domain and task. For this purpose, we can utilize libraries of reusable ontologies available on the web and in scholarly articles.
3. Listing out critical terms: Creating a list of terms that we aim to discuss or explain is beneficial. This involves identifying the terms of interest, understanding their properties, and determining what specific information or statements we want to convey about these terms.
4. Defining classes and their hierarchies: Top-down, bottom-up, or a combination of both can be used for defining the class hierarchy. The top-down development approach initiates with the identification of general concepts in the domain, which are then incrementally narrowed down. Conversely, the bottom-up process starts with the most specific classes at the bottom of the hierarchy and groups them into general concepts. The combination approach initially identifies the most prominent concepts and then appropriately generalizes and refines them.
5. Defining class properties (slots): Merely defining classes is insufficient; It's essential to also detail the internal structure of these concepts once they have been defined.
6. Specifying slot facets such as cardinality, value type, and domain- range: Slots can possess various facets detailing aspects like the type of value, permissible values, cardinality, and other characteristics of potential slot values. The cardinality of a slot specifies the number of values it can have. The value-type facet of a slot indicates the kinds of values it can contain, such as String, Number, Boolean, Enumerated, and Instance. Table 4.1 describes slot-value types. For slots of the Instance type, the allowed classes are often referred to as the slot's range, while the classes associated with a slot or those that a slot describes are known as the slot's domain.
7. Creating instances: The final step is generating specific instances of classes within the hierarchy and this includes selecting a class, creating a unique instance of that class, and populating the slot values for that instance.

Ontology concerning software development refers to the use of formal, structured representations of knowledge and information to enhance the development, management, and understanding of software systems. An ontology provides a common vocabulary for researchers seeking to collaborate within a specific domain, offering machine-understandable definitions of core concepts and their interrelationships

Table 4.1 Slot-value types in an ontology

Slot-value type	Description
String	The most basic value type used for slots like name
Number	Slots that hold numerical values
Boolean	Basic yes or no flags
Enumerated	Specify a list of values that are permitted for the slot
Instance-type	Allow defining relationships among individuals. Slots with value type Instance must also specify a list of permitted classes from which the instances can come

[12]. The reasons for employing ontologies include promoting a shared understanding of structures of information among individuals and software agents, facilitating the reuse of domain knowledge, clarifying domain assumptions, separating domain knowledge from operational knowledge, and enabling the analysis of domain knowledge.

To cope with the ever-changing needs of the industry, new methods are being integrated into software development. In recent years, ontologies have been integrated into the software development process. Ontologies have relevance across different phases of software development, including software analysis and design, implementation, deployment and runtime, and maintenance.

Ontology, in the context of RE, provides a structured and formal representation of the requirements of a software system. Ontology in RE creates a semantic model that captures the domain's entities, relationships, and constraints. This semantic model, also known as a domain ontology, serves as a shared vocabulary for development teams and stakeholders to communicate unambiguously. Key benefits of using ontology in RE include:

1. Clear and Consistent Communication: Ontologies provide a standardized vocabulary that helps bridge the communication gap between stakeholders with diverse backgrounds, such as business analysts, domain experts, and technical developers. This leads to a more accurate and shared understanding of requirements.
2. Requirement Reusability: Ontologies enable the modularization of requirements by breaking them down into well-defined concepts and relationships. This facilitates requirement reuse across different projects and helps maintain consistency.
3. Traceability and Impact Analysis: With an ontology-based approach, it becomes easier to trace the relationships between different requirements, ensuring that changes to one requirement are properly understood in the context of the entire system.
4. Adaptability to Change: As requirements evolve over time, ontologies offer a flexible framework that can accommodate changes by updating or extending existing concepts and relationships.

Table 4.2 Major studies on ontology

Ontology—definitions	[11–13]
General applications	[14–17]
Applications in SE	[18–23]
Applications in RE	[24–26]

5. Adaptability to Change: As requirements evolve over time, ontologies offer a flexible framework that can accommodate changes by updating or extending existing concepts and relationships. Enhanced Requirement Analysis: Ontologies enable automated reasoning and inference, which can be used to analyze requirements for inconsistencies, conflicts, or missing information.

Ontology's entire body of work (Table 4.2) can be organized into four main divisions: (1) concerning definitions, (2) encompassing general applications, (3) oriented towards software engineering, and (4) dedicated to applications in requirement engineering. There are primarily three pieces of work related to the definitions.

The work [11] begins by defining why an ontology is needed and what is it. The paper describes the methodology for creating an ontology with an example ontology OASys (Ontology for Autonomous Systems). Ontology can be used to share a common understanding of a domain and related information, and by using an ontology, anybody can reuse the knowledge generated. Ontology can be used to analyze the domain knowledge and separate domain knowledge from operational.

Noy [12] defines steps for creating an ontology with an example ontology of wines. Ontology development starts by defining its domain and scope. Existing ontologies if any, can be reused. The next step is to identify important terms in the ontology. This may include the terms that we would like to talk and their properties. Then classes, class hierarchy, properties of classes (called slots), facets o slots are defined and finally, instances are created.

A conceptualization is a simplified, abstract representation of the world that we intend to use for a specific goal [13]. A conceptualization (extensional relational structure) is defined by Genesereth and Nilsson as a tuple (D, R), where D is a set known as the universe of discourse and R is a set of relations on D. Guarino states that a triple $C = (D, W, R)$, where W is a set of feasible worlds and R is a set of conceptual relations on the domain space $< D, W >$, is a conceptualization (intentional relational structure). In this context, a world is a completely ordered set of world states that reflects the system's progression over time. A world state for a system is a distinct assignment of values to all the observable variables that define the system.

The following works showcase how ontology is applied to software engineering.

Oberle [20]. Introduces the Core Software Ontology that formalizes the shared concepts in the field of software engineering. Since the domain of the software is sufficiently complex, and stable, to prevent modeling from scratch, the stable core can be captured in an ontology. The core concepts of the software domain, such as software, data, classes, and methods, are introduced in this ontology.

Happel [18] showcases instances of applying ontologies across various stages of the Software development process, investigates the advantages of using ontologies in each scenario, and presents a structure for categorizing ontological utilization in Software Engineering. The significance of ontologies in software development is explored within the domains of analysis and design, implementation, deployment, runtime, and maintenance phases. Table 4.3 illustrates how ontologies are applied in these phases of software development.

ODYSSEY [19], an ontology for the Software Development Life Cycle (SDLC) domain, captures the knowledge included in the SDLCs and then, formalizes and implement the SDLC major concepts and their properties. This ODYSSEY ontology has been created by the Enterprise Ontology (EO) approach and implemented using Descriptive Logic (DL).

Gavsevic [21] emphasizes the significance of ontologies in the field of software engineering, defines software engineering as an application context for ontologies

Table 4.3 Instances of applying ontologies across various stages of the software engineering lifecycle by Happel [18]

Software analysis and design	Requirements engineering: To delineate SRS documents and formalize the representation requirements knowledge Component reuse: To depict the component's functionality using a knowledge representation formalism facilitating more convenient and robust querying
Implementation	Integration with Software Modelling Languages: The integration of ontology languages like RDF and OWL can enhance software modeling languages and methodologies Ontology as Domain Object Model: Utilizing ontology tools capable of developing an Application Programming Interfaces directly from the ontology Coding Support: Enhancing Application Programming Interfaces by incorporating semantic information Code Documentation: Ontologies create a unified frame work for representing both the problem domain and source code, thereby simplifying cross-references between the two
Deployment and run time	Semantic Middleware: Ontologies serve as a mechanism to capture knowledge about the problem domain. Semantic tools can then create an information space to store this knowledge Business Rules: Ontology-based methods in Software Engineering handle frequently changing declarative knowledge through specialized middleware Semantic Web Services: Overlay a semantic layer onto the existing web service framework
Maintenance	Project Support: In software maintenance workflows, various types of interconnected data coexist without explicit connections. Ontologies integrates data from various sources into a cohesive semantic model Updating: Ensuring that one's application remains current requires investing time in tasks such as monitoring for updates, downloading them, and installing them Testing: Employing ontologies can support the development of essential test cases by encoding domain knowledge into a format that machines can interpret

and describes how ontology finds relevance in the Analysis, Design, Implementation, Integration, Maintenance, and Retirement phases of software development.

Bhatia [27] suggests a tool-based approach for automating software documentation using ontologies. The suggested multi-stage software development frame work based on ontology multi-phase framework, aimed at automating documentation, consists of six phases: domain ontology, software requirements specification ontology, design phase, source code phase, source code listing ontology, and test cases ontology.

Akerman [23] presents a software development approach centered on architectural decisions, employing ontology. In this method, architecture is encapsulated within an ontology instance, which comprises four key elements: architecture assets, architectural decisions, stakeholder concerns, and an architectural roadmap. The method is exemplified by means of a case study featuring a real-time credit-approval system.

Saiyd [28] introduces a structured ontology framework for the context of the software design lifecycle. This framework addresses issues related to ambiguity in terminology, particularly among designers and users with diverse backgrounds and varying levels of knowledge. It also aims to elucidate the rationale behind designers' choices among various design alternatives.

The following works illustrate how ontology is applied to RE.

Mavin arranges the various RE approaches using an ontology-based framework [24]. Since the selection of RE approaches for software or systems development projects are complex due to numerous options; this study employs two scales: Lifecycle phases and technology readiness levels (TRLs) to measure and relate RE approaches. This study positions each established RE approach on a two-dimensional chart, considering the RE lifecycle phase and the'maturity' of the approach to clearly illustrate the connections between these approaches.

GUITAR, a tool developed by Nguyen, automatically identifies incorrectness, incompleteness, and inconsistency among artifacts [25]. GUITAR stands for Goal Use case Integration Tool for Analysis of Requirements. The tool comprises modules for modeling and analysis of requirements. GUITAR automates the analysis of goals and use cases and empowers the generation of problem explanations and alternative resolutions.

Castaneda conducts a thorough review, outlining how ontologies are applied in RE. [26]. This review highlights the challenges that need to be addressed during RE phase. In RE, ontologies find applications in the following ways: describing requirement documents, providing a formal structure of requirements, and formally representing knowledge about the application domain.

The literature reviewed shows that generative AI models, typically trained on general-purpose code, struggle with domain-specific constraints and dependencies, limiting their ability to produce semantically accurate and contextually relevant code. Generative AI can create precise code for specific domains by incorporating domain-specific ontologies. Additionally, generative AI faces challenges in scaling and generalizing across complex systems. The novelty of our approach lies in integrating ontological reasoning into generative AI, which enhances both the scalability and adaptability of AI models for more extensive and diverse coding tasks. Unlike existing methodologies, our approach enables the generation of more semantically accurate, contextually relevant code while addressing key limitations in scalability and adaptability. This combination of ontology-driven knowledge and generative coding uniquely contributes to our study.

To apply our approach, we assume that domain-specific ontologies are available and well-defined, as the effectiveness of our method depends on the quality of these ontologies. A limitation is that our approach may be less practical in domains where ontologies are underdeveloped or unavailable. Additionally, the method may require significant computational resources to manage complex ontologies and large-scale systems, which could impact its efficiency in specific scenarios.

4.6 Requirements to Ontology

Typically expressed in natural language, software requirements lack structure, com plicating the automation of transitioning from requirements to executable code. Consistency and standardization in representing software requirements are needed to ensure certainty in requirement engineering. Consequently, transforming natural language requirement text into an ontology structure offers a solution. Ontologies provide a semantic model that facilitates information storage, retrieval, and inference. Formalizing requirements ensures that all stakeholders share a common understanding. Organizing requirements in a structured manner makes them easily analyzable by humans and machines.

Reqtology [29] outlines a structured process, taking requirements as user stories and extracting relevant information based on predefined rules. Subsequently, this extracted information is categorized to facilitate its utilization in forming ontologies. The work flow comprises six distinct steps, commencing with input provided through user stories and culminating in generating entities suitable for ontology formation.

Figure 4.2 illustrates the workflow of Reqtology. This system accepts requirements as input provided through user stories.

Example requirement. "As a teacher, I want to login, so that I can enter marks."

Categories of Information in User Stories (Fig. 4.3):

1. Actors: Actors are entities that execute specific functions to accomplish tasks or enable certain functionalities. In the above user story, teacher is an actor who wants login functionality.

Fig. 4.2 Reqtology architecture [29]

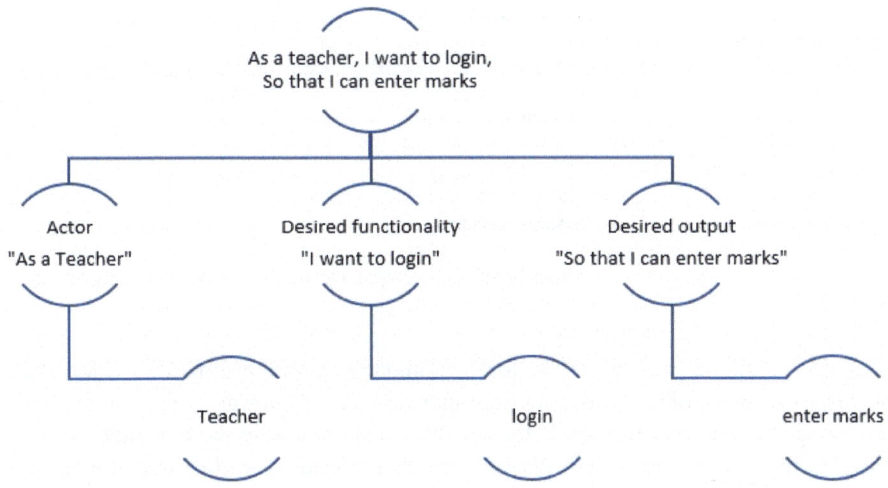

Fig. 4.3 Categories of information in user stories [29]

2. Desired functionality: The "desired functionality" or "action" represents the functionalities that actors require or seek to accomplish. The actor, identified as the teacher in this scenario, requires the functionality of login.
3. Output of desired functionality: This category encompasses entities that serve as the desired outcomes or outputs described within user stories. In the above user story, enter marks is the desired outcome.

4.6.1 Input Text

Example: As a user, I want to login, so that I can enter marks.

The first step of the technique entails ensuring that the input text meets specific criteria. This includes verifying that the text comprises either a single user story or, if multiple, a dot separates each story. Furthermore, the user stories must adhere to the appropriate format and comply with the rules specified in the INCOSE software requirement standard. The text is effectively prepared for subsequent processing by ensuring these conditions are met. If the input text aligns with all the specified conditions, it progresses to the next phase, sentence segmentation.

4.6.2 Sentence Segmentation

The initial phase involves breaking down the input text into smaller segments, called sentence segmentation, which consists of identifying and marking the boundaries

Fig. 4.4 Sentence segmentation [29]

Input Text:

As a teacher, I want to login, so that I can enter marks.

Delimiter Rule:

["."+" "+[A-Z]]

Output Text:

As a teacher, I want to login, so that I can enter marks

between individual sentences. The text's boundaries are determined by a delimiter. In this case, we utilize the dot (.) delimiter and the occurrence of a capital letter following the dot (.) delimiter. If the conditions specified by the delimiter rule are not met, the output comprises only one sentence. Figure 4.4 illustrates the process of sentence segmentation.

4.6.3 Tokenization

The tokenization partitions the text into individual words, treating them as discrete entities for tagging by the POS tagger in the subsequent step. In this phase, the input consists of a single sentence segmented in the prior step. Tokenization employs delimiters to separate the sentence into words, using the space delimiter for this purpose. Figure 4.5 illustrates the process of tokenization.

Tokenization Output:

T1	As
T2	a
T3	teacher
T4	I
T5	want
T6	to
T7	login
T8	so
T9	that
T10	I
T11	can
T12	enter
T13	marks

Fig. 4.5 Tokenization [29]

POS tagging :

Input:

T1	As
T2	a
T3	teacher
T4	I
T5	want
T6	to
T7	login
T8	so
T9	that
T10	I
T11	can
T12	enter
T13	marks

Output:

As/IN a/DT teacher/NN I/PRP want/VBP to/TO login/VB so/IN that/IN I/PRP can/MD enter/VB marks/NN

Fig. 4.6 POS tagging [29]

4.6.4 *POS Tagging*

In the subsequent step, the part-of-speech tagging process assigns a grammatical category to each tokenized entity derived from the prior steps. Each word receives a part-of-speech tag based on its role within the grammatical structure. Figure 4.6 illustrates the process of POS tagging.

4.6.5 *Entity Detection*

Entity detection is the key step in the information extraction process. Entities are detected after the POS tagger has tagged them. Within entity detection, tagged entities are categorized into classes and sub-classes. Figure 4.7 illustrates the process of entity detection.

The entity is selected only when both Pre-condition and post-condition are met. The pre-condition is met if the trigger token tag is detected. Upon satisfying the Pre-condition tag, the subsequent entity is disregarded, proceeding directly to the post-condition. In the post-condition phase, if the token tag corresponds to the tag specified in the post-condition, the entity between the pre-condition and post-condition is chosen, signifying the fulfillment of both conditions. The entity is incorporated into the pattern once both Pre-condition and post-condition are met. Following extraction,

Entity Detection:

As/IN a/DT teacher/NN I/PRP want/VBP to/TO login/VB so/IN that/IN I/PRP can/MD enter/VB marks/NN

Rules:

1. **Actor detection**

   Pre-condition:
   POS tags = {IN, DT} matches => "As a"
   Condition:
   POS tags = {NN} matches =>" teacher"
   Post-Condition:
   POS tags = {PRP} matches => "I"

2. **Action detection**

   Pre-condition:
   POS tags = {TO} matches => "to"
   Condition:
   POS tags = {VB} matches => "login"
   Post-Condition:
   POS tags = {IN} matches => "so"

3. **Action detection**

   Pre-condition:
   POS tags = {PRP, MD, VB} matches => "I can enter"
   Condition:
   POS tags = {NN} matches => "marks"
   Post-Condition:
   NONE

   Result:

   Detected entities:
   [{"teacher", class = Actor}, {"login", class = Action}, {"enter marks", class=desired outcome}]

Fig. 4.7 Entity detection [29]

the entity compares with words stored in the domain-specific dictionary. This dictionary encompasses the definitions and category definitions of words. If the extracted entity aligns with any word in the domain dictionary, it is categorized accordingly based on its definition.

4.6.6 Relationship Detection

Relationship detection plays a key role in ontology development, enabling the representation of connections and associations between entities detected. This process involves identifying and defining relationships between concepts or individuals, thereby enriching the semantic model of the ontology. Figure 4.8 illustrates the process of relation detection.

Various methods are employed to identify relationships among entities, with one prevalent approach involving using trigger words. For instance, consider the sentence "John attended the workshop". In this example, the relationship can be

Relation Detection:

Input:

Detected Entities: [{"teacher", class = Actor}, {"login", class = Action}, {"enter marks", class=desired outcome}]

Relationship types:

	Relation Name	Example
1	Student/Required Functionality	Student, register
2	Student/Desired Outcome	Student, access
3	Teacher/ Required Functionality	Teacher, login
4	Teacher/ Desired Outcome	Teacher, Access
5	Action/ Desired Outcome	Login, access
6	Action/Subject	Fill, details

Relationship rules:

Rule1:
[Entity E1 from Actor class = "teacher" + "want to" + Entity E2 from class Required Functionality = "login"]
Rule2:
[Entity E1 from Actor class = "teacher" + "want to" + Entity E2 from class Desired Outcome = "enter marks"]

Result:

"Teacher" want to "login"
"Teacher" want to "enter marks"

Fig. 4.8 Relation detection [29]

readily extracted using the trigger phrase "attended." However, extracting relationships from the text can pose challenges, mainly when trigger words are not explicitly present or when dealing with text-encompassing terms from diverse domains.

4.6.7 Ontology Mapping

The diagram 9 depicts the classification of "person" as a subclass of the "actor" category. Furthermore, the entities "student" and "teacher", derived from user stories, are classified as subclasses of the "person" category (Fig. 4.9).

In Fig. 4.10, the term "login" is derived from the user story and is categorized as a subclass within the "action" class. The elements depicted in the diagram are predefined as components of the "login" entity within the domain dictionary. Upon matching the term "login" from the user story with the entry in the domain dictionary, the components of the login entity specified within the domain dictionary are assigned as instances of the "login" subclass within the class hierarchy.

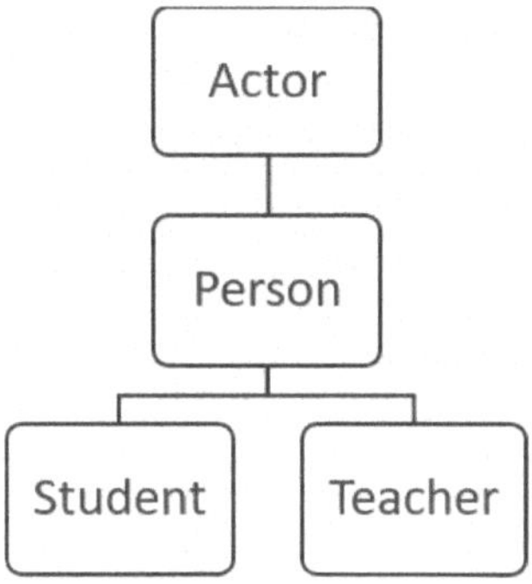

Fig. 4.9 Entities (actors) [29]

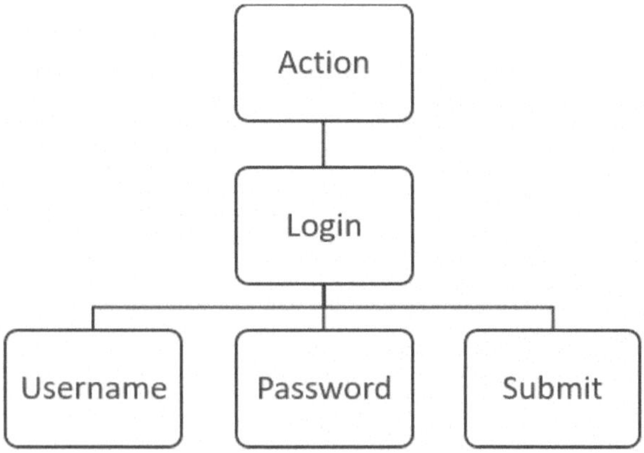

Fig. 4.10 Entities (desired functionality) [29]

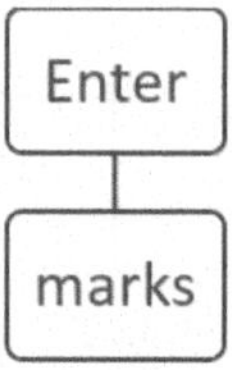

Fig. 4.11 Entities (output of desired functionality) [29]

In the diagram 11, the term "marks" is extracted from the user story text and subsequently categorized as a subclass of "enter" following its identification in the domain dictionary (Fig. 4.11).

Transforming software requirements into ontologies offers a significant leap forward in software engineering by systematically converting informal requirements into a structured, formal representation. This transformation aids in mitigating ambiguities and inconsistencies, which are common challenges in traditional requirements engineering. Moreover, the ontology-based representation of requirements contributes to higher quality and more efficient software development. The ontology provides a detailed and precise specification of the concepts, relationships, and rules within a domain, which can be interpreted by automated tools to produce the corresponding code. This automation streamlines the development process, reduces the likelihood of human error, and ensures consistency between the requirements and the implemented software.

4.7 Ontology to Code Generation

Automatic code generation represents a paradigm shift in software engineering, aiming to streamline the development process by automating the creation of executable code from higher-level specifications or models. The rationale for automatic code generation stems from the need to improve software development efficiency, reduce manual effort, and mitigate the risk of human error. By automating the translation of abstract models or specifications into executable code, developers can accelerate the development cycle, enhance code quality, and ensure consistency across the software system. At the core of automatic code generation are models that serve as input for the code generation process. These models can take various forms, including ontologies, formal specifications, Domain-Specific Languages (DSLs), and graphical models such as UML diagrams. Models capture the software system's essential structure, behavior, and functionality, providing a high-level representation that can be translated into code.

Ontology-to-code generation (Fig. 4.12) automatically converts information stored in ontologies, representing knowledge within a particular domain, into executable code in a programming language. This process is essential for linking high-level domain specifications, typically expressed in formal languages like OWL (Web Ontology Language), with the practical implementation of software systems.

Fig. 4.12 Code generation from ontology

There are three works primarily dedicated to the generation of code from ontologies.

OnToCode [30] is a tool designed to facilitate code generation in various programming languages from ontologies. It enables the generation of code in multiple languages from a single primary model (ontology) by utilizing formal specifications as the foundation for all models in software prototypes. Given that modern software applications often incorporate components written in different programming languages, OnToCode simplifies the process by generating platform-specific models (derivatives) from a single platform-independent domain model. Developed as a Python package, OnToCode leverages Owlready2 for ontology loading and querying, offering a fixed API for these tasks. Additionally, it provides an extensible API for result processing and template instantiation, along with built-in support for Jinja2-based templates.

The construction of the PIDESO ontology [31] involves three main stages: Conceptualization, Operationalization, and Ontologization. Conceptualization entails identifying concepts and relationships from raw data, where concepts describe cognitive entities informally within the domain. Ontologization involves partial formalization to facilitate representation in a formal language, ultimately resulting in a fully operational ontology. Operationalization aims to establish a formal structure of concepts and relations, represented as OWL classes. Process-oriented discrete event simulation models (PODESO) are depicted as OWL instances to transform them into another form known as XPISM instances. These XPISM instances are an intermediate representation for the executable model, which is converted into Java code via an XSLT stylesheet. This stylesheet contains transformation rules executed by an XSLT processor, utilizing XPath to identify nodes in the source document (OWL instance) and generate the result document (XPISM instance).

The model in [32] comprises three layers: analyzer, solution finder (reasoner), and converter. In the first layer, each database schema is analyzed to extract relationships between tables, determine field meanings, and analyze user stories to identify functions performed by each software user role. The second layer deduces new functions based on the findings of the first layer and extracts knowledge containing inferred solutions. WordNet and a Backend Ontology, constructed from scratch, serve as the knowledge bases. In the third layer, solutions are transformed into source code using templates extracted from the knowledge and configured, which are then applied to the templates.

Source code can be automatically generated from formal models such as UML models also. The ultimate goal of software engineering has long been the automatic generation of code from requirements. This involves transforming requirements into a formally equivalent model, the foundation for code generation. The work R2D2C ("Requirements to Design to Code") [33] aims to develop a methodology for generating code directly from requirements, ensuring that the generated code is a provably correct implementation of those requirements. Automating code production based on customer requirements would further minimize the risk of developer-inserted errors. While various tools for automatic code generation exist in the market, they often produce code sections that remain unexecuted or lack justification from the requirements or the model. Furthermore, these tools fail to build a provable equivalence among the original requirements and the developed code.

In R2D2C, engineers have the option to express requirement specifications as scenarios in constrained natural language or different formats such as UML use cases. These scenarios are the basis for deriving a formal model, ensuring equivalence with the initial requirements. The R2D2C approach involves five phases:

1. Scenario Capture: Clients use constrained natural language to document scenarios describing intended system operations.
2. Traces Generation: Based on the scenarios that have been documented, traces and sequences of atomic events are generated.
3. Model Inference: An automatic theorem prover infers a formal model or specification, typically expressed in Communicating Sequential Processes (CSP).
4. Analysis: Various analyses are conducted based on the formal model, utilizing existing public domain tools as well as specialized tools planned for development.
5. Code Generation: The generated code may be low-level code for electromechanical devices, high-level programming language code, or business procedures and instructions in NL.

Automatic code generation from ontologies represents a paradigm shift in software development, offering a novel approach to translating high-level specifications into executable code. By harnessing the structured knowledge captured in ontologies, developers can automate and streamline many aspects of software engineering, from requirements analysis to code implementation. It accelerates the development process and enhances the quality and maintainability of the resulting software systems. With the ability to encode domain knowledge and semantic relationships into the codebase, automatic code generation from ontologies holds immense potential for advancing software engineering practices. As research and adoption in this field continue to grow, it promises to reshape how software is developed, ultimately leading to more efficient, robust, and intelligent software solutions.

Compared to existing methods that rely on general-purpose models for generative coding, our approach integrates domain-specific ontological reasoning, offering several advantages. Traditional generative models often struggle with domain-specific constraints and dependencies, leading to code that may be syntactically correct but lacks semantic depth and contextual relevance. Our methodology addresses this by embedding ontologies into the generative process, ensuring more accurate and context-aware code generation.

Additionally, while many current models face challenges in scaling and adapting to complex systems, our approach enhances scalability by leveraging structured ontological knowledge, allowing the system to manage dependencies more effectively. However, it's important to note that relying on well-defined ontologies means our approach is best suited to domains where such ontologies are available and comprehensive.

4.8 Conclusion

Integrating ontologies for automatic code generation from requirement documents represents a significant advancement in software engineering methodologies. Combining ontological knowledge structures with generative AI provides a systematic and semantic foundation for AI-driven code generation. It enhances the generation of context-aware code, improving the structure and consistency of the output. The practical applications include automatically generating high-quality code, reducing manual coding in routine tasks, and contributing to developing more intelligent Inte grated Development Environments (IDEs) and AI-driven programming assistants. These tools can offer precise suggestions, code completions, and debugging insights by understanding the intent and context behind the code, thus increasing accuracy and efficiency.

Future research could focus on creating domain-specific ontologies and integrating these with advanced large language models like GPT. Using structured guidelines would explore how ontologies can automatically verify and validate generated code for correctness, security, and performance. Additionally, scaling these ontological frameworks to handle more extensive and complex software projects will be crucial.

Integrating ontologies in automatic code generation can revolutionize software development practices, leading to more efficient and accurate development processes. With ongoing advancements and innovations, this approach promises to enhance the quality and maintainability of software systems, empowering developers to create high-quality solutions more effectively than ever before.

References

1. L. Banh, G. Strobel, Generative artificial intelligence. Electron. Mark. **33** (2023). https://doi.org/10.1007/s12525-023-00680-1
2. E. Brynjolfsson, D. Li, L.R. Raymond, Generative AI at work, in *Working Paper 31161, National Bureau of Economic Research* (2023). https://doi.org/10.3386/w31161. http://www.nber.org/papers/w31161
3. H.S. Sætra, Generative AI: here to stay, but for good? Technol. Soc. **75**, 102372 (2023). https://doi.org/10.1016/j.techsoc.2023.102372

4. A. Bandi, P.V. Adapa, Y.E. Kuchi, The power of Generative AI: a review of requirements, models, input–output formats, evaluation metrics, and challenges. Future Internet **15**, 260 (2023). https://doi.org/10.3390/fi15080260
5. S. Feuerriegel, J. Hartmann, C. Janiesch, P. Zschech, Generative AI. Bus. Inf. Syst. Eng. **66**, 111–126 (2023). https://doi.org/10.1007/s12599-023-00834-7
6. C. Ebert, P. Louridas, Generative AI for software practitioners. IEEE Softw. **40**, 30–38 (2023). https://doi.org/10.1109/ms.2023.3265877
7. B. Nuseibeh, S. Easterbrook, Requirements engineering. Proc. Conf. Futur. Soft-Ware Eng. (2000). https://doi.org/10.1145/336512.336523
8. M. Faisal, G.F. Issa, I. Ayub et al., (2022) How automate requirements engineering system effects and support requirement engineering. Int. Conf. Bus. Anal. Technol. Secur. (ICBATS) **18**, 1–3 (2022). https://doi.org/10.1109/icbats54253.2022.9758997
9. G. Kotonya, I. Sommerville, *Requirements Engineering: Processes and Techniques*, 1st edn (Wiley Publishing, 1998)
10. T. Clancy, *The Standish Group Report*. Chaos Report (1995)
11. J. Bermejo, A simplified guide to create an ontology. Madrid University, 1–12 (2007)
12. N. Noy, Ontology development 101: a guide to creating your first ontology (2001). https://api.semanticscholar.org/CorpusID:500106
13. N. Guarino, D. Oberle, S. Staab, What is an ontology? in *Handbook on Ontologies* (2009), pp. 1–17
14. J. An, Y.B. Park, Methodology for automatic ontology generation using database schema information. Mob. Inf. Syst. **2018**, 1–13 (2018). https://doi.org/10.1155/2018/1359174
15. D.U. Vidanagama, A.T.P. Silva, A.S. Karunananda, Ontology based sentiment analysis for fake review detection. Expert Syst. Appl. **206**, 117869 (2022). https://doi.org/10.1016/j.eswa.2022.117869
16. S. Elnagar, V. Yoon, M. Thomas, An automatic ontology generation framework with an organizational perspective. Proc. Annu. Hawaii Int. Confer-Ence Syst. Sci. (2020). https://doi.org/10.24251/hicss.2020.597
17. X. Xue, Q. Huang, Generative adversarial learning for optimizing ontology alignment. Expert. Syst. (2022). https://doi.org/10.1111/exsy.12936
18. H.-J. Happel, S. Seedorf, Applications of ontologies in software engineering (2006). https://api.semanticscholar.org/CorpusID:16634666
19. J.I. Olszewska, I. Allison, Odyssey: software development life cycle ontology (2018)
20. D. Oberle, S. Grimm, S. Staab, An ontology for software, in *Handbook on Ontologies* (Springer, 2009), pp. 383–402
21. D. Gašević, N. Kaviani, M. Milanović, Ontologies and software engineering, in *Handbook on Ontologies* (2009), pp. 593–615
22. T.S. Dillon, E. Chang, P. Wongthongtham, Ontology-based software engineering—Software engineering 2.0, in *19th Australian Conference on Software Engineering (ASWEC 2008)* (IEEE, 2008) , pp. 13–23
23. A. Akerman, J. Tyree, Using ontology to support the development of software architectures. IBM Syst. J. **45**(4), 813–825 (2006)
24. A. Mavin, S. Mavin, B. Penzenstadler, C.C. Venters, Towards an ontology of requirements engineering approaches, in *2019 IEEE 27th International Requirements Engineering Conference (RE)*. (IEEE, 2019), pp. 514–515
25. T.H. Nguyen, J. Grundy, M. Almorsy, Guitar: an ontology-based automated requirements analysis tool, in *2014 IEEE 22nd International Requirements Engineering Conference (RE)* (IEEE, 2014), pp. 315–316
26. V. Castañeda, L. Ballejos, M.L. Caliusco, M.R. Galli, The use of ontologies in requirements engineering. Glob. J. Res. Eng. **10**(6), 2–8 (2010)
27. M.P.S. Bhatia, A. Kumar, R. Beniwal, Ontology-driven software development for automated documentation. Webology **15**(2) (2018)
28. N. Al Saiyd, I. Al Said, A. Al Neaimi, Towards an ontological concept for domain-driven software design, in *2009 First International Conference on Networked Digital Technologies* (IEEE, 2009), pp. 127–131

29. S. Ahmed, B. Ahmad, Transforming requirements to ontologies (2020)
30. P. Schäfer, OnToCode: template-based code-generation from ontologies. J. Open Source Softw. **4**, 1513 (2019). https://doi.org/10.21105/joss.01513
31. Y. Gheraibia, A. Bourouis, Ontology and automatic code generation on modeling and simulation, in *2012 6th International Conference on Sciences of Electronics, Technologies of Information and Telecommunications (SETIT)*. (IEEE, 2012), pp. 69–73
32. alokla anas, W. Gad, M. Aref, A.-B. Salem, Source code generation-based on NLP and ontology. Int. J. Intell. Comput. In-Form. Sci. 1–12 (2022). https://doi.org/10.21608/ijicis.2022.117905.1160
33. M.G. Hinchey, J.L. Rash, C.A. Rouff, Requirements to design to code: towards a fully formal approach to automatic code generation. Technical report (2005)
34. J. Sauvola, S. Tarkoma, M. Klemettinen et al., Future of software development with Generative AI. Autom. Softw. Eng. (2024). https://doi.org/10.1007/s10515-024-00426-z

Chapter 5
Case Studies: Machine Learning Approaches for Software Development Effort Estimation

Sarika Mustyala, Pravali Manchala, and Manjubala Bisi

Abstract Software is crucial in current life cycle management, impacting social, political, financial, healthcare, and military sectors. As software becomes more complex, accurately estimating the effort required for development has become increasingly challenging. Software development effort estimation (SDEE) involves predicting the effort needed to develop software based on its specifications, measured in person-months or person-hours. Accurate effort estimation is essential during the early stages of a project, particularly in planning and requirement assessment, as it aids in project planning, budgeting, monitoring, scheduling, and resource allocation. Effective estimation helps manage large and complex projects and improves application quality by addressing issues of overestimation and underestimation. Effort estimation methodologies are broadly categorized into expert judgment, algorithmic models, and machine learning processes. Algorithmic models rely on statistical analysis of project input data to estimate effort, using mathematical formulas derived from numerical inputs of one or more projects. Expert judgment, on the other hand, depends on the experience of domain experts in providing effort estimates. Still, this approach can lack objectivity and is challenging to fine-tune, especially when data or numerical expertise is scarce. Machine learning techniques have emerged as a promising alternative, often matching or surpassing the accuracy of algorithmic methods while offering improved understandability and ease of application. This chapter contains different software effort estimation projects that can be solved using machine learning techniques. This chapter will briefly present a case study of applications of each machine learning model for various tasks in software development effort estimation.

S. Mustyala · P. Manchala · M. Bisi (✉)
National Institute of Technology, Warangal, India
e-mail: manjubalabisi@nitw.ac.in

S. Mustyala
e-mail: ms23csr1r02@student.nitw.ac.in

P. Manchala
e-mail: mpravali@student.nitw.ac.in

T. R. Benala et al. (eds.), *Boosting Software Development Using Machine Learning*,
Artificial Intelligence-Enhanced Software and Systems Engineering 7,
https://doi.org/10.1007/978-3-031-88188-6_5

Keywords Software development effort estimation · Machine learning models · Expert judgment · Algorithmic models

5.1 Overview of Software Development Effort Estimation

Accurate software development effort estimation is difficult for effective project planning, budgeting, and bidding. Poor estimation can lead to significant financial losses or missed business opportunities. Studies, including one by the Standish Group, emphasize the importance of precise estimation as a key factor in software project success [1]. Despite efforts over the past 20 years, software cost estimation accuracy has seen limited improvement. Estimation remains a significant challenge, as many methods developed by researchers and managers still lack precision. Software development effort estimation (SDEE) involves predicting the work required, measured in person-hours or person-months, based on functional and non-functional requirements [2].

For software development to be successful, a more precise prediction of work is necessary; otherwise, several problems could occur, including delays in project delivery, overspending, and other problems that lower the quality of the end product [3]. Furthermore, software developers and managers are more likely to see project failures due to inaccurate effort estimations than they are due to proper methodology in software development. However, when the effort assessment is precise, the resources can be employed effectively and maintained sensibly, and the result will be of excellent quality and satisfy user expectations. Several techniques have been developed for estimating software effort. Specific methods relied on expert judgment, while others relied on algorithmic and parametric approaches like Use Case Point (UCP) and Function Point Analysis (FPA). Still more employed machine learning techniques. Unfortunately, not all of these methods produced reliable findings. Thus, new methods are still needed to generate more precise estimates of the efforts needed for software projects. For example, most effort estimating models use project functional size to estimate effort. These models work well with waterfall, spiral, and incremental techniques, among other classic lifecycle developments. Nevertheless, to finish the work estimation process, agile software development needs various data, like narrative points.

In software engineering, accurate software effort estimation is crucial because it helps to address the problems of overestimation and underestimating [4, 5]. Overestimating could lead to resource misallocation, which would go over budget and negatively impact the progress of other important initiatives. On the other hand, underestimating could lead to overspending and delivering poor-quality projects, budget overruns, or software that falls short of client expectations. According to an analysis of software development, an average of 30–40% more work is required for 60–80% of projects. Therefore, it is possible to increase software production and improve software quality by having a thorough awareness and control over the software development process. Six categories are used to classify software effort estimating methodologies [6]. These categories are then further divided into 3 groups:

- Expert judgment
- Algorithmic models
- Machine learning models.

Expert Judgement

Expert judgment is a popular and valued technique for software effort estimation because of its simplicity and reliance on experienced professionals. Experts use their knowledge of similar projects to estimate effort, aiding project planning and management. This approach is advantageous for its speed, ease of use, and minimal need for complex models or extensive historical data, making it useful when detailed information is scarce early in a project [7]. Expert judgment leverages intuitive insights from past experiences, allowing quick and flexible estimation. Though less structured than data-driven methods, it remains effective, especially in the project's initial stages when information is limited. A project's ability to proceed without significant delays can be ensured by the experts' prompt provision of approximate estimates, which aid in early project planning and resource allocation.

Using their experience and knowledge of the proposed project, a group of experts or an individual expert is consulted to estimate the project's cost using expert judgment procedures. Formal procedures such as Delphi are frequently employed to coordinate divergent ideas among estimators. There are several variants of the Delphi technique. Wideband Delphi encourages participants to talk among themselves about the issue. The steps listed below are used to apply this technique [7]:

1. The coordinator will get to know each expert with the project parameters and estimation methodology.
2. The coordinator will organize a specialist conference to discuss the problems around the presented value.
3. Every expert will fill out the questionnaire on their own, on their own time.
4. The coordinator will schedule an additional meeting to review the estimates.
5. Experts will fill up the paperwork once more.
6. The abovementioned steps will be repeated until a consensus is established.

Algorithmic Models

Software development projects can be estimated for effort using structured, systematic methodologies called algorithmic models. Depending on various project criteria, these models use statistical methods and mathematical formulas to forecast the time, resources, and effort required. The primary goal is to find relationships between various project characteristics and effort; these relationships are frequently found in empirical research and historical project data. Functional Point Analysis (FPA) and the Constructive Cost Model (COCOMO) are well-known instances of algorithmic models in software effort estimation.

One of the key advantages of algorithmic models is their ability to generate objective, trustworthy, and repeatable estimations. Algorithmic models employ predetermined calculations and parameters that yield the same outcome given the same input, in contrast to expert judgment, which can be subject to individual preferences and subjective viewpoints. Because of their consistency, they are beneficial in large businesses where project management uniformity and established procedures are crucial. Furthermore, these models may take into consideration a variety of elements that affect project effort because they make use of statistical analysis and historical data, which could result in estimations that are more precise and reliable.

Constructive Cost Model (COCOMO)

COCOMO is a widely used parametric model to estimate software development effort based on project size, typically in KLOC, and factors like hardware limitations, team experience, and product complexity. It has three versions—Basic, Intermediate, and Detailed—providing increasing accuracy by incorporating more variables. Although calibration with historical data is essential for reliability, COCOMO offers a systematic approach for managing large software projects [8].

Function Point Analysis (FPA)

FPA estimates software effort by measuring the functionality provided to users, regardless of technology or programming language. It assigns weights to components based on complexity and quantifies usefulness through inputs, outputs, user interactions, files, and external interfaces [9]. FPA is useful for evaluating software size and difficulty from a functional perspective, offering consistent metrics for comparisons and being adaptable to various projects.

Software Life Cycle Management (SLIM)

SLIM is an estimation methodology generated from Quantitative Software Management (QSM) that forecasts software effort, duration, and cost using mathematical models and historical data [10]. It employs a probabilistic approach to account for uncertainties and compares new projects with a database of completed ones. SLIM integrates with project management systems for real-time tracking and forecasting, providing strategic planning and comprehensive project control through historical data and advanced analytics.

Machine Learning Models

Machine learning models are increasingly used for software effort estimation because of their ability to learn from past project data and identify complex patterns that traditional methods may miss. These models, such as decision trees, neural networks, and support vector machines, can adapt to new data and continuously improve their predictions. Unlike traditional parametric models, they don't require explicit descriptions of variable relationships, making them flexible and powerful tools for estimating effort in dynamic software projects. Machine learning models are being adopted more frequently for software effort estimation because they can:

- Handle Complicated Relationships: They can model non-linear interactions between variables that traditional methods may overlook.
- Adapt to New Data: Their accuracy improves over time by continuously learning from new data.
- Process Huge Datasets: They can manage large amounts of data, incorporating various project attributes for more accurate estimates.
- Minimize Human Bias: Unlike expert judgment-based methods, they rely on data, reducing the impact of arbitrary biases in estimation.

Uses of different machine learning models

- The amount of effort required for software development is frequently estimated using regression models, such as ridge, lasso, and linear regression, based on a range of project variables. These models perform admirably when there is a clear relationship between the effort and the input factors (such as size and complexity). For example, estimating the man-hours needed for a project by counting the function points or lines of code.
- To estimate the effort for new software projects, Case-Based Reasoning (CBR) models reference a database of similar, previously completed projects. The effort for the new project is then inferred from the efforts of the closest matches. For example, this approach estimates effort by finding and adapting information from past projects that resemble the current project in terms of scale, scope, and complexity.
- The use of Artificial Neural Networks (ANNs) in estimating software effort comes from their capacity to represent complicated non-linear correlations between project variables and effort. To make better forecasts, they might gain knowledge from past experiences. An illustration of this would be estimating the effort required for large, complicated projects in which there may be non-linear interactions and a complex relationship between input factors and effort.
- Fuzzy-based neural networks are useful for managing uncertainty and imprecision in software effort estimation because they include the benefits of fuzzy logic and neural networks. They come in especially handy when the input data is unclear or imprecise. Consider an example of estimating work when project needs are not completely specified or when expert opinions are required, and project data is inaccurate.
- Extreme Learning Machines (ELM) are a kind of feedforward neural network that performs well in generalization and offers quick learning rates. They speed up the processing of large data sets and provide precise forecasts in software effort estimation. For instance, estimating work quickly for projects in situations where accuracy and speed are crucial in real-time.
- The use of Ensemble Learning Models is to increase accuracy and robustness. Ensemble learning models—like Random Forests and Gradient Boosting Machines—combine the predictions of several models. When estimating software effort, they work incredibly well with complicated and varied datasets. As an illustration, consider combining many models to more precisely estimate effort and utilize the advantages of each model to handle various data points.

5.2 Overview of Machine Learning Models

Machine learning (ML) techniques have demonstrated accuracy with algorithmic methods, and they may also provide improved application ease and comprehensibility. The machine learning algorithms create "rules" that suit a predetermined training dataset and can also be logically applied to data that has never been seen before. The majority of machine learning (ML) models used in effort estimate can be classified into several types, including ensemble learning, extreme learning, artificial neural networks (ANN), fuzzy-based neural networks, regression models, and case-based reasoning (CBR) models.

I. **Regression Models**

Regression analysis is a statistical technique used to explore the relationship between variables. It examines how changing one independent variable while keeping others constant affects a dependent variable [11]. This technique is used by data scientists and mathematicians for forecasting and prediction, selecting the best model to fit data. Regression analysis helps in understanding relationships within data sets and is essential for extracting meaningful results from complex data. Different regression models are used to analyze various types of data with multiple relationships.

- Linear regression
- Polynomial regression
- Ridge regression
- Lasso regression.

Applications of Regression Models in effort estimation

1. *Linear Regression*:

Linear regression is a commonly used method for software effort estimation, establishing a linear relationship between the dependent variable (effort) and independent variables like lines of code, project complexity, or team experience. Its simplicity makes it a useful tool for interpreting and applying effort estimates based on past data, especially when the relationship between variables is linear. However, its effectiveness diminishes when there are complex interactions or non-linear relationships between effort and project parameters.

2. *Polynomial Regression*:

Polynomial regression builds on linear regression by fitting a polynomial equation to the data, enabling the modeling of non-linear relationships between effort and project attributes. Software effort estimation can capture complex interactions between parameters, such as when effort increases more rapidly with project size or complexity. While it enhances accuracy when linear regression falls short, polynomial regression requires careful management to avoid overfitting.

3. *Ridge Regression*:

In software effort estimation, multi-collinearity arises when independent variables have a high degree of correlation. This issue can be addressed by ridge regression. This can happen when there is a relationship between several project parameters, like project duration and team size. Ridge regression adds a regularization component to the linear regression model to restrict large coefficients and also helps to prevent overfitting. While estimating effort for new projects that differ from the past data, ridge regression improves the model's capacity, making it more reliable by shrinking the coefficients.

4. *Lasso Regression*:

Lasso regression, a regularization technique similar to ridge regression, is used in software effort estimation with the added benefit of feature selection. It penalizes large coefficients, reducing some to zero, which is particularly useful when there are many potential predictors. By focusing on the most important components, lasso regression produces more interpretable and effective effort estimates.

Regression techniques range from simple linear relationships to more complicated models that address non-linearity, multicollinearity, and feature selection, and each one has a distinct advantage for software effort estimation [11]. The choice of a regression model depends on the specific characteristics of the software project and the data available.

II. **Case-Based Reasoning (CBR)**

The field of cognitive science gave rise to CBR, primarily through the work of Yale University's Roger Schank and his students. Riesbeck and Schank [12] developed the basic definition of CBR.

> A case-based reasoner solves problems by using or adapting solutions to old problems.

Case-based reasoning (CBR) in effort estimation creates solutions based on the concept that similar problems have similar solutions, using past successful projects to address new ones. It applies similarity measures, such as Euclidean or Manhattan distance, to find matches and set a learning pattern that improves as more solutions are stored. Key parameters include feature selection, feature weighting, similarity measures, and the k-nearest neighbor algorithm, which recognizes the closest past cases. Careful parameter tuning is essential to minimize estimation error and enhance accuracy [13].

Selection of feature is one of the CBR technique's that is used to pick an input variable subset which can lower estimation errors and sufficiently explain the input data. Three categories exist for feature selection techniques: approaches that are filter-based, embedded-based, and wrapper-based [14].

- Filter-based approaches determine the impact of features using statistical metrics or parameters instead of incorporating machine learning algorithms based on the fundamental properties of the training data.

- The embedded-based techniques combine wrapper and filter-based techniques [14].
- Wrapper-based techniques estimate the relevance of features by explicitly using machine learning algorithms. These techniques typically produce results that are more accurate than the last one.

In Case-Based Reasoning (CBR), feature weighting assigns an ideal weight to each feature, highlighting its significance by distinguishing important from irrelevant features. Feature weighting methods include filter-based and wrapper-based approaches. Filter-based methods calculate weights based on overall data characteristics, while wrapper-based methods determine weights using a specific machine learning algorithm as a fitness function [14]. Research suggests that feature weighting tends to yield higher accuracy than feature selection, as feature selection is binary (features are either chosen or not) and considered a subset of feature weighting, which assigns weights in the interval [0, 1].

Within the framework of the CBR approach, the similarity measure is one of the key performance factors. It alludes to a range of ratings and measurements employed to analyse the distinctions between various kinds of data. This parameter is crucial since it identifies the neighbours with the new project's attributes nearest to one another or most comparable to it. Furthermore, a common parameter in the retrieval stage of CBR is the k nearest neighbour. Even though this parameter is easily implemented, the CBR technique's performance will be impacted by the k-value selection, which is entirely subjective [11].

The steps involved in using the CBR approach are as follows [13] and as shown in Fig. 5.1.

1. Retrieve
2. Reuse
3. Revise
4. Retain.

Retrieval Phase: Identifying a group of previous projects (cases) comparable to the present project is the first stage in the CBR process. Usually, factors including project size, complexity, team experience, and technology utilized are considered to establish similarity.

Reuse Phase: After identifying the most comparable situations, the effort related to these earlier projects is adapted to meet the current project's needs. This could entail modifying the effort in light of variations in the project's attributes.

Revise Phase: Evaluating the efforts of the retrieved cases yields the estimated effort for the new project. This can be achieved by using more advanced approaches to weight the cases according to their significance or by averaging the efforts of the most similar situations.

Retention Phase: After a project is completed, estimated and actual efforts are compared, and the project data is saved to the case library for future use. The algorithm

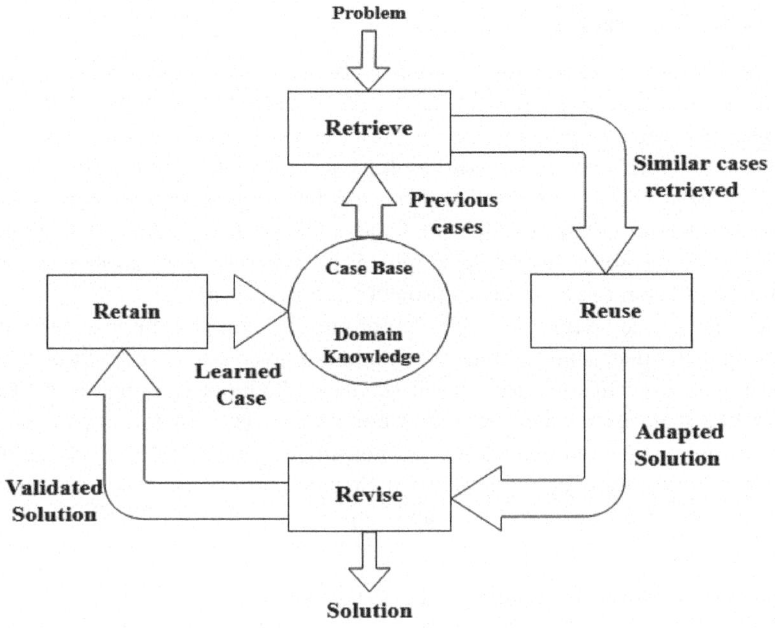

Fig. 5.1 Phases of CBR

learns from the success or failure of previous estimates, improving over time. Case adaptation is then used to modify the retrieved solution to better fit new problems, adding intelligence to the CBR process. Adaptation is crucial for CBR's effectiveness, transforming it from a basic pattern-matching approach into a more sophisticated, adaptive system.

There are numerous methods for modifying CBR initiatives [11]:

In Case-Based Reasoning (CBR), a retrieved solution can be applied to a new problem either as is or with modifications if it doesn't fully suit the current situation. Prior methods can be re-implemented, with or without adjustments, based on their relevance to the new context. When multiple cases are recalled, they may produce either a single solution or several alternative solutions for consideration.

Example of CBR Usage

Imagine a software development firm is responsible for estimating the effort needed for a new web application project. The firm has access to a case library filled with data from similar past projects. The firm identifies the most comparable cases by applying CBR, such as previous web applications with similar features and team sizes. By examining the effort involved in these past projects and adjusting for any differences, the firm can more accurately estimate the effort required for the new project.

III. **Artificial Neural Network (ANN)**

The ability of Artificial Neural Networks (ANNs) to simulate intricate, non-linear interactions seen in software projects has made ANNs more and more valuable in the assessment of software projects. ANNs may use vast datasets to discover complex patterns and interactions between different project attributes and the effort required, in contrast to standard estimation approaches that frequently rely on professional opinion or oversimplified assumptions. ANNs are an effective tool for estimating the effort required for software development because of their capacity to manage complexity and adapt to a variety of project conditions [15].

Artificial Neural Networks (ANNs) mimic the human brain's information processing using interconnected neurons: an input layer, one or more hidden layers, and an output layer. In software effort estimation, the input layer represents project features like team experience, size, and complexity. Hidden layers process these inputs through weighted connections and non-linear activation functions, capturing complex relationships. The output layer then generates an effort estimate, enabling ANNs to learn from historical data and improve estimations based on identified patterns.

Structure of ANN for Estimating Software Effort

Input Layer

To estimate software effort, an ANN's input layer includes neurons representing key project features. These inputs, such as project size (LOC or FP), complexity (cyclomatic complexity, module count), team expertise, project type, and development environment, form the foundation for the estimation process. Each feature is transformed through hidden layers, enabling the network to capture complex relationships among variables. Ultimately, this produces an accurate effort prediction at the output layer.

Hidden Layers

In an ANN, inputs from the input layer are processed by neurons in the hidden layers, where each neuron calculates a weighted sum of its inputs, adds a bias term, and applies an activation function to introduce non-linearity. This non-linearity enables the ANN to detect complex patterns and correlations between input features and the target output (effort estimation). Hyperparameters, such as the number of hidden layers and neurons per layer, can be adjusted to optimize model performance, allowing the ANN to model intricate relationships between project attributes and required effort.

Output Layer

An ANN's output layer typically contains a single neuron that represents the estimated effort, expressed in person-hours, person-days, or person-months. After processing the input through hidden layers, this neuron produces a final prediction useful for resource allocation and project planning. While appearing simple, the

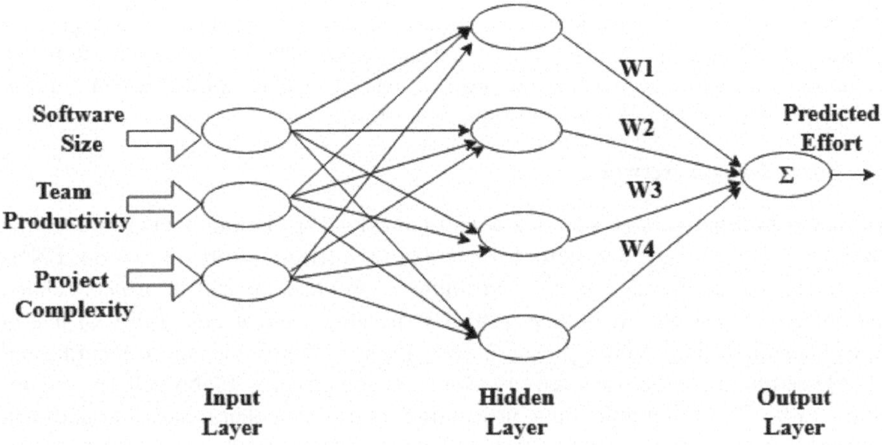

Fig. 5.2 Architecture of ANN

output layer encapsulates the complex interactions learned by the hidden layers, translating them into a concrete estimation of the effort required.

The input layer collects essential project data, the hidden layers process and model the data, and the output layer produces an ultimate estimate that may be utilized for decision-making. Each level is essential to the ANN's ability to correctly anticipate software effort [16].

The architecture of ANN in software effort estimation [17] is shown in Fig. 5.2. Let's look at an example dataset, such as the NASA93 dataset, which contains a history of software project data with numerous attributes pertinent to effort estimation, to demonstrate how ANNs work.

The ANN is trained on the NASA93 dataset by comparing its effort predictions to actual recorded efforts. A loss function calculates the error, and backpropagation adjusts weights and biases to capture underlying patterns. Over multiple iterations, the ANN learns to make accurate predictions for new projects by analyzing these patterns.

- The ANN begins with an input layer made up of neurons that reflect important project features in the framework of the NASA93 dataset. The project size (measured in lines of code), team experience (estimated in years of experience), and complexity metrics (such cyclomatic complexity) could be examples of input features for this dataset. Data obtained from the NASA93 dataset is supplied into each neuron in this layer, which correlates to one of these attributes.
- In the hidden layer the network can learn these complicated associations from the NASA93 dataset thanks to the hidden layers, which allow the relationship between project size and complexity to be non-linear.
- The final prediction, in this case the anticipated effort needed for the project, is provided by the output layer of the artificial neural network. This could be expressed in person-hours or person-months for the NASA93 dataset. Following

processing across the hidden layers, an estimate is generated by the output layer using the relationships and patterns that it has learned. During training, the network modifies its weights and biases based on past data from the NASA93 dataset, which enhances its accuracy in predicting effort.

IV. **Fuzzy Neural Network**

A fuzzy neural network (FNN) is a potent hybrid ML technique that combines the strengths of fuzzy logic and neural networks. In software effort estimation, FNNs effectively handle the uncertainty and imprecision often present in project-related data, which is typically qualitative [18]. By merging the learning ability of neural networks with the interpretive power of fuzzy logic, FNNs provide more accurate and reliable estimates of the work required for software projects. FNNs achieve precise estimates by fuzzifying input data, processing it through a neural network, and then defuzzifying the output to yield a final effort estimate.

Input Fuzzification: The crisp input variables (such as "Lines of Code," "Team Experience," and "Project Complexity") are converted into fuzzy values during the fuzzification process. Membership functions are used in this process to classify the inputs into fuzzy sets like "Low," "Medium," and "High." If "Team Experience" is used as input, a team with moderate experience may partially belong to both the "Medium" and "High" groups, with membership values of 0.6 and 0.4, respectively [18].

Neural Network Processing: Next, a neural network receives the inputs that have been fuzzified. Usually, this neural network consists of several layers containing hidden layers whereby learning occurs. The network minimizes the discrepancy between the estimated and actual effort values by learning from past project data and modifying the weights and biases of the connections between neurons [18]. The resilience of the model is increased by the fuzzy logic component, which enables the network to handle imprecise inputs more effectively.

Defuzzification of Outputs: The neural network generates an output but retains its fuzzy form after processing the inputs [18]. This output is transformed back into a crisp value in the defuzzification stage, which represents the estimated work in a particular unit (person-months, for example). This phase usually entails applying a technique similar to the centroid calculation, at which the fuzzy set's "center of gravity" is used to determine the final result.

Figure 5.3 illustrates the steps needed in estimating software effort using FNN [18]. The NASA93 dataset, well-known in the software engineering community, will be examined to understand better how FNNs function in software effort estimation. There are ninety-three software projects with different attributes in the NASA93 dataset, such as KLOC (thousands of lines of code), Complexity, Language, and Effort.

- The NASA93 dataset has fuzzified every attribute. For instance, established membership methods could be used to fuzzify KLOC into groups such as "Small,"

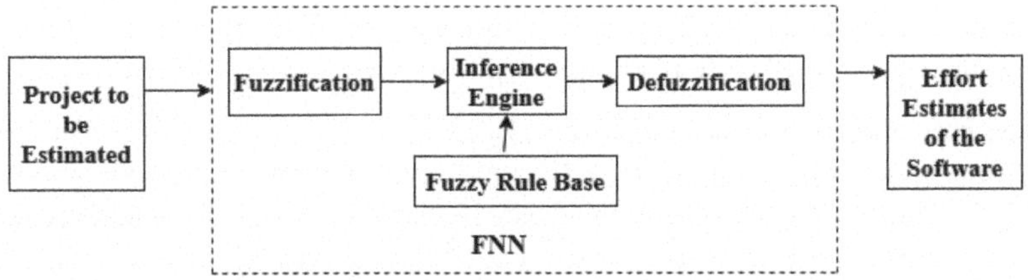

Fig. 5.3 Estimation of efforts using FNN

"Medium," and "Large". Similar fuzzification of the complexity could be applied with classifications such as "Low," "Moderate," and "High."

- Inputs that have been fuzzified, such as membership values for KLOC and Complexity, are passed into the neural network. With weights adjusted to minimize the discrepancy between estimated and real efforts, the network is trained on a subset of the NASA93 dataset. Because of the imprecision of the inputs, the model accounts for these by learning the correlations between various input characteristics and the effort required during training.
- The FNN can be used to project effort for new projects once it has been trained. For instance, the fuzzification process may identify a new project as "Medium" with a membership value of 0.7 in "Medium" and 0.3 in "Large" if its KLOC is 50. These variables are then processed by the neural network to estimate the effort. The defuzzification process, which comes last, transforms the fuzzy output into a precise effort estimate, like 1200 person-hours.

Fuzzy neural networks (FNNs) provide an advanced, reliable approach to software effort estimation by combining fuzzy logic reasoning with neural network learning. Their complexity enables them to deliver accurate, resilient estimates, making FNNs valuable for software project management.

V. **Ensemble Learning**

Ensemble learning integrates multiple models to improve prediction accuracy and robustness, making it valuable in software effort estimation. By merging outputs from diverse models such as neural networks, decision trees, and regression models, ensemble learning captures different data aspects and compensates for individual model biases. This approach provides more reliable estimates than single models, addressing the complexity of software projects [19]. Ensemble methods are generally categorized into three main types.

1. Bagging
2. Boosting
3. Stacking.

Bagging in Software Effort Estimation: Bagging improves software effort estimation by training multiple models, like decision trees or neural networks, on different subsets of data (e.g., NASA93 dataset). Each model provides an effort estimate, and the final prediction is the average of all estimates [19]. This approach helps reduce variance and overfitting, increasing prediction stability and reliability.

Boosting in Software Effort Estimation: Boosting enhances software effort estimation by combining simpler models, each focused on correcting the errors of its predecessor. For example, it may start with a basic linear regression model and progress to more complex models like decision trees or neural networks. Each new model targets projects where previous estimates were less accurate, improving overall prediction accuracy [19]. This technique is particularly effective for datasets with complex, non-linear relationships, significantly boosting the precision of effort estimates.

Stacking in Software Effort Estimation: Stacking combines predictions from multiple base models (e.g., neural networks, decision trees, linear regression) using a meta-model, which could be another neural network or regression model. In software effort estimation, base models process different project data elements, and the meta-model integrates these predictions into a final estimate. Stacking is particularly effective when base models have complementary strengths, resulting in a more reliable and accurate overall prediction [19].

The method of ensemble learning in software effort estimation is depicted in Fig. 5.4. The process begins with an input layer where data from software projects, such as their size and complexity, are fed into a variety of base models, such as decision trees, neural networks, and linear regression. In order to create a single, more precise Final Output that represents the estimated software effort, the results from different models are then pooled in an Aggregation process.

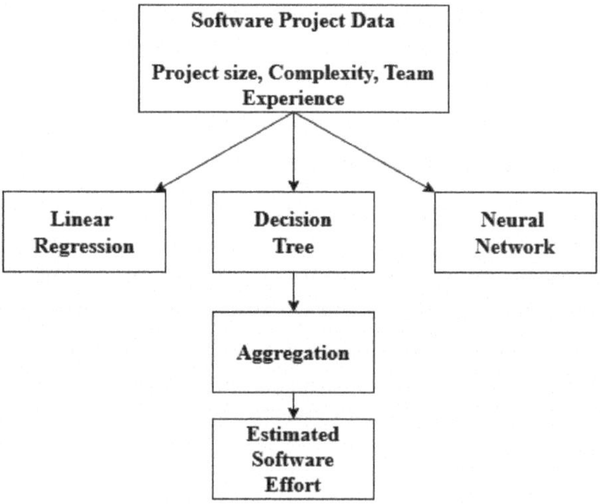

Fig. 5.4 Ensemble learning for software effort estimation

5.3 Challenges in Effort Estimation

Software effort estimation presents unique issues for both traditional & machine learning techniques. While machine learning models struggle with issues of interpretability, computational complexity, overfitting, and data needs, traditional models suffer from subjectivity, rigidity of assumptions, and data limits. The project's unique context, the data's accessibility, and the requirements of interpretability and scalability all play a role in selecting the best model.

Traditional Models

- COCOMO assumes a linear or semi-linear relationship between software size and effort, which may not hold for complex projects with non-linear growth. Its accuracy depends on extensive historical data for parameter calibration, and many organizations struggle with incorrect estimates due to insufficient data or improper calibration. Additionally, COCOMO is less adaptable to dynamic environments like Agile, as it struggles with changing project conditions or requirements. Estimating program size and other input characteristics often involves subjectivity, leading to potential inaccuracies in the estimation process.
- Function point analysis (FPA) requires a deep understanding of system functionality, which can be time-consuming and prone to misinterpretation. It relies on well-defined criteria, making it difficult to apply to projects with evolving or unclear requirements. FPA primarily focuses on functional requirements, potentially overlooking non-functional factors like performance, security, or maintainability. Additionally, the subjective assignment of weights to different functionalities can lead to inconsistent estimates.
- Expert opinion is highly subjective and prone to biases, such as optimism bias, where professionals overestimate effort due to deadline pressure or overconfidence. The accuracy of expert judgment depends on the estimator's experience and knowledge, with overconfident or inexperienced estimators likely to produce inaccurate estimates. Different experts may provide varying estimates for the same project, causing discrepancies that can affect resource allocation and planning. For large or complex projects, expert judgment becomes less scalable as individual experience may not capture the full scope of work required.

Machine Learning Models

Software effort estimation can be done in a data-driven manner using machine learning models, such as ensemble approaches, decision trees, and neural networks. Although they have some advantages over conventional models, they also have particular challenges.

- ANNs require more high-quality, labeled data for practical training, but many organizations lack sufficient historical data. Inadequate data or overly complex networks can lead to overfitting, reducing performance on new applications. Additionally, ANNs are often considered "black-box" models, making it difficult to interpret their reasoning, which can hinder adoption in situations requiring transparency. Additionally, training ANNs can be time-consuming and computationally demanding, mainly when dealing with complicated network architectures or big datasets.

- Training multiple models in parallel, as in ensemble methods like boosting or bagging, can be computationally expensive and time-consuming, especially with large datasets. While ensemble methods often provide more accurate estimates than single models, they can make the estimation process harder to explain to stakeholders. Booster techniques, in particular, are prone to overfitting if not properly controlled, especially with noisy or high-dimensional data. Additionally, ensemble methods may be difficult to apply and require extensive adjustments to hyperparameters, which can be challenging for practitioners with limited machine-learning experience.
- Designing a fuzzy neural network (FNN) requires expertise in both fuzzy logic and neural networks. Developing appropriate fuzzy rules and membership functions adds complexity, and while fuzzy logic aims to improve interpretability, the combined model may still be difficult to understand. FNNs are more computationally demanding than standard neural networks due to additional fuzzy logic operations, making them slower to train, especially with large datasets. FNNs also require tuning multiple parameters for both the neural network and fuzzy system, often through trial and error. Like other machine learning models, FNNs need high-quality, complete data for accurate effort estimation.
- Linear regression assumes a linear relationship between effort and input factors, which can be inaccurate for software projects due to the frequent presence of non-linear relationships [20]. Non-linear regression models may overfit the training data, especially with many input variables, leading to poor generalization for new projects [21]. Multicollinearity, or high correlation between independent variables, can cause instability in regression models and make it hard to assess the impact of each variable. Regression models are also sensitive to the quality of input data, with outliers or missing data potentially skewing estimates. Handling categorical data can be challenging unless properly encoded, and simple regression models may overlook complex relationships better captured by non-linear or ensemble methods.

CBR's effectiveness is based on the case quality of the case base, as outdated or incomplete cases can limit the model's ability to identify relevant analogies. Retrieving pertinent cases from a large case base can be computationally demanding, and as the number of cases increases, the process may slow down [22]. The adaptation phase, which adjusts solutions to fit new contexts, can become subjective and complex, leading to inaccurate estimates. Maintaining the case base becomes more difficult as it grows, requiring continuous updates to ensure cases are relevant. Overreliance on historical data can cause overfitting, particularly when cases are too similar, and subjectivity in selecting features can affect estimation consistency. CBR struggles when no close match exists in the case base, leading to less accurate estimates.

5.4 Case Studies of Machine Learning Models for Effort Estimation

In this section, case studies were performed using linear and ridge regression, bagging and boosting ensemble models, and CBR, ANN, and ANFIS models as effort estimators. Data collection, performance measures, and case studies are explained throughout the section.

5.4.1 *Data Collection*

The quality of the dataset used in effort estimation models is crucial for accurate predictions in software development. To minimize bias and improve reliability, multiple data sources and machine learning techniques were employed. Five benchmark datasets—Albrecht, China, Cocomo81, Kemerer, and Maxwell—were selected for evaluating machine learning models. These datasets typically contain metrics like project size, complexity, functional points, tool skills, and effort, which are essential for estimating software development efforts. Table 5.1 shows a detailed summary of these five datasets, outlining the project count (instances), the features count (attributes, metrics), and the effort units (Person-Months (PM) and Person-Hours (PH)). Effort estimation involves predicting the work required for software development despite challenges such as incomplete, uncertain, and noisy data. The five selected datasets vary in project size, with Albrecht and Kemerer featuring fewer projects, while China, Cocomo81, and Maxwell include larger datasets. These datasets exhibit common challenges like multi-collinearity, outliers, and uneven effort distribution, which complicates accurate effort estimation.

(1) **Albrecht dataset**: The Albrecht dataset consists of 24 software programs developed using languages such as PL1, COBOL and other database management languages. It features six independent numerical attributes and one dependent numerical attribute. Detailed description of the Albrecht dataset is provided in Table 5.2.

Table 5.1 Summary of software effort estimation datasets

S. No	Datasets	#Projects	#Features	Effort Unit
1	Albrecht	24	8	PM
2	Cocomo81	63	17	PM
3	China	499	18	PH
4	Kemerer	15	7	PM
5	Maxwell	62	27	PH

Table 5.2 Albrecht dataset description

S. No.	Features	Description
1	Input	The total number of inputs a program processes
2	Output	The total number of outputs generated by a program
3	Inquiry	The number of inquiries or queries a program must handle
4	File	The total number of files a program reads from or writes to
5	FPAdj	Used to adjust raw function points called Function Point Adjustment Factor
6	RawFPcounts	The raw function points, calculated using Function Point Metrics
7	AdjFP	The adjusted function points = FPAdj * RawFPcounts
8	Effort	Effort needed to develop software

(2) **China dataset**: The China dataset, used for predicting software effort, includes 19 attributes and encompasses a total of 499 project instances. Table 5.3 provides descriptive statistics for the China.

(3) **COCOMO81 dataset**: The COCOMO81 dataset has been commonly utilized to validate various effort estimation methods. This dataset includes 63 software projects, each characterized by 18 attributes and a specific effort measurement.

Table 5.3 China dataset description

S. No.	Features	Description
1	AFP	Adjusted function points
2	Input	Input associated function points
3	Output	External output's function points
4	Enquiry	F unction points for inquiries related to external outputs
5	File	Internal logical files function points
6	Interface	External interface function points
7	Added	Function points for newly added functions
8	Changed	Function points of functions that have been modified
9	Deleted	Function points for functions that have been removed
10	PDR_AFP	The Productivity delivery rate of adjusted function points
11	PDR_UFP	The Productivity delivery rate of Unadjusted function points
12	NPDR_AFP	Normalized PDR of adjusted function points
13	NPDR_UFP	Normalized PDR of Unadjusted function points
14	Resource	Type of development team
15	Dev. type	Type of development process
16	Duration	Total time taken to complete the project
17	N_effort	Effort level adjusted for normalization
18	Effort	Summary of the total work effort

Effort in the COCOMO81 dataset is quantified in person-months. Table 5.4 offers a comprehensive description of the COCOMO81 dataset. To reduce the effort required for project development, increase a few features and decrease a few feature values.

(4) **Kemerer dataset**: The Kemerer dataset described using six metrics and one effort metric quantified in "man-months" and a total of 15 software projects included in Kemerer dataset. The dataset features two categories and four numerical attributes that together encompass the six qualities. Table 5.5 provides a detailed description of the Kemerer.

Table 5.4 Cocomo81 dataset description

S. No.	Features	Description
1	Rely	The level of software reliability required
2	Data	The size of the database
3	Cplx	The complexity of the product
4	Time	Constraints related to execution time
5	Stor	Constraints on main storage capacity
6	Virt	The volatility of the virtual machine
7	Turn	The required turnaround time
8	Acap	The capability of the analyst
9	Aexp	Experience of applications
10	Pcap	The capability of the programmer
11	Vexp	Experience of Virtual machine
12	Lexp	Experience with the programming language
13	Modp	The use of modern programming practices
14	Tool	The use of software tools
15	Sced	The Required schedule for development
16	Loc	The number of Lines of code
17	Actual	The needed effort expended in person-months

Table 5.5 Kemerer dataset description

S. No.	Metrics	Description
1	Language	The employed programming language for the project
2	Hardware	The utilized hardware type for the project
3	Duration	The project's duration measured in months
4	KSLOC	The estimated size of the project is in the thousands of source lines of code
5	AdjFP	Function points after adjustment
6	RAWFP	Function points before adjustment
7	EffortMM	Required Effort

Table 5.6 Maxwell dataset description

S. No.	Metrics	Description
1	Syear	The year the project began
2	App	Application name
3	Har	Used hardware platform
4	Dba	Database management system
5	Ifc	The technology used for the user interface
6	Source	The source code management system
7	Telonuse	Whether or not the project is using IBM Telon
8	Nlan	The number of programming languages utilized in the project
9	T01	Level of customer involvement
10	T02	Adequacy of the development environment
11	T03	Availability of staff
12	T04	Standards
13	T05	Methods
14	T06	Tools
15	T07	Logical complexity of the software
16	T08	Volatility
17	T09	Quality
18	T10	Efficiency
19	T11	Installation
20	T12	Skills of the staff in analysis
21	T13	Knowledge of the staff regarding the application
22	T14	Skills of the staff with tools
23	T15	Team skills of the staff
24	Duration	The duration of the project in months
25	Size	The project size, measured in lines of code
26	Time	The total time spent on the project, measured in person-months
27	Effort	The total amount of effort expended on the project, in person months

(5) **Maxwell dataset** The Maxwell dataset, sourced from a major commercial bank in Finland, includes 62 projects, each characterized by 23 attributes. The only numerical attribute is Project Size in Function Points. Table 5.6 provides a detailed summary of the Maxwell.

5.4.2 Evaluation Measures

This section outlines several performance measures, including RMSE (Root Mean Squared Error), MMRE (Mean Magnitude of Relative Error), BMMRE (Balanced

Mean Magnitude of Relative Error), MAE (Mean Absolute Error), and PRED (0.25) (Prediction), used to assess the effectiveness of effort estimation models. Due to the asymmetric distribution of MRE often leads to claims of bias in measures based on MRE, these indicators alone cannot be used to select the best model. However, despite these limitations, this case study includes them for comparative analysis, as they have been widely used in previous research. These performance assessment measures are defined as follows:

RMSE quantifies the difference between predicted values and actual values by computing the square root of the average squared differences. The RMSE is calculated as follows:

$$\mathrm{RMSE} = \sqrt{\frac{1}{n}\sum_{i=1}^{n}(\mathrm{AE}_i - \mathrm{EE}_i)^2}$$

MAE measures the average magnitude of the errors in a set of predictions, without considering their direction (i.e., it treats overestimates and underestimates equally). Taking the absolute difference between predicted and the actual values ensure that errors are always positive. The MAE is calculated as follows:

$$\mathrm{MAE} = \frac{1}{n}\sum_{i=1}^{n}|\mathrm{AE}_i - \mathrm{EE}_i|$$

MMRE measures how close the predicted values are to the actual values in terms of magnitude. It is particularly useful in cases where the scale of the predictions and actual values can vary significantly. MMRE measures the average magnitude of the prediction errors relative to the actual values. The MMRE is calculated as follows:

$$\mathrm{MRE} = \frac{|\mathrm{AE}_i - \mathrm{EE}_i|}{\mathrm{AE}_i}$$

$$\mathrm{MMRE} = \frac{1}{n}\sum_{i=1}^{n}\mathrm{MRE}_i$$

BMMRE is a variation of the MMRE, used primarily in software engineering and effort estimation. The BMMRE adjusts the MMRE by adding a small constant (Threshold) to the actual values in the denominator. This adjustment helps to prevent extremely high error values when actual values are very small and ensures that the measure is more stable and less sensitive to outliers, which can otherwise skew the results. The BMMRE is calculated as follows:

$$\text{BMMRE} = \frac{1}{n}\sum_{i=1}^{n}\left|\frac{\text{AE}_i - \text{EE}_i}{\min(\text{AE}_i,\ \text{EE}_i)}\right|$$

1. **PRED (0.25)** measures the proportion of predictions that fall within a certain relative error threshold, in this case, 25%. PRED (0.25) represents the percentage of predictions for which the relative error is within 25% of the actual value. The formula for PRED (0.25) is:

$$\text{PRED} = \frac{A}{n}$$

$$A = \sum_{i=1}^{n}\begin{cases} 1, & \textit{if } \text{MRE}_i \geq 0.25 \\ 0, & \textit{otherwise} \end{cases}$$

Here 'AEi' represent actual efforts and 'EEi' represent estimated efforts, while 'n' and 'm' represent the number of projects and the number of independent features in the dataset. Lower values of RMSE, MAE, MMRE, and BMMRE indicate that the model's predictions are, closer to the actual values, reflecting better model performance with relatively small errors. Conversely, a high PRED (0.25) value suggests that a significant proportion of the predictions fall within the specified error margin of the actual values, indicating strong performance in terms of relative accuracy.

Making decisions amidst high uncertainty poses a major challenge in software engineering. In this study, we created a software effort estimation model to forecast effort levels using traditional machine learning techniques, ensemble learners, Case-Based Reasoning (CBR), Adaptive Neuro-Fuzzy Inference Systems (ANFIS), and Artificial Neural Networks (ANN).

5.4.3 Case Study: Traditional Machine Learning Models for Effort Estimation

The case study performed using linear and ridge regression as estimation models over above mentioned five distinct software datasets. During the validation process, 70 and 30% splitting for training and testing data considered. The results have been reported in Tables 5.7 and 5.8, in terms of RMSE, MAE, MMRE, BMMRE, and PRED in addition to the bar plots shown in Figs. 5.5 and 5.6.

Maxwell has the highest RMSE of 7599.745, suggesting significant errors in predictions compared to the other datasets. Lower MAE values suggest better accuracy. The China dataset has the highest MAE (391.043), indicating larger average errors compared to datasets like Albrecht with a lower MAE (10.369). The China dataset has the lowest MMRE (0.217), suggesting that predictions are relatively close to actual values when normalized by the actual values. In contrast, Cocomo81 has a very high MMRE (15.433), indicating poor prediction accuracy relative to its actual

Table 5.7 LR estimation model performance measures

Datasets	RMSE	MAE	MMRE	BMMRE	PRED
Albrecht	13.720	10.369	1.353	0.772	0.275
China	930.658	391.043	0.217	0.252	0.76
Cocomo81	1800.638	1001.79	15.433	5.296	0.089
Kemerer	289.649	231.487	2.204	0.587	0.16
Maxwell	7599.745	5693.837	1.735	0.694	0.205

Table 5.8 RR estimation model performance measures

Datasets	RMSE	MAE	MMRE	BMMRE	PRED
Albrecht	18.669	11.187	0.844	0.602	0.263
China	859.024	375.310	0.226	0.22	0.741
Cocomo81	1124.368	544.517	8.094	2.887	0.174
Kemerer	276.504	210.072	1.662	0.704	0.160
Maxwell	7270.310	4795.037	1.267	0.244	0.300

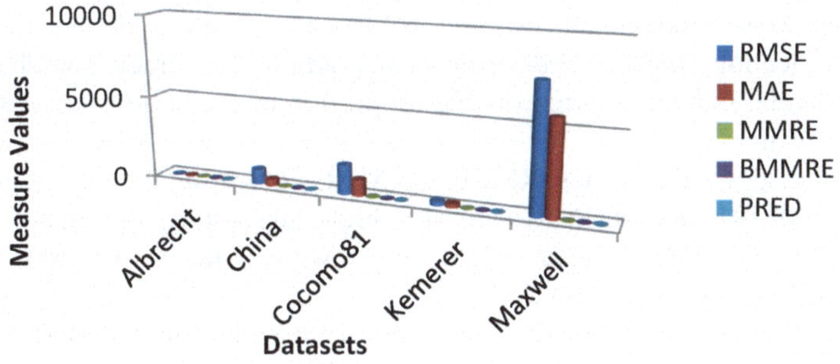

Fig. 5.5 Performance of LR model

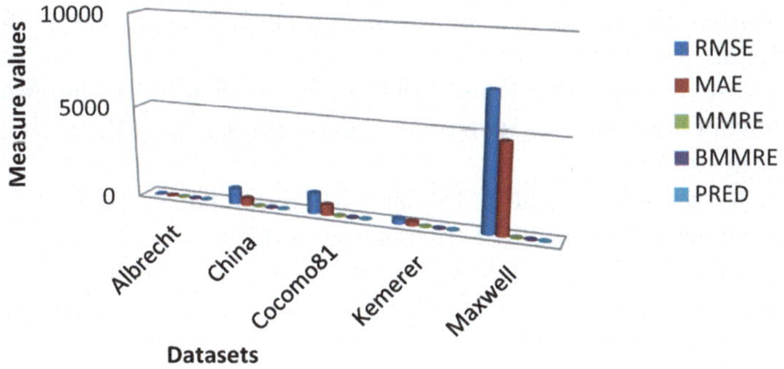

Fig. 5.6 Performance of RR model

values. China has the lowest BMMRE (0.252), suggesting that predictions are relatively balanced and accurate across different scales. Cocomo81 shows a much higher BMMRE (5.296), indicating less accurate predictions when scaled. China has the highest PRED (0.760), meaning a higher percentage of predictions are within 25% of the actual values. Cocomo81 has the lowest PRED (0.089), showing that only a small fraction of predictions are within the 25% error margin.

In summary linear regression model over Albrecht dataset shows relatively low RMSE and MAE, indicating reasonably accurate predictions. The MMRE and BMMRE are also moderate, suggesting a balanced prediction performance. The PRED value is moderate, showing that a fair proportion of predictions are within 25% of the actual values.

Over china dataset exhibits very high RMSE and MAE, but the lowest MMRE and BMMRE, indicating relatively accurate predictions in terms of relative error, despite the large absolute error. The high PRED value suggests that many predictions are within the acceptable error margin.

Cocomo has extremely high RMSE and MAE, and very high MMRE and BMMRE, indicating poor prediction accuracy. The low PRED value shows that very few predictions are within the 25% error threshold.

Kemerer shows moderate RMSE and MAE values, with moderate MMRE and BMMRE, reflecting a balanced performance in prediction accuracy. The PRED value is also moderate, indicating a reasonable proportion of predictions are within 25% of the actual values.

Maxwell displays the highest RMSE and MAE, indicating the largest prediction errors. The MMRE and BMMRE are also high, suggesting significant errors in relative terms. The PRED value is relatively low, showing fewer predictions are within the 25% error range (Table 5.7).

Ridge regression over Albrecht dataset has increased RMSE and MAE compared to LR, suggesting larger errors. MMRE and BMMRE are lower, indicating improved relative error performance. PRED is slightly lower (26.3%) compared to LR.

Over China datasets, RMSE compared to LR model, but MAE is slightly higher. MMRE and BMMRE are similar, indicating consistent relative error performance. PRED is slightly lower (74.1%).

Over Cocomo dataset, RMSE and MAE are lower than in LR, but MMRE and BMMRE remain high, showing persistent relative error issues. PRED improves to 17.4%, though still low.

RR with Kemerer Improved RMSE and MAE compared to LR, with similar MMRE and higher BMMRE. PRED remains low (16%).

Slightly reduced RMSE and MAE compared to LR, with lower MMRE and BMMRE. PRED is slightly higher (30%), showing improved performance within the 25% error margin (Table 5.8).

Overall, each dataset presents a different level of prediction performance. The RR estimation model generally shows improved performance in relative error metrics (MMRE and BMMRE) for most datasets, while RMSE and MAE are often higher compared to the LR model. PRED values are similar or slightly improved in the RR model, indicating better proportionate accuracy in some cases.

5.4.4 Case Study: CBR Model for Effort Estimation

This section empirically assesses the performance of the CBR-based estimation model using a series of five evaluation metrics across multiple datasets, each reflecting typical characteristics of software effort. Each dataset is analyzed independently due to differences in attribute domains, preventing them from being combined into a single, larger dataset. The results are summarized in Table 5.9 **and depicted as box plot in** Fig. 5.7.

Albrecht: Shows moderate RMSE and MAE, indicating reasonable absolute prediction errors. The MMRE and BMMRE values are moderate, suggesting relatively balanced accuracy in proportion to actual values. The PRED value of 38.8% indicates that a significant proportion of predictions (around 39%) are within 25% of the actual values, reflecting good prediction performance.

China: Exhibits very high RMSE and MAE, indicating large absolute errors. However, it has the lowest MMRE and BMMRE, suggesting good performance relative to actual values. The very high PRED value of 95.8% means that most predictions (nearly 96%) fall within 25% of the actual values, demonstrating excellent relative prediction accuracy.

Cocomo81: Shows high RMSE and MAE, pointing to significant absolute errors. The MMRE and BMMRE are high, reflecting poor relative error performance. The

Table 5.9 CBR estimation model performance measures

Datasets	RMSE	MAE	MMRE	BMMRE	PRED
Albrecht	13.3	8.218	0.636	0.83	0.388
China	1227.363	410.103	0.113	0.12	0.958
Cocomo81	966.222	446.060	1.509	2.041	0.132
Kemerer	230.763	140.937	0.594	0.947	0.34
Maxwell	4747.699	3318.335	0.632	0.764	0.274

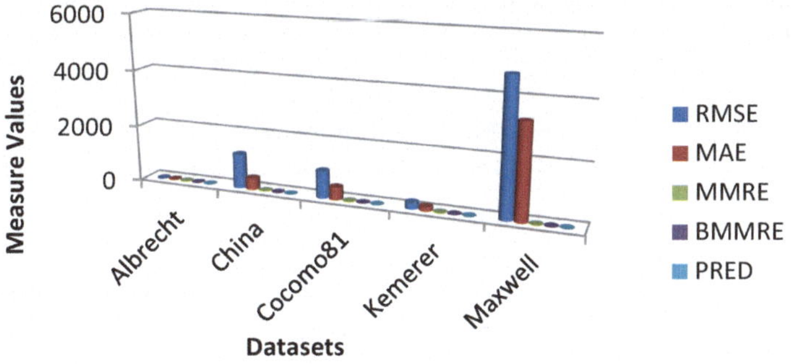

Fig. 5.7 Performance of CBR model

low PRED value of 13.2% suggests that only a small fraction of predictions are within 25% of the actual values, indicating lower prediction accuracy.

Kemerer: Displays moderate RMSE and MAE, indicating decent absolute prediction accuracy. The MMRE is moderate, while the BMMRE is relatively high, suggesting some issues with relative error. The PRED value of 34.0% shows that about a third of the predictions are within 25% of the actual values, which is reasonable but not as high as in China.

Maxwell: Shows very high RMSE and MAE, indicating substantial absolute prediction errors. The MMRE and BMMRE are moderate, suggesting relatively balanced performance in proportion to actual values. The PRED value of 27.4% indicates that a smaller proportion of predictions (about 27%) are within 25% of the actual values.

In summary China demonstrates the best relative prediction accuracy with the highest PRED value, despite having large absolute errors (RMSE and MAE). Cocomo81 and Maxwell have high absolute errors and lower prediction accuracy within the 25% error margin. Albrecht and Kemerer show moderate results across most metrics, with reasonable performance but not as strong as China in relative terms.

5.4.5 Case Study: Ensemble Model for Effort Estimation

This section presented and analyzed the results from experiments evaluating the performance of ensemble methods for effort estimation—specifically bagging and boosting. The objective is to determine whether these ensemble techniques offer improvements over traditional estimation methods. The performance of these methods are summarized in Tables 5.10 and 5.11 **and depicted as box plot in** Figs. 5.8 and 5.9.

Albrecht: Boosting shows better performance with lower RMSE (13.105 vs. 15.694) and MAE (9.047 vs. 10.218) compared to bagging. However, bagging has a slightly lower MMRE (0.554 vs. 0.681) and BMMRE (0.880 vs. 0.820), indicating that boosting may have a higher proportion of relative errors but more accurate overall predictions.

Table 5.10 Bagging estimation model performance measures

Datasets	RMSE	MAE	MMRE	BMMRE	PRED
Albrecht	15.694	10.218	0.554	0.88	0.362
China	1277.244	413.472	0.101	0.107	0.964
Cocomo81	976.942	456.400	1.969	2.223	0.184
Kemerer	160.962	10.81	0.624	0.803	0.460
Maxwell	5132.702	2918.112	0.597	0.647	0.437

Table 5.11 Boosting estimation model performance measures

Datasets	RMSE	MAE	MMRE	BMMRE	PRED
Albrecht	13.105	9.047	0.681	0.820	0.388
China	1241.450	387.300	0.117	0.122	0.939
Cocomo81	967.818	435.488	1.997	2.109	0.153
Kemerer	162.092	108.192	0.846	1.061	0.300
Maxwell	5459.27	3068.598	0.539	0.651	0.442

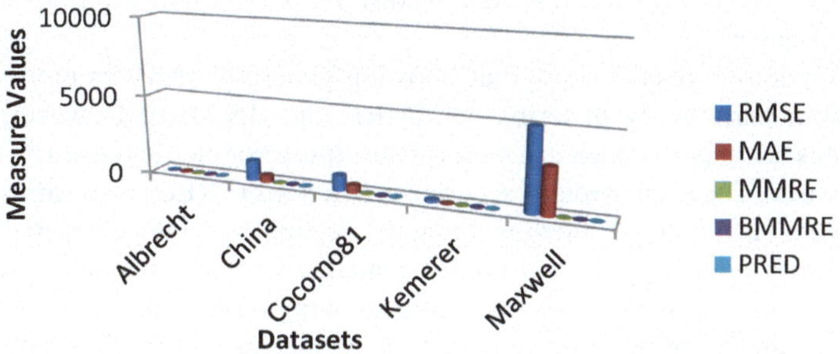

Fig. 5.8 Performance of bagging model

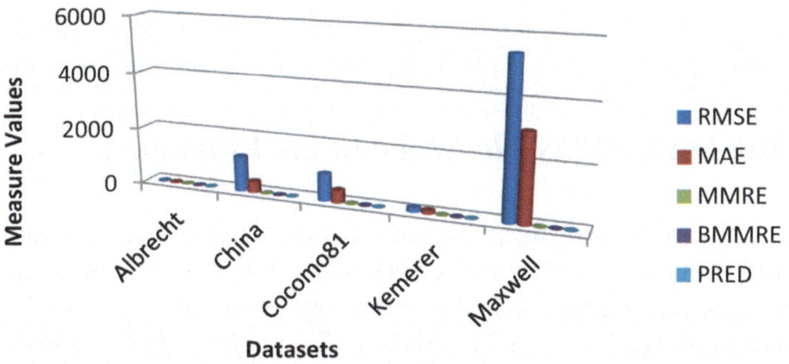

Fig. 5.9 Performance of boosting model

China: Both methods show similar performance, but boosting has slightly higher RMSE and MAE compared to bagging. However, boosting still performs better in terms of MMRE (0.117 vs. 0.101) and BMMRE (0.122 vs. 0.107), and the PRED value is slightly lower for boosting (93.9% vs. 96.4% for bagging), indicating similar predictive accuracy with marginal trade-offs.

Cocomo81: Bagging performs better in terms of RMSE (976.942 vs. 967.818) and MAE (456.400 vs. 435.488). Both methods show high MMRE and BMMRE values, but bagging has a higher PRED value (18.4% vs. 15.3%), suggesting better performance in predicting effort within the 25% margin.

Kemerer: Bagging achieves lower RMSE (160.962 vs. 162.092) and MAE (10.810 vs. 108.192), with a better PRED (46.0 vs. 30.0%). However, boosting has a lower MMRE (0.846 vs. 0.624) and BMMRE (1.061 vs. 0.803), reflecting more balanced error distribution.

Maxwell: Bagging has lower RMSE (5132.702 vs. 5459.270) and MAE (2918.112 vs. 3068.598), with a better PRED (43.7% vs. 44.2%). Boosting, however, has lower MMRE (0.539 vs. 0.597) and BMMRE (0.651 vs. 0.647), indicating better relative error performance.

The comparative results show that boosting generally provides more balanced error metrics, particularly in terms of MMRE and BMMRE. However, bagging demonstrates better performance in terms of absolute error metrics (RMSE and MAE) in several cases. The proportion of accurate predictions (PRED) also varies between methods, with each method showing strengths depending on the dataset.

In summary, both bagging and boosting methods offer valuable improvements in software effort estimation, with boosting showing more consistent relative error performance, and bagging offering better absolute error metrics. The effectiveness of each method can vary significantly depending on the dataset characteristics. So choosing the appropriate method depends on the specific needs of the software development project and the characteristics of the available data (Tables 5.10 and 5.11).

5.4.6 Case Study: ANN Model for Effort Estimation

In this case study, we evaluate the performance of an Artificial Neural Network model for estimating software development effort. The ANN model focuses on accuracy and reliability in effort estimation. The results are summarized in Table 5.12 **and depicted as box plot in** Fig. 5.10. The ANN model employed in this study comprised three layers: an input layer with nodes representing the features, a hidden layer, and an output layer dedicated to effort estimation. The hidden layer utilized the ReLU (Rectified Linear Unit) activation function, while the output layer applied a linear activation function. The model was trained using the backpropagation algorithm with a learning rate of 0.01 and a batch size of 32. Training was conducted over 50–100 epochs with early stopping to prevent overfitting.

Albrecht: The ANN model exhibits an RMSE of 14.627 and MAE of 10.931. The MMRE (1.297) and BMMRE (1.345) are higher, indicating relatively higher errors compared to other models. The PRED value of 25.0% shows that a quarter of the predictions fall within 25% of the actual values.

China: The model performs exceptionally well with an RMSE of 829.417 and MAE of 311.446. The low MMRE (0.099) and BMMRE (0.103) suggest very accurate

Table 5.12 ANN estimation model performance measures

Datasets	RMSE	MAE	MMRE	BMMRE	PRED
Albrecht	14.627	10.931	1.297	1.345	0.250
China	829.417	311.446	0.099	0.103	0.965
Cocomo81	1372.891	519.934	1.207	2.105	0.153
Kemerer	161.996	113.684	0.753	0.852	0.400
Maxwell	6276.224	3640.179	0.473	0.893	0.279

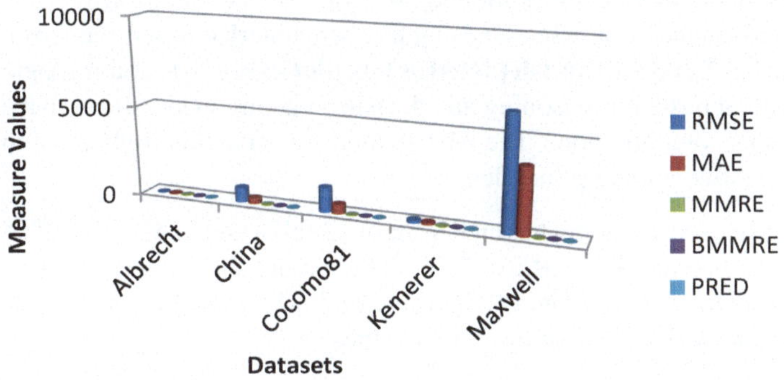

Fig. 5.10 Performance of ANN model

relative error performance. The PRED value of 96.5% indicates that the model makes accurate predictions within 25% of the actual values in most cases.

Cocomo81: The ANN model shows high RMSE (1372.891) and MAE (519.934), reflecting substantial absolute errors. The MMRE (1.207) and BMMRE (2.105) are also high, indicating significant relative errors. The PRED value of 15.3% suggests fewer predictions within the acceptable margin.

Kemerer: The ANN model has a RMSE of 161.996 and MAE of 113.684, which are relatively competitive. The MMRE (0.753) and BMMRE (0.852) indicate moderate relative errors. The PRED value of 40.0% reflects a reasonable proportion of accurate predictions.

Maxwell: The model's RMSE (6276.224) and MAE (3640.179) are high, showing considerable absolute errors. The MMRE (0.473) and BMMRE (0.893) are lower in relative terms, indicating better relative error performance. The PRED value of 27.9% shows a moderate accuracy level.

The ANN model demonstrates variable performance across different datasets. It excels with the China dataset, showing low relative errors and high accuracy in practical predictions. While it performs well on some datasets, there is room for improvement, especially in handling datasets like Cocomo81 and Maxwell, where it exhibits high absolute errors and lower prediction accuracy. The results suggest that further model refinement and hyperparameter tuning are necessary to improve performance. Additionally, the ANN model's effectiveness can vary significantly with different datasets (Table 5.12).

5.4.7 *Case Study: ANFIS Model for Effort Estimation*

This section details the outcomes of using the Adaptive Neuro-Fuzzy Inference System (ANFIS) model for estimating software development effort. The ANFIS model's performance was assessed using five benchmark datasets, and the results are summarized in Table 5.13 **and depicted as box plot in** Fig. 5.11. Fuzzy logic provides a mathematical basis for reasoning and decision-making in situations that are uncertain, imprecise, or ambiguous. The ANFIS model used in this study combines neural networks and fuzzy logic principles.

Albrecht: The ANFIS model has an RMSE of 13.6974 and MAE of 11.9508. With an MMRE of 1.3668 and BMMRE of 0.7672, the model shows moderate performance in terms of relative errors. The PRED value of 25.0% indicates that a quarter of the predictions are within 25% of the actual values.

Table 5.13 ANFIS estimation model performance measures

Datasets	RMSE	MAE	MMRE	BMMRE	PRED
Albrecht	13.6974	11.9508	1.3668	0.7672	0.25
China	1045.53	355.262	0.0986	0.0991	0.9879
Cocomo81	1192.39	432.44	1.2945	1.7504	0.2063
Kemerer	140.146	95.1225	0.4712	0.3357	0.3333
Maxwell	5275.59	3450.29	0.5453	0.7848	0.3045

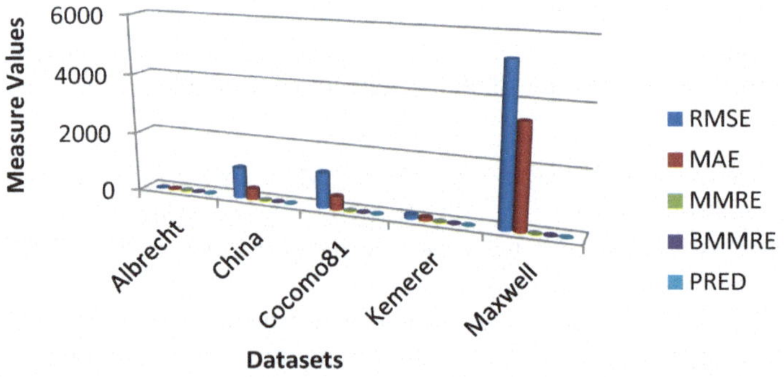

Fig. 5.11 Performance of ANFIS model

China: The model performs very well with an RMSE of 1045.53 and MAE of 355.262. The MMRE (0.0986) and BMMRE (0.0991) are among the lowest, demonstrating excellent relative error performance. The high PRED value of 98.79% shows that nearly all predictions fall within 25% of the actual values.

Cocomo81: With an RMSE of 1192.39 and MAE of 432.44, the ANFIS model displays significant absolute errors. The MMRE (1.2945) and BMMRE (1.7504) are high, indicating substantial relative errors. The PRED value of 20.63% is relatively low, suggesting fewer accurate predictions within the 25% margin.

Kemerer: The model has an RMSE of 140.146 and MAE of 95.1225, showing competitive performance in absolute terms. The MMRE (0.4712) and BMMRE (0.3357) indicate lower relative errors, and the PRED value is 33.33%.

Maxwell: The RMSE (5275.59) and MAE (3450.29) are high, indicating significant absolute errors. However, the MMRE (0.5453) and BMMRE (0.7848) are lower in relative terms. The PRED value of 30.45% reflects a moderate accurate prediction.

The ANFIS model demonstrates strong performance on the China dataset with very low relative errors and high prediction accuracy. However, the model performs less favorably on datasets like Cocomo81 and Maxwell, with higher absolute errors and lower prediction accuracy, which indicates a need for further refinement of the fuzzy rules and membership functions. Additionally, the variability in performance across different datasets suggests that dataset-specific adjustments are necessary (Table 5.13).

References

1. K. Molokken, M. Jorgensen, A review of software surveys on software effort estimation, in *2003 International Symposium on Empirical Software Engineering, 2003. ISESE 2003. Proceedings* (IEEE, 2003), pp. 223–230
2. J. Wen, S. Li, Z. Lin, Y. Hu, C. Huang, Systematic literature review of machine learning based software development effort estimation models. Inf. Softw. Technol. **54**(1), 41–59 (2012)
3. F. Ferrucci, C. Gravino, R. Oliveto, F. Sarro, Genetic programming for effort estimation: an analysis of the impact of different fitness functions, in *2nd International Symposium on Search Based Software Engineering* (IEEE, 2010) , pp. 89–98; F. Author, Article title. Journal **2**(5), 99–110 (2016)
4. K. El Emam, A. Günes Koru, A replicated survey of IT software project failures. IEEE software **25**(5), 84–90 (2008)
5. P. Manchala, M. Bisi, TSoptEE: two-stage optimization technique for software development effort estimation. Clust. Comput. 1–20 (2024)
6. B. Boehm, C. Abts, S. Chulani, Software development cost estimation approaches—a survey. Ann. Softw. Eng. **10**(1), 177–205 (2000)
7. S. Shilpi, International journal of application or innovation in engineering & management. Face HR Pract. Today'S Scenar. Indian Banks **2**(1)
8. Boehm, B.W, Software engineering economics. Softw. Eng. Barry W. Boehm's Lifetime Contrib. Softw. Dev. Manag. Res. **69**, 117 (2007)
9. A.J. Albrecht, J.E. Gaffney, Software function, source lines of code, and development effort prediction: a software science validation. IEEE Trans. Software Eng. **6**, 639–648 (1983)

10. L.H. Putnam, A general empirical solution to the macro software sizing and estimating problem. IEEE Trans. Softw. Eng. **4**, 345–361 (1978)
11. M. Jørgensen, Regression models of software development effort estimation accuracy and bias. Empir. Softw. Eng. **9**(4), 297–314 (2004)
12. C.K. Riesbeck, R.C. Schank, *Inside Case-Based Reasoning* (Psychology Press, 2013)
13. A. Trendowicz, R. Jeffery, Software project effort estimation. Found. Best Pract. Guid. Success Constr. Cost Model–COCOMO **12**, 277–293 (2014)
14. W. Ali, S. Malebary, Particle swarm optimization-based feature weighting for improving intelligent phishing website detection. IEEE Access **8**, 116766–116780 (2020)
15. P. Rijwani, S. Jain, Enhanced software effort estimation using multi layered feed forward artificial neural network technique. Procedia Comput. Sci. **89**, 307–312 (2016)
16. A. Heiat, Comparison of artificial neural network and regression models for estimating software development effort. Inf. Softw. Technol. **44**(15), 911–922 (2002)
17. A.B. Nassif, L.F. Capretz, D. Ho, Estimating software effort using an ANN model based on use case points, in *2012 11th International Conference on machine learning and applications*, vol. 2. (IEEE, 2012), pp. 42–47
18. S.-J. Huang, N.-H. Chiu, Applying fuzzy neural network to estimate software development effort. Appl. Intell. **30**, 73–83 (2009)
19. F. Matloob, T.M. Ghazal, N. Taleb, S. Aftab, M. Ahmad, M.A. Khan, S. Abbas, T.R. Soomro, Software defect prediction using ensemble learning: a systematic literature review. IEEE Access **9**, 98754–98771 (2021)
20. A. Gupta, A. Sharma, A. Goel, Review of regression analysis models. Int. J. Eng. Res. Technol **6**(08), 58–61 (2017)
21. R. Tibshirani, Regression shrinkage and selection via the lasso. J. R. Stat. Soc. Ser. B Stat Methodol. **58**(1), 267–288 (1996)
22. S. Hameed, Y. Elsheikh, M. Azzeh, An optimized case-based software project effort estimation using genetic algorithm. Inf. Softw. Technol. **153**, 107088 (2023)

Chapter 6
Hybridizing Metaheuristics and Analogy-Based Methods with Ensemble Learning for Improved Software Cost Estimation

Anupama Kaushik, Kalpana Yadav, Prabhjot Kaur, Kavita Sheoran, Nikhil Bhutani, Ritvik Kapur, and Bhavesh Singh

Abstract For a considerable time, the software development life cycle has been obstructed by the volatile factors associated with software cost estimation. The stakeholders involved have been severely impacted by the inaccuracies they cause during the estimating process. This can be alleviated by estimating the cost using machine learning algorithms, which significantly reduce process volatility and produce more accurate outcomes. Thus, implementing analogy-based estimation along with stacking using SVR, Ridge Regressor, K-nearest neighbours and Linear Regression in level-0 and SGD Regressor in level-1, on various datasets has given highly accurate results. The Metaheuristic and Analogy-based approach, for software cost estimation using ensemble learning, built with metaheuristic algorithm-based hyperparameter tuning that we propose is substantiated on 4 datasets Nasa93, China, Maxwell and ISBSG and evaluated using MAE, RMSE, R^2, PRED, MMRE evaluation metrics. We find competent performance being displayed by our proposed model with ISBSG and China datasets displaying the most promising results.

Keywords Software cost estimation · Software effort estimation · Stacked generalization · Ensemble learning · Random forest · K-nearest neighbours · Lightgbm

---

A. Kaushik (✉) · P. Kaur · R. Kapur
Department of IT, Maharaja Surajmal Institute of Technology, New Delhi, India
e-mail: anupama@msit.in

K. Yadav
Department of IT, Indira Gandhi Delhi Technical University for Women, Delhi, India

K. Sheoran · N. Bhutani · B. Singh
Department of CSE, Maharaja Surajmal Institute of Technology, New Delhi, India

T. R. Benala et al. (eds.), *Boosting Software Development Using Machine Learning*, Artificial Intelligence-Enhanced Software and Systems Engineering 7,
https://doi.org/10.1007/978-3-031-88188-6_6

6.1 Introduction

The design, development, operation, and maintenance of software systems are studied and approached using a methodical, quantitative, and disciplined approach known as software engineering. Estimation of efforts for a software is the technique of evaluating accurately, the time and financial resources required to complete a software-based project. It is a crucial initial phase in Software Development Life Cycle.

In the absence of an accurate cost estimate, the resources necessary will either be overestimated, which will be costly for both the software firm and the customer, or underestimated, that will take longer, involve fewer developers, and result in insufficient analysis and training. Consequently, either of these scenarios can lead to project failure. Thus, software effort estimation is a challenging process, having various models and techniques to estimate the effort [1–6].

In this paper, analogy-based estimation, metaheuristic algorithms and ensemble learning techniques have been used for better accuracy and promising performance.

Analogy based estimation (ABE) is similar to case-based reasoning and has been used for effort estimation by many researchers [3, 7]. In ABE, a new project is compared with the historical projects present in the database to retrieve similar kind of projects using various similarity functions. The effort of this new project is calculated based upon the effort values of the retrieved similar projects. ABE is simpler than expert judgement or parametric model as it relies on historical data and experience rather than mathematical formulas.

When confronted with imperfect or incomplete information or restricted computing power, a metaheuristic is an efficient higher-level procedure designed to find, generate, or choose a heuristic (partial search algorithm) that might give an adequate solution to an optimization or machine learning problem. Ensemble learning merges various prediction models and mitigate the errors present in individual models which leads to enhancement of accuracy of prediction models.

Several methods and models were synthesized using ensemble approaches. Many researchers have used ensemble models for software effort estimation [8–10]. This work is an attempt to enhance accuracy and produce superior outcomes in software effort estimation using analogy approach. The study uses stacking model configuration with linear regression, K nearest neighbours, support vector regressor and ridge regression as base models; and Stochastic Gradient Descent (SGD) regressor as a meta model. The use of diverse models strengthens the model, leading to good estimations. The paper is organised as follows: Sect. 6.1 describes the literature review; Sect. 6.2 discusses various background approaches; Sect. 6.3.

6.2 Literature Review

Many previous studies have been conducted with the aim of assessing software costs, and many machine learning techniques have also been proposed. Resmi and Vijayalakshmi [11] analyse different techniques such as, including correlation coefficient analysis, classification accuracy, clustering accuracy, the MMRE, and prediction accuracy. The article ends by utilizing a firefly algorithm inspired by nature and applying it to the datasets using fuzzy logic. The findings indicate that deep-structured multilayer perceptron effort estimation outperforms multivariate linear regression effort estimation. Probabilistic model-based EM clusters offer more accurate predictions and lower MMRE values compared to vector quantized k-means clusters. When compared to two existing methods, MDELP and COA-FIS, the proposed method demonstrates a significant improvement in the accuracy of effort estimation.

Dashti et al. [12] propose using the LEM algorithm to optimize feature weighting in the analogy-based approach. The proposed approach was tested on two datasets, Maxwell and Desharnais, using the MMRE, MdMRE, and PRED (0.25) metrics for evaluation. Various combinations of key ABE parameters—such as KNN, similarity functions, and solution functions—were explored to find the optimal configuration. The best result was achieved with Euclidean distance and k-nearest neighbors paired with the median solution function, providing the most effective and responsive model across both datasets. Overall, the proposed model produced more accurate estimates during testing than other methods, with the worst results coming from MLR.

In their publication, Sakhrawi et al. [13] utilized projects based on the Scrum framework. The first model incorporated three machine learning techniques: Linear Support Vector Regression (LinearSVR), Decision Tree Regression (DTRegr), and Random Forest Regression (RFR). The second model used a Stacking Regressor from the scikit-learn library. The models were evaluated using the RMSE, MAE, and MSE metrics, yielding the following results: RMSE = 0.595, MAE = 0.206, and MSE = 0.406.

Sakhrawi et al. [14] employed Gradient Boosting Regression (GB regr) and Random Forest Regression as base learners, with Linear Support Vector Regression as the second-level classifier. However, the results showed an improvement, with a Root Mean Square Error (RMSE) of 0.1973 and a Mean Absolute Error (MAE) of 0.0383, demonstrating better software EME estimation. ul Hassan and Khan [15] explore different methods for software cost estimation, categorized into algorithmic and non-algorithmic approaches. The hybrid grey wolf optimization algorithm (HACO-BA) is employed to find optimal initial weights for network training. Their proposed deep neural network (DNN) model demonstrates improved performance and reduced execution time compared to other models such as WNN-FA-MORLET, Deep-MNN, and ECS-DBN. The experimental results indicate that HACO-BA outperforms both ACO and BA when tuning COCOMO II. Additionally, HACO-BA achieves better optimization of the DNN in terms of accuracy and speed. Khan and Jabeen [16] investigate meta-heuristic algorithms for software effort estimation, focusing on Deep Neural Network (DNN) models. The study compares

Grey Wolf Optimizer (GWO) and Strawberry (SB) algorithms with other methods, revealing that GWO provides superior accuracy. They propose a new DNN model, GWDNNSB, which utilizes GWO and SB for optimizing the learning rate and initial weights. This model shows better results in software effort estimation compared to previously existing DNN models. Phannachitta [17] examines the Analogy-Based Software Effort Estimation (ABE) technique. The study compares the performance of ABE models using both heuristic-based and machine learning-based effort adaptation techniques and also investigates the potential of combining these methods. The results indicate that the top-performing ABE model, with a combined effort adaptor, could become a candidate for a new state-of-the-art model-based software effort estimator. The article provides a detailed evaluation, implementation, and analysis of these effort adaptors and their effectiveness. Przemysław Pospieszny, Beata Czarnacka-Chrobot, and Andrzej Kobylinski [18] applied Support Vector Machines, Neural Networks, and Generalized Linear Models in their research on software project effort and duration estimation. Their findings revealed that these methods had a limited impact on improving software estimation techniques. Despite advances in parametric approaches, the research showed that estimation still heavily relies on expert judgment, involves complex calculations, and often leads to overly optimistic assumptions. Mohammad Azzeh and Ali Bou Nassif [19] tackle the challenge of identifying the optimal set of analogies for each specific project within analogy-based effort estimation (ABE) for software effort estimation. Their proposed method seeks to enhance understanding of dataset characteristics and automatically determine the best analogies for each test project. By utilizing the bisecting k-medoids algorithm, the authors can uncover the dataset's structure and pinpoint the most relevant analogies while discarding irrelevant ones. This dynamic selection approach allows for flexibility in choosing analogies, moving away from a fixed number that may not be suitable for individual projects or varying datasets.

Harish Kumar and Srinivas [20] propose a methodology known as ANOVA Convolutional Neural Network (A-CNN) to enhance accuracy and decrease training time in effort estimation (EE). To cluster similar values based on specific features, they utilize the Modified K-Harmonic Means (MKHM) algorithm. The clustered data is then input into the A-CNN algorithm. In their experimental evaluation, the proposed methodology is compared to existing research methodologies using metrics such as accuracy, MMRE, PRED, and execution time. The results indicate that the proposed methodology surpasses existing approaches, achieving excellent performance in both accuracy and estimation. Sripada Rama Sree and Chatla Prasada Rao [21] utilized Neural Networks (NN), Adaptive Neuro Fuzzy Inference System (ANFIS), Random Forests, Support Vector Machines (SVM), and Fuzzy Logic in their research, using VAF, BRE, Pred, MARE, and VARE as evaluation metrics. Their findings revealed that the ANFIS model achieved outstanding results, boasting a VAF of 99.9% and a Pred of 94%. Omar Hidmi and Betul Erdogdu Sakar [22] applied their algorithm to the Desharnais and Maxwell datasets, achieving an accuracy of 85.48% on the Maxwell dataset and 91.35% on the Desharnais dataset, using k-nearest neighbors and support vector machines. To enhance analogy-based estimation prediction, the SABE model proposed by Kaushik et al. [3] introduces an analogy-based estimation

(ABE) method that employs a stacking ensemble technique. It is evaluated using several metrics, including standard accuracy (SA) and median magnitude of relative error (MdMRE), among others. Harish Kumar and Srinivas [23] proposed the Hyperbolic Tangent Recurrent Radial-Recurrent Neural Networks (HTRR-RNN) algorithm for software effort estimation. They utilized the fuzzy Mamdani technique to generate rules based on extracted features and incorporated the Levy Flight-Rock Hyraxes Swarm (LH-RHS) optimization algorithm for relevant feature extraction. The datasets used include MAXWELL, China, and COCOMO81, and the authors found that their approach outperformed published techniques. They emphasized that adjustment and adaptation through case-based reasoning (CBR) are vital steps for achieving accuracy. The authors Hameed et al. [24] employed a Genetic Algorithm (GA) to identify the optimal set of parameters for Case-Based Reasoning (CBR). The proposed CBR-GA technique enhanced the accuracy of CBR. This technique was evaluated using the publicly available PROMISE data repository.

6.3 Techniques Used

6.3.1 Analogy Based Estimation

Analogy-based estimation is a technique used to estimate the required effort for a new project by associating it to related previous projects, falling under the case-based reasoning (CBR) category. The process involves the following steps:

- Identify the new project along with all its relevant features.
- Apply a similarity function to compare the new project with projects in the historical database.
- Calculate the effort for the new project based on the effort values of the most similar projects retrieved, typically selecting up to five similar projects.

So, the main components of analogy-based estimation are similarity function, K nearest neighbours and the solution function. The similarity function used in this work is Correlation Similarity Function.

Correlation similarity is bounded between − 1 and 1 and it is the cosine similarity between centered versions of a and b.

$$Corr(a, b) = \frac{\langle a - \bar{a}, b - \bar{b} \rangle}{\|a - \bar{a}\| \|b - \bar{b}\|}$$

Cosine similarity is linked to correlation if the vectored angles are taken as paired samples.

The K nearest neighbours are the most similar projects retrieved after applying the similarity function. Here, K = 5 is used in the evaluations.

Once the K most similar projects are selected, the effort for the new project is calculated using specific statistics derived from these similar projects, a process known as the solution function. In the experimental analysis, the Random Forest algorithm is used as the foundation for the solution function. The core concept is to aggregate multiple decision trees to determine the final output, rather than depending on individual decision trees alone, thus enhancing the accuracy and reliability of the estimation.

6.3.2 *Stacking Model*

The architecture of a stacking model incorporates two or more base models, called level-0 models, and a meta-model that collects the predictions of base model, called level-1 models [18].

The study uses the following base models:

6.3.2.1 Ridge Regressor

Ridge regression is a linear regression technique used in machine learning to minimize overfitting, particularly when variables are closely correlated. It prevents overfitting by penalizing the coefficients, which means adding a penalty term to the error function that reduces the size of the coefficients. Ridge regression, unlike standard least squares regression, ensures that the coefficients do not get excessively large. This approach is effective for data with a lot of noise since it minimizes the model's sensitivity to individual data points. Ridge regression is frequently combined with other approaches, like as cross-validation, to reduce overfitting and enhance generalization.

Ridge regression works by including a penalty element in the cost function that is proportional to the total of the squared coefficients. This penalty phrase is referred to as the L2 norm. As a result, the optimization issue gets simpler to solve, and the coefficients shrink in size. The cost function for ridge regression has two components: the first is the same cost function as in linear regression, which ensures that the model fits the training data well. The second component is the L2 penalty or regularization term, which is intended to keep the parameters minimal and avoid overfitting.

$$RSS_{Ridge}(w, b) = \sum_{i=1}^{n} (y_i - (w_i x_i + b))^2 + a \sum_{j=1}^{p} w_j^2$$

6.3.2.2 Linear Regression

The linear regression algorithm posits a linear correlation between a variable that is dependent (y) and more than one independent variables (x), which is why it is called linear regression. It describes how the value of the dependent variable fluctuates in response to fluctuations in the independent variable(s).

The linear regression model generates a gradient straight line to visually delineate the relationship between these variables. Linear regression is represented as follows:

$$Y = a_0X + a_1X + \in Y$$

In linear In linear regression, Y represents the dependent or target variable.

- X is the independent or predictive variable.
- The intercept (a_0) adds flexibility to the line.
- The linear regression coefficient (a_1) determines the scaling factor for each input value.
- $\in$ represents random error, which accounts for the difference between expected and actual values.

6.3.2.3 Support Vector Regressor

This regressor supports both linear and non-linear regression and is based on Support Vector Machine (SVM) principles. SVR functions similarly to SVM, with a few notable distinctions. Instead of employing a curve as a decision boundary, SVR searches for the best match between the data points and the curve's position. Support vectors play an important part in determining the closest match between data points and the function expressing them. SVR uses a variety of kernels, which are functions that quantify the similarity of input vectors, as well as a number of hyperparameters that influence model behavior.

6.3.2.4 K-Nearest Neighbours

This algorithm identifies the neighbourhood of an unknown input by determining its range or distance from known data points, along with other relevant parameters. It operates on the principle of "information gain," aiming to determine what is most appropriate for predicting an unknown value. The K-Nearest Neighbours (KNN) algorithm can be applied to classification and regression problems, making it versatile in different predictive tasks.

Meta Model used in study is:

In this study, the Stochastic Gradient Descent (SGD) regressor is implement as a meta-model to integrate the base models' predictions in most appropriate manner. It is well-suited for large-scale regression tasks due to its ability to handle high-dimensional datasets and its fast training time. The SGD Regressor updates model weights iteratively using small random subsets of the training data instead of the entire dataset, making it computationally efficient for large datasets. It includes several tuneable hyperparameters, such as the learning rate, regularization term, and number of iterations, to optimize performance.

The SGD Regressor utilizes stochastic gradient descent for optimization. This iterative algorithm updates the model's parameters in small batches, using the gradient of the cost function with respect to the parameters to guide each update.

6.4 Proposed Work

Figure 6.1 showcases the entire flow of the proposed approach. The steps of implementation consists of:

- Importing libraries and datasets.
- Performing Exploratory Data Analysis and Data Visualisation.
- Performing Data Normalisation and removal of Skewness.
- Feature selection calculating feature importance with the help of random forest regressor.
- Analogy-Based Estimation
- Hyperparameter Tuning.
- Model configuration and evaluation.

6.4.1 Importing Libraries and Dataset

The proposed technique is implemented using python. The libraries used are: Pandas; for data manipulation and analysis, SkLearn; for various machine learning frameworks; Matplotlib and Seaborn; for data visualisation, Skopt; for hyperparameter optimization especially bayesian optimization and SciPy; for providing utility functions for optimization and stats. Four effort estimation datasets are utilized in this study: NASA 93, ISBSG, China, and Maxwell. Exploratory Data Analysis (EDA) involves interpreting raw data by examining its properties, identifying abnormalities, and conducting basic statistical and graphical analysis. Here, EDA starts by checking for null or redundant values in the dataset. The df.describe() function is employed to extract statistical details. Feature correlation is visualized using a heatmap, with darker shades indicating positive correlations and lighter shades showing negative correlations. This analysis helps identify features that can be dropped to improve the model's efficiency (Fig. 6.2).

Normalization aims to transform data so that it becomes dimensionless and/or has similar distributions. In normalization (or min–max normalization), all values are scaled within a specified range, typically between 0 and 1. Skewness is measured

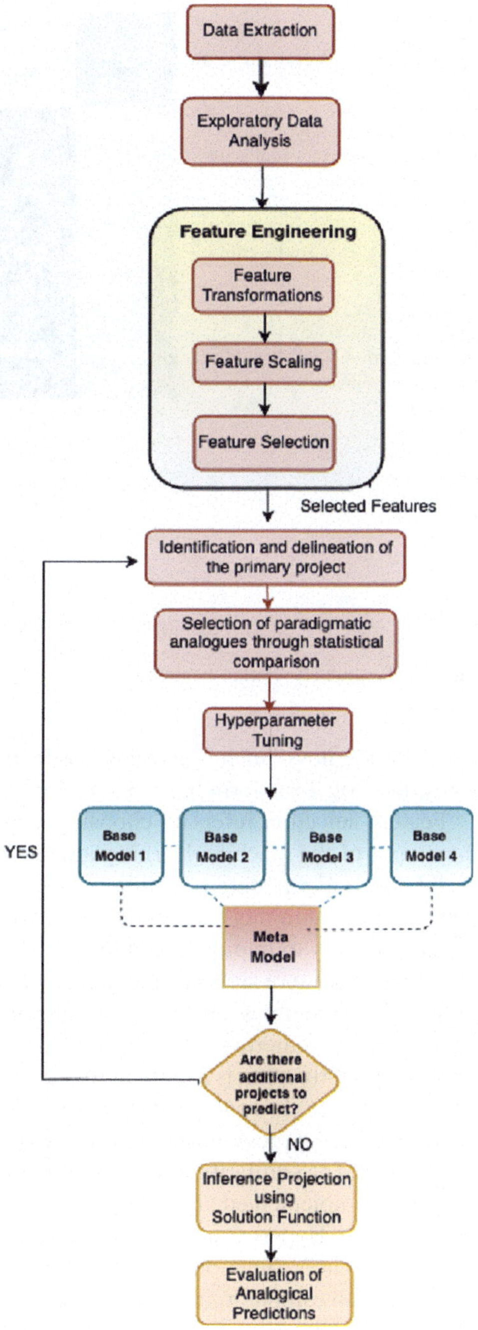

Fig. 6.1 Proposed work

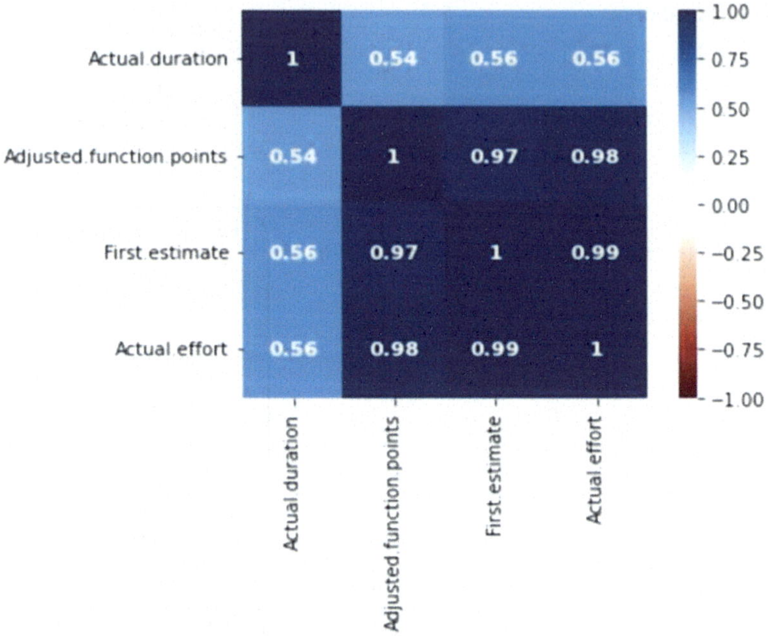

Fig. 6.2 Feature correlation-heatmap

to determine how much a distribution deviates from a normal distribution. In this study, the Log Transform has been applied to address skewness.

Feature selection refers to choosing only the features that significantly impact the output of the machine learning model, while discarding those with minimal or no influence. For feature selection in this project, Random Forest has been utilized. Apiece tree within this Random Forest determines the relevance of a feature based on its capability to improve upon the purity of the leaves. The higher the increase in leaf purity, the more significant the trait. This calculation is performed for each tree, averaged over all trees, and then normalized to 1 to ensure that the Random Forest's sum of feature importance scores equals one.

Analogy estimation is a method for calculating parameters specific for future jobs using prior data. In analogy-based effort estimation, the effort of a new project is determined using previous development experiences of similar projects, as described in Sect. 3.1. The similarity function used is a correlation function, while the solution function is implemented via a random forest regressor, highlighting the robustness of this approach. Hyperparameter tuning is conducted using a genetic algorithm through the GASearchCV function, which applies a "fit" and "score" method. If supported by the chosen estimator, it also offers additional functionalities like "predict," "predict_proba," "decision_function," and "predict_log_proba". The parameters of the estimator are enhanced via a cross-validated search across various settings.

The framework introduced in this study differs notably from other stacking-based effort estimation approaches, primarily due to its distinct choice of level-0 algorithms in the stacking model configuration, along with the incorporation of analogy-based estimation and hyperparameter tuning via genetic algorithms (GA). In this approach, base models include Support Vector Regression (SVR), K-Nearest Neighbours, Ridge Regression, and Linear Regression, while the meta-model is built using the Stochastic Gradient Descent (SGD) Regressor. A thorough explanation of these methods is provided in Sect. 3.2.

6.5 Results and Discussion

The experiment is conducted on four effort estimation datasets Maxwell, China, Nasa93 and ISBSG. The final model is evaluated on the basis of the following evaluation metrics:

- **Mean Absolute Error (MAE)**: The average difference between projected and actual values indicates how near predictions are to true values.
- **Root Mean Square Error (RMSE)**: It measures how far predictions deviate from actual values using Euclidean distance. RMSE is computed by taking the residuals (differences between predicted and actual values), finding the norm of each residual, averaging them, and then taking the square root of this mean.
- **R-Squared ($\mathbf{R^2}$)**: This statistic represents how well a regression model fits the data. The ideal R^2 value is 1, with values closer to 1 indicating a better fit.
- **Magnitude of Relative Error (MRE)**: MRE quantifies the error as a percentage of the actual effort, measuring the relative size of the prediction error.
- **PRED**: The predictor output is the predictions' percentage that lie within a specified range of the actual value, showing the model's predictive accuracy.

$$D_i = \begin{cases} 1 & \textit{if } MRE_i \leq 0.25 \\ 0 & \textit{otherwise} \end{cases}$$

$$PRED(X) = \frac{100}{n} \sum_{i=1}^{n} D_i$$

- **MMRE**: For the n number of projects, it is the average of MRE.

Among these evaluation criteria, **RMSE, MMRE**, and **MAE** act as loss indicators, evaluating the prediction errors. Meanwhile, **PRED** and $\mathbf{R^2}$ are accuracy indicators, demonstrating how well the model fits the data. The experimental results of the datasets, using these jumbled evaluation criteria, are displayed in Table 6.1.

The table indicates that the **ISBSG** dataset yields the best results, achieving the best $\mathbf{R^2}$ score, followed closely along the lines of the **China** dataset. Both the ISBSG and China datasets exhibit remarkably low values for **MAE**, **MMRE**, and **RMSE**,

Table 6.1 Experimental Results

Dataset	MAE	RMSE	MMRE	R2	PRED (%)
China	0.492716	0.352827936	0.063413518	0.9689	100
Maxwell	0.366511	0.195886966	0.046932302	0.8012	99.67
Nasa93	0.770650	0.779810000	0.118663128	0.8226	100
ISBSG	0.204509	0.057238846	0.025519410	0.9897	100

highlighting the effectiveness of the model configuration used in this study. Additionally, ISBSG, China, and **NASA 93** demonstrate a perfect accuracy of 100% in **PRED**, while **Maxwell** achieves over 95% accuracy. Among the evaluation metrics, $\mathbf{R^2}$ is deemed the most critical as it reflects the extent of variability in the dependent variable in relation to the independent variable, indicating the model's fitting efficiency. In this context, the China and ISBSG datasets show high $\mathbf{R^2}$ scores, whereas Maxwell and NASA 93 have lower scores, underscoring the model's superior performance with the China and ISBSG datasets.

Figure 6.3 shows the results in a graphical format for better visualization. In Figs. 6.4, 6.5, 6.6 and 6.7 ISBSG, China, Nasa93 and Maxwell dataset results have been depicted, respectively, with negligible deviation in predicted values as compared to the actual values. There is diminutive deviation, if any, from the actual values.

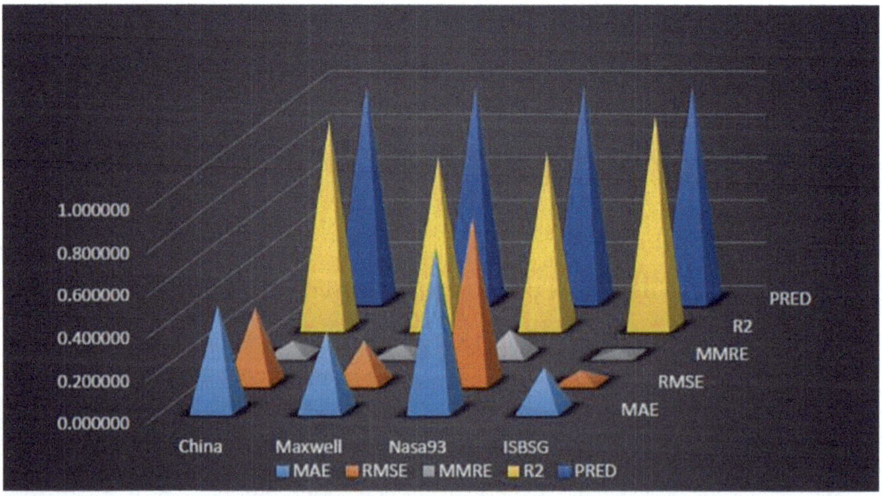

Fig. 6.3 Graphical representation of predicted results

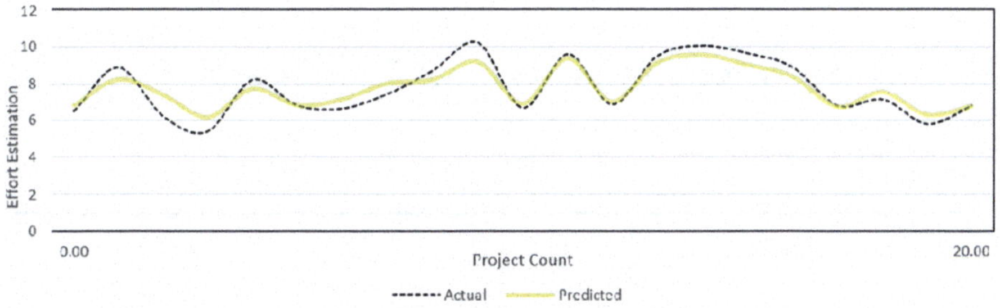

Fig. 6.4 Representation of China dataset—actual versus predicted values

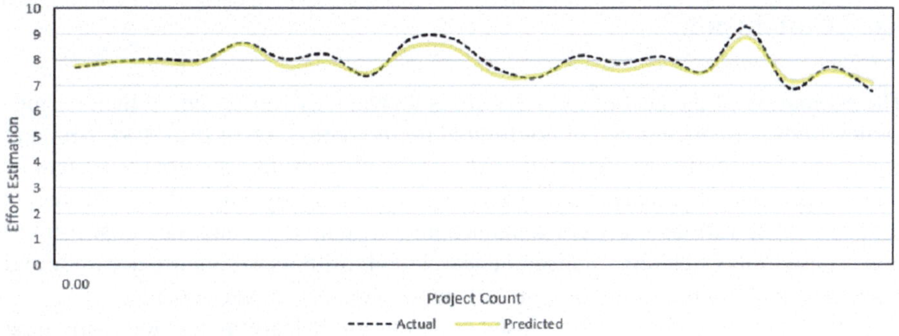

Fig. 6.5 Representation of ISBSG dataset—actual versus predicted values

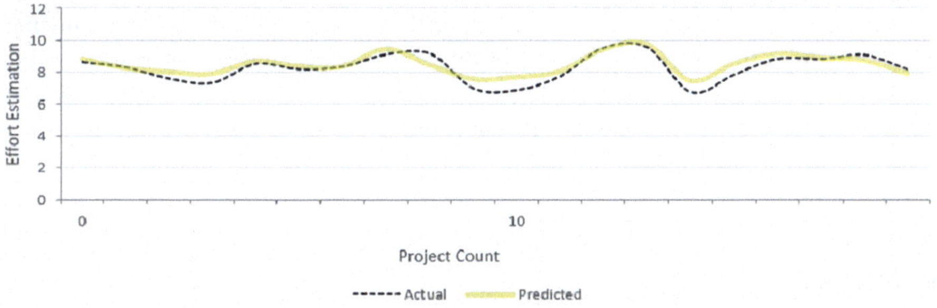

Fig. 6.6 Representation of Maxwell dataset—actual versus predicted values

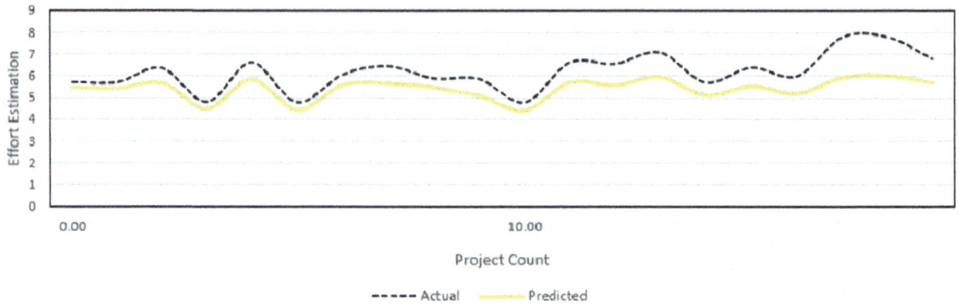

Fig. 6.7 Representation of Nasa93 dataset—actual versus predicted values

6.6 Conclusion

The estimation of the cost of the software is a vivacious constituent of the development of software, including but not limited to, the prediction of the effort, size, and cost required to fully, appropriately and effectively complete a software system or project. Traditionally, this process was carried out manually, but with advancements in technology, various software tools are now available to automate cost estimation. These tools often rely on machine learning algorithms to enhance accuracy and efficiency, making the estimation process more streamlined and reliable.

In this study, stacking has been used as an ensemble learning method combining various models for a better result. How better those results are, depends on how the stacking model is configured and other machine learning techniques utilized alongside it such as analogy-based estimation of hyperparameter tuning. Opining a versatile model with various distinct algorithms to compile the most promising results has been the primary objective of this study in which it has been successful.

However, portending to the future, with further advent of technology, new and efficient algorithms might be developed and show results even more accurately than those being presented today.

References

1. F.J. Heemstra, Software cost estimation. Inf. Softw. Technol. **34**, 627–639 (1992)
2. B. Boehm, C. Abts, S. Chulani, Ann. Softw. Eng. **10**, 177–205 (2000)
3. A. Kaushik, P. Kaur, N. Choudhary et al., Stacking regularization in analogy-based software effort estimation. Soft. Comput. **26**, 1197–1216 (2022)
4. A. Kaushik, D.K. Tayal, K. Yadav, A comparative analysis on effort estimation for agile and non-agile software projects using DBN-ALO. Arab. J. Sci. Eng. **45**, 2605–2618 (2019)
5. A. Kaushik, N. Singal, M. Prasad, Incorporating whale optimization algorithm with deep belief network for software development effort estimation. Int. J. Syst. Assur. Eng. Manag. **13**, 1637–1651 (2022)
6. A. Kaushik, D.K. Tayal, K. Yadav, A. Kaur, Integrating firefly algorithm in artificial neural network models for accurate software cost predictions. J. Softw.: Evol. Process **28**(8), 665–688 (2016)

7. T.R. Benala, R. Mall, Dabe: differential evolution in analogy-based software development effort estimation. Swarm Evol. Comput. **38**, 158–172 (2018)
8. A. Idri, M. Hosni, A. Abran, Improved estimation of software development effort using classical and fuzzy analogy ensembles. Appl. Soft Comput. **49**, 990–1019 (2016)
9. J.T.H.D.A. Cabral, A.L. Oliveira, Ensemble effort estimation using dynamic selection. J. Syst. Softw. **175**, 110904 (2021)
10. M. Azzeh, AliBou Nassif, L.L. Minku, An empirical evaluation of ensemble adjustment methods for analogy-based effort estimation. J. Syst. Softw. **103**, 36–52 (2015)
11. V. Resmi, S. Vijayalakshmi, *Analogy-Based Approaches to Improve Software Project Effort Estimation Accuracy*. De Gruyter (2020). https://doi.org/10.1515/jisys-2019-0023
12. M. Dashti, T.J. Gandomani, D.H. Adeh, H. Zulzalil, A.B.M. Sultan, LEMABE: a novel framework to improve analogy-based software cost estimation using learnable evolution model. PubMed. Central (PMC) (2022). https://doi.org/10.7717/peerj-cs.800
13. Z. Sakhrawi, A. Sellami, N. Bouassida, Software enhancement effort estimation using stacking ensemble model within the scrum projects: a proposed web interface, in *Proceedings of the 17th International Conference on Software Technologies—Volume 1: ICSOFT*, pp. 91–100 (2022). ISBN 978-989-758-588-3. https://doi.org/10.5220/0011321000003266
14. Z. Sakhrawi, A. Sellami, N. Bouassida, Software enhancement effort estimation using correlation-based feature selection and stacking ensemble method. Cluster Comput. **25**(4), 2779–2792 (2021) (Springer Science and Business Media LLC). https://doi.org/10.1007/s10586-021-03447-5
15. C.A. ul Hassan, M.S. Khan, R. Irfan, J. Iqbal, S. Hussain, S. Sajid Ullah, R. Alroobaea, F. Umar, Optimizing deep learning model for software cost estimation using hybrid meta-heuristic algorithmic approach. Optim. Deep. Learn. Model. Softw. Cost Estim. Using Hybrid Meta-Heuristic Algorithmic Approach (2022). https://doi.org/10.1155/2022/3145956
16. M.S. Khan, F. Jabeen, S. Ghouzali, Z. Rehman, S. Naz, W. Abdul, Metaheuristic algorithms in optimizing deep neural network model for software effort estimation. IEEE Access **9**, 60309–60327 (2021). https://doi.org/10.1109/ACCESS.2021.3072380. (keywords: {Software; Estimation; Software algorithms; Optimization; Genetic algorithms; Analytical models; Market research; Software effort estimation; meta-heuristic algorithms; deep neural networks; deep learning)
17. P. Phannachitta, On an optimal analogy-based software effort estimation. **125**, 106330 (2020). https://doi.org/10.1016/j.infsof.2020.106330
18. P. Pospieszny, et al., An effective approach for software project effort and duration estimation with machine learning algorithms. J. Syst. Softw. **137**, 184–196 (2018). https://doi.org/10.1016/j.jss.2017.11.066
19. A.B. Nassif, A.B. Nassif, Analogy-based effort estimation: a new method to discover set of analogies from dataset characteristics. IET Software **9**(2); in *Institution of Engineering and Technology* (2015), pp. 39–50, https://doi.org/10.1049/iet-sen.2013.0165
20. K. Harish Kumar, K. Srinivas, Efficient analogy-based software effort estimation using ANOVA convolutional neural network in software project management, in *Efficient Analogy-based Software Effort Estimation Using ANOVA Convolutional Neural Network in Software Project Management* (SpringerLink, 2022). https://doi.org/10.1007/978-981-16-9669-5_35
21. S.R. Sree, C.P. Rao, A study on application of soft computing techniques for software effort estimation. A Journey Bio-Inspired Tech. Softw. Eng. 141–165 (2020)
22. O.H. Alhazmi, M.Z. Khan, Software effort prediction using ensemble learning methods. J. Softw. Eng. Appl. **13**(07), 143–160 (2020). https://doi.org/10.4236/jsea.2020.137010
23. K. Harish Kumar, K. Srinivas, An improved analogy-rule based software effort estimation using HTRR-RNN in software project management. Expert. Syst. Appl. **251**, 124107 (2024)
24. S. Hameed, Y. Elsheikh, M. Azzeh, An optimized case-based software project effort estimation using genetic algorithm. Inf. Softw. Technol. **153**, 107088 (2023)
25. D. Zherebtsov, Stacking 101: all you need to know. Medium (2021). https://danilzherebtsov.medium.com/stacking-101-all-you-need-to-know-3c72fe9171a0

Chapter 7
A Review on Detection of Design Pattern in Source Code Using Machine Learning Techniques

Sourav Biswas, Meghavarshini Senthilkumar, Jyoti Prakash Meher, and Rajib Mall

Abstract Design patterns are integral to software development, offering solutions to recurring design challenges. Detecting these patterns within code enhances comprehension, documentation, and maintenance. We review reported design pattern detection techniques based on supervised learning, feature-based analysis, clustering, graph-based methods, dynamic approaches, and ontology integration. We investigate the effectiveness of each technique and their strengths and limitations.

Keywords Design pattern · Machine learning · Review · Survey

7.1 Introduction

A design pattern in software development refers to a reusable solution for addressing common problems within a specific context [1]. Design patterns offer a template to solve recurring issues in software design, promoting efficient and effective solutions. Design patterns are valuable because they provide a structured way to handle challenges that often arise during software development. They encapsulate tried-and-true solutions, enabling developers to avoid redundant work and adhere to established best practices. Additionally, design patterns foster clear communication among

S. Biswas (✉)
Centre for Internet of Things (CIOT), Siksha 'O' Anusandhan (Deemed to be) University, Bhubaneswar, Odisha, India
e-mail: bsws.sourav@gmail.com

M. Senthilkumar
Department of Computer Science and Engineering, Mepco Schlenk Engineering College, Sivakasi, Tamil Nadu, India

J. P. Meher
Department of Computer Science and Engineering, Indian Institute of Technology, Kharagpur, India, West Bengal, India

R. Mall
Department of Computer Science and Engineering, Shiv Nadar Institution of Eminence, Greater Noida, Uttar Pradesh, India

T. R. Benala et al. (eds.), *Boosting Software Development Using Machine Learning*, Artificial Intelligence-Enhanced Software and Systems Engineering 7,
https://doi.org/10.1007/978-3-031-88188-6_7

software professionals by providing a shared language to discuss prevalent design issues and their solutions. Ultimately, design patterns guide the creation of well-organized and resilient software systems [2]. Identifying design patterns can be a challenging task, especially when it comes to reverse engineering, as this process relies on understanding the existing codebase, which serves as the input for pattern detection. The automated detection of design patterns holds significant importance within the field of software engineering, as it facilitates the recognition of repetitive design solutions within code [3]. Also, the automated identification of design patterns has proven advantageous in aiding software developers to rapidly and accurately grasp and manage unfamiliar source code [4].

An important objective of our review is to systematically evaluate the effectiveness of these techniques and to identify tools that exhibit proficiency in efficient pattern detection within software systems. We investigate the effectiveness of detection techniques and identify tools capable of performing extensive and efficient pattern detection. To accomplish this, we conducted a systematic mapping study, reviewing primary studies published from 2006 to 2021 and comparing their findings and heuristics to gain a deeper understanding of the landscape.

The rest of this paper is structured as follows. Section 7.2 discusses basic terminologies and concepts that are essential for a comprehensive understanding of the review. Section 7.3 presents our method for the selection of research papers for our review. Section 7.4 presents the methodologies being used for design pattern detection including supervised learning, feature-based analysis, clustering, and graph-based techniques, ontology-based detection for pattern detection and categorization. Section 7.5 discuss the results and analysis of the reported works. Finally, Sect. 7.6 concludes this paper.

7.2 Basic Concepts

Though many design patterns have been proposed, possibly the GoF patterns are most widely discussed and used [5]. There are 23 well-established design patterns categorized into three primary types: Creational, Structural, and Behavioral. In this section, we review a few important concepts for understanding the background of the related work presented in this survey paper.

Creational design pattern: Creational design patterns [6] deal with how classes are created and objects are instantiated. These patterns include techniques for creating classes or generating object instances. Builder, prototype, Abstract Factory, Singleton, and Factory Method are a few examples of creational design patterns.

Structural design pattern: Structural design patterns [6] pertain to patterns oriented towards the organization and composition of classes and objects within larger software structures. Adapter, Bridge, Composite, Decorator, Facade, Flyweight, and Proxy are examples of structural design patterns.

Behavioural design pattern: These patterns concentrate on how objects interact and communicate with each other [6]. Chain of Responsibility, Command, Interpreter, Iterator, Mediator, Memento, Observer, State, Strategy, Template Method, and Visitor are a few examples of behavioral design patterns.

A few machine learning techniques have gained prominence in the area of detecting design patterns from source code. We provide an overview of those techniques in the following.

Supervised learning-based detection: Supervised learning-based detection is a widely used for automatically recognizing common design patterns [7] within codebases. In this approach, a geometric model is employed to represent the structural characteristics of design patterns, facilitating their identification. Moreover, a supervised machine learning classifiers meticulously trained on a labeled dataset. In this dataset, each instance is associated with a specific design pattern label, enabling the classifier to acquire knowledge and make generalizations based on the patterns provided. The labeled dataset serves as a foundation for the supervised classifier to discern and classify design patterns effectively.

Feature-based detection: This strategy [8], focuses on identifying plagiarism by extracting particular traits or features from the documents. To find similarities in documents, it entails analyzing multiple elements like grammar, vocabulary, and structure.

Clustering-based detection: In this detection approach, similar documents are grouped together using clustering algorithms [9], which makes it simpler to spot possible instances of patterns. Similar content is probably present in documents that are part of the same cluster.

Graph-based detection: In this technique [10], documents are represented as graphs, and instances is found by contrasting the structures of the graphs. Graph algorithms [10] can be used to find relationships and patterns in a set of texts.

Pattern Pre-Detection: It is a unique method for identifying design patterns through static analysis of source code before the actual application of those patterns. This approach utilizes the source code before any design patterns are applied, allowing it to differentiate between design patterns that may have identical structures but serve different purposes or address distinct issues. This approach not only aids in early pattern detection but also significantly enhances the efficiency of the software development process.

This technique helps developers anticipate design considerations and make informed decisions during the coding process.

Dynamic Pattern Detection: Dynamic pattern detection approach [11, 12] dynamically examines the patterns and behavior of documents over time to find alterations that might point to plagiarism. This detection method can be helpful for figuring out when content has been changed or altered. Ontology-based detection: This technique [13] uses structured knowledge representations known as ontologies to find semantic similarities between documents. Instead of only examining textual similarity, it focuses on comprehending the meaning and context of the information.

7.3 Research Method

To conduct our comprehensive review, we have implemented a systematic and structured methodology, which encompasses the following key steps.

7.3.1 Selection of Literature

In the process of selecting literature, we aim to ensure the trustworthiness and quality of the data included in our review. To accomplish this, we have opted for specific repositories that are known for their reliability and relevance to our research topic. These repositories include (a) IEEE Xplore, (b) Science Direct, (c) Orielly Media, (d) Springer, and (e) ACM.

7.3.2 Determination of Research String

To effectively identify studies that employ diverse detection processes for design patterns, we have meticulously crafted our search strings. These search strings act as keywords that guide our search within the selected repositories. The chosen search strings are: "design pattern detection," which broadly encompasses the topic of detecting design patterns in software, "object-oriented design pattern detection," which narrows the focus to detection methods specific to object-oriented programming, and "software design pattern survey," which helps us locate surveys and comprehensive overviews of software design patterns. These carefully chosen search strings will facilitate the identification of relevant research articles in our review process.

7.4 Overview of Detection Process

7.4.1 Supervised Learning-Based Detection

Chaturvedi et al. [14] proposed a methodology to use machine learning to find design patterns in Java code. There are four main steps in the method. First, Source Monitor[1] was used to look at 71 Java projects and get information about the code, like number of statements, percentage of branch statements, and number of classes or interfaces are there. Only four of these features was used: Lines of Code, Statements, Branch Statements Percentage, and Classes/Interfaces.

[1] https://www.derpaul.net/SourceMonitor.

In the second step, Principal Component Analysis (PCA) [14] was employed to simplify the data. This process involved an examination of how different segments of the data resembled one another, facilitating the categorization of the data into Class A through Class M. The method also showed that one of the groups of data, called Principal Component 1 (PC1), was really important because it explained a lot of the differences in the data. This helps us understand the patterns and trends in the data better.

Moving to the third step, four distinct machine learning methods (Linear Regression, Polynomial Regression, Support Vector Regression, and Artificial Neural Networks) were applied to the data to determine the optimal approach for identifying design patterns in Java code. Root Mean Square Error (RMSE) used to decide the most accurate method. Support Vector Regression had the lowest RMSE, which means it was really good at finding design patterns. It made predictions that were very close to the actual values, showing that it was very accurate. It results that among all the ML methods, Support Vector Regression was the best, especially when checking it using cross-validation tests (threefold, fivefold, and tenfold), where it always had the highest accuracy.

However, this method does have its limitations. Firstly, it explored only four machine learning methods, potentially overlooking more effective alternatives. Secondly, it utilized only 40% of the available data features, possibly missing crucial information related to design patterns. Thirdly, the assessment relied solely on accuracy and RMSE as performance metrics, neglecting other valuable metrics like recall, precision, and the F1 score. Additionally, the data collection process involving code analysis using a tool may be time-consuming. Lastly, certain elements like class names and comments were not considered during the pattern identification process.

7.4.2 Feature Based Detection

Zheng et al. [8] introduced an approach called DPDF (Feature-Based Design Pattern Detection) for design pattern detection, which relies on a feature-based approach. Feature-based approach, means looking at specific aspects of the code to identify patterns.

The approach begins by gathering a significant amount of Java code from the Pattern-like Micro-Architecture Repository (PMART). However, this repository contains varying numbers of examples for each pattern. To standardize, 100 examples were selected for each pattern, resulting in a final set of 1300 code files after excluding irrelevant ones.

Three individuals were involved in labeling the code using a dedicated tool to help the system learn about patterns. Interestingly, they mostly agreed on the labels, with a score of 0.74 showing good agreement. They looked at different parts of the code, like attributes in classes and methods. They also changed the code into plain text so computers could understand it better. This change involved making a clear map that explains how different parts of the code are connected.

A computer algorithm known as word2vec is employed to construct a model called JEM (Java Embedded Model). This algorithm aids in computer understanding of word meanings within the code, utilizing the CBOW (Continuous Bag-of-Words) [15] method and for code comprehension, a machine learning approach employing random decision trees is used, complemented by the SAMM-R algorithm to further improve results. The system's evaluation involves cross-validation, a technique that partitions the data into segments for rigorous testing.

The system's performance is evaluated using precision, recall, and F1-Score measures. The results indicate a favorable performance, with precision consistently exceeding 80% and recall consistently at 79% or higher. Various design patterns are examined, each with its level of detectability. For instance, the system excels in identifying the Visitor pattern with nearly 97% precision, but encounters challenges in the case of the Façade pattern, where it achieves only 68% precision. The difference in performance could be because the Façade and Proxy patterns are complex and used in various ways. When aggregating the results for all patterns, the system demonstrates strong overall performance, boasting an average precision exceeding 80% and a recall rate of 79%.

This approach has internal and external validity threats. The study recognizes threats to internal validity, including differences in counting methodologies and unavailability of a project, but mitigates them through labelling and comparison. Potential bugs in code and classifiers have been addressed through debugging efforts. Disagreement among labellers during data labelling is reduced by employing multiple labellers. The study acknowledges that the reference set may not represent all relevant Java source code and plans to increase its size and diversity in future research. The study takes measures to address these threats, improving the reliability of the findings.

7.4.3 Clustering Based Detection

Uchiyama et al. [16] introduced a novel object-oriented pattern detection approach. This method is based on source code metrics (e.g., number of methods—NOM) which constitutes a numerical measurement of software attribute and machine learning techniques for focusing on role-based pattern recognition without necessitating class structural data and similar information. For instance, consider the source code metric "number of methods" (NOM), which quantifies the count of methods within a class [17]. The machine learning-based detection process comprises five steps, divided into two phases. The initial three steps constitute the learning phases, while the remaining two steps form the detection phases. These five steps encompass: Pattern detection, Metric determination, Application of machine learning techniques, Candidate role assessment, and Pattern identification.

The patterns are classified based on three terms, namely creational, behavioural, and structural patterns, in which five of three types are detected using this approach, namely singleton adapter, template, state, strategy pattern, and 12 roles. 12 role refers to a specific responsibility or behavior that a class or object within a software system plays in the implementation of a design pattern. Pattern specialists employ the GQM (Goal Question Metric Method) to assess roles within design patterns by formulating questions and goals related to attributes and operations within those roles, culminating in the identification of relevant metrics.

Clustering-based detection is typically not applied to metrics. Metrics are primarily used to evaluate the effectiveness and quality of design pattern implementations, rather than for clustering purposes. Instead, clustering is applied for the assignment of candidate roles. To fine-tune the weight adjustments, a backward propagation technique [18] is employed. Instead of using a step function, a sigmoid function [18, 19] is implemented to achieve smoother transitions.

Relationships are distinguished in terms of pattern structure and roles (e.g., inheritance, aggregation, interface relations), an agreement value is computed. Furthermore, a pattern agreement value is derived from the role and relation agreement values. Evaluation is done by 5K fold test for candidate role This proposed methodology uses Goal Question Metric Method [9] used by pattern specialists to judge role within design patterns. This method is used to define goals and questions related to it, which is qualitative and sometimes may not capture all relevant aspects.

The comparison with TSAN [20] showed that the proposed method is effective for singleton pattern if we add some special implementation which utilises a Boolean variable to proposed method, but TSAN or DIET [21] does not detect the singleton pattern. However, there is a limitation with false positives in the proposed technique. Accurate role judgment is crucial for high recall in pattern detection. Template Method and Adapter patterns posed challenges due to lower role judgment accuracy. Overall, pattern detection accuracy relied heavily on role judgment accuracy, offering insights for refining the technique, especially in large-scale programs and specific pattern detection.

7.4.4 Graph Based Detection

Zanoni et al. [22] employed a hybrid approach combining graph matching and machine learning approaches, known as the MARPLE DPD (Metrics and Architecture Recognition Plug-in for Eclipse Design Pattern Detection) tool. This tool consists of three modules: Micro-structures detector, Joiner, and Classifier. The Joiner module identifies instances that match precise rules, prioritizing high recall despite lower precision due to its focus on fundamental traits. The Classifier module then submits a limited number of candidates for classification, specifically targeting significant true instances. The Microstructures detector employs static analysis to identify microstructures supported by a set of visitors. This detection process unfolds in two steps: matching and merging. The matching phase produces a set of role mappings, while merging entails combining role mappings into pattern instances. The classification process encompasses in two steps:

1. Transition from specific occurrences of patterns to the mapping of roles.
2. Transition from role mappings to creating instances based on patterns.

A noteworthy proposition is representation of pattern instances through role mappings [22] and micro-structures [22] as features. This involves disassembling pattern instances into role mappings, translating mappings into feature vectors, clustering these vectors, and finally expressing pattern instances as Boolean vectors based on these clusters. This representation facilitates training and classification through supervised classification algorithms. This methodology accommodates varying pattern instance sizes and capitalizes on micro-structures.

The experimentation carried out within the framework of MARPLE-DPD focused on assessing the accuracy of diverse machine-learning models in detecting design pattern instances. The primary goals of the experimental studies were to show that machine learning models outperformed a baseline model, to determine which models performed the best, and to compare different methods for pre-processing data, with a focus on clustering algorithms. The experimentation involved datasets representing five design patterns: Singleton, Adapter, Composite, Decorator, and Factory Method. Their frequency of occurrence guided the choice of these patterns in specialized literature. The datasets were constructed by extracting instances using the Joiner module from projects, which were subsequently manually classified.

Multiple machine learning models were applied in the experiments, including the ZeroR (Zero Rule) [23] algorithm, OneR (One Rule) [24, 25], Naïve Bayes (a classification and probabilistic modeling algorithm), JRip (JRules Induction algorithm for producing RIPPER rules) [25], Random Forest [26], C4.5 (an extension of Quinlan's earlier ID3 algorithm) [27], SVMs (Support Vector Machines with different kernel functions) [28], SimpleKMeans [29], and CLOPE (Clustering with sLOPE) [30], a clustering algorithm used in data mining and machine learning.

The approach involved optimizing these models by adjusting their settings using three key performance metrics: accuracy, F1-measure, and the area under the receiver operating characteristic curve (AUC). Genetic algorithms sourced from the JGAP library (Java Genetic Algorithms and Genetic Programming)[2] were specifically applied to single-level patterns such as Singleton and Adapter. Additionally, a proprietary clustering algorithm was implemented to represent pattern instances as feature vectors.

For single-level patterns, each instance was associated with just one mapping, resulting in feature vectors containing only a single authentic feature. Evaluation of instances followed precise criteria for consistency. The methodology utilized a tenfold cross-validation strategy, and the evaluated pattern cases can be accessed through the Design Pattern Detection Tools Benchmark Platform (DPB). The primary objective of this work was to gauge the effectiveness of various machine learning models in detecting design pattern instances and enhancing their performance by adjusting model parameters. The outcomes underscored the efficacy of machine

[2] https://sourceforge.net/projects/jgap/.

learning in pattern detection than baseline classifier [22], particularly for single-level patterns and for multi-level patterns, the choice of clustering algorithms and dataset dimensions played a pivotal role.

Measuring recall, which helps determine if all instances are found, was challenging because there are no standard methods. This highlights the need for more ways to measure and compare their technique with existing tools. The dataset they used came from only ten projects and may not represent all patterns and software systems, so the results may not apply universally. Additionally, there's no agreed-upon standard for determining if a pattern was correctly identified, which can lead to disagreements. Lastly, they limited themselves to classifier and clustering algorithms available in the Weka software package,[3] which may not cover all possibilities and could impact its effectiveness.

Oruc et al. [31] introduce an innovative approach named DesPaD which stands for Design Pattern Detection to detect design patterns within object-oriented software projects using a graph mining strategy [32]. The method involves three core steps: constructing a model graph, generating design pattern templates, and executing pattern detection. The model graph is formed by representing the software as a directed graph with labels, where vertices denote classes, abstract classes, templates, and interfaces, and edges signify relationships like inheritance and aggregation. Design pattern templates are derived from analysing class and sequence diagrams of well-known Gang of Four (GoF) design patterns. These templates are subsequently used to seek corresponding sub-graphs within the model graph through an isomorphic sub-graph mining algorithm.

Experimental evaluation is conducted across multiple open-source projects, including JUnit 3.8, JUnit 4.1[4] and Java AWT 1.3.[5] The accuracy of the tool's pattern detection is gauged using precision and recall metrics. The approach showcases an average precision of 80% and an average recall of 88%, highlighting its effectiveness in precisely identifying design patterns. Despite slight variations in precision and recall based on project complexity, the tool consistently delivers remarkable performance, particularly evident in smaller projects like JUnit 3.8, where precision and recall are nearly flawless. A comparative analysis with PINOT (Pattern INference and recOvery Tool) [33], a similar tool, underscores DesPaD's superior ability to detect a higher number of design pattern instances, translating into improved recall values.

Additionally, the research delves into DesPaD's computational efficiency in terms of runtime. The runtime evaluations reveal that the tool's efficiency depends on project size and complexity. While runtime remains reasonable for smaller projects. For larger projects, such as Java AWT 1.3, demand more time due to the NP-complete nature of isomorphic sub-graph mining. This observation accentuates the potential for optimization to elevate DesPaD's performance. Overall, the study effectively showcases DesPaD's adeptness in accurately detecting design patterns.

---

[3] http://www.cs.waikato.ac.nz/ml/weka/.

[4] http://www.junit.org/.

[5] http://docs.oracle.com/javase/7/docs/api/java/awt/.

The comprehensive evaluation sheds light on its strengths and offers insights into areas with potential for enhancement.

Gupta et al. [6] introduced a matrix-based approach that enhances the visualization of relationships graphs and associated matrices. They created a structured method to find design patterns in system designs. The methodology relies on calculation called Normalized cross-correlation (NCC) values [22], which measure how similar design patterns are to system design matrices. Link matrices from link graphs are made with the help of UML diagrams [34]. However, this process is complex and takes a lot of time because of the many steps to calculate the link matrices.

Pradhan et al. [34] presented a novel approach that involves representing a system's class structure and the design patterns are identified through analysis of directed graphs, wherein nodes signify classes, and edges denote their relationships. The study highlights two techniques: Graph Isomorphism and Normalized Cross Correlation (NCC) [35, 36]. Graph Isomorphism identifies instances of the design pattern within the system graph by identifying structural similarities. On the other hand, NCC quantifies the degree of pattern presence within the system graph, offering a percentage value that reflects the likeness between the pattern and the system. A significant aspect of this methodology involves the Candidate Filtering Algorithm, which plays a pivotal role in refining the pool of candidate classes for more accurate pattern matching. Through iterative assessment of candidate classes, this algorithm enhances the dependability of the detection process. The proposed methodology has been implemented using the programming language C++ and verified on four open-source software tools. The study gives results that demonstrate the number of occurrences of varied design patterns that are fully matched and partially matched. The needs to handle the possibility of false positives or negatives and the scalability issues.

Tsantalis et al. [20] present a novel approach for the automated detection of design patterns in software systems. This methodology tackles challenges found in existing pattern detection techniques [37], including identifying modified pattern versions, managing the explosion of search space in large systems, and accommodating new patterns. The approach's core lies in a similarity scoring algorithm that draws inspiration from link analysis algorithms [38] and graph matching techniques. This algorithm facilitates the comparison of graph vertices, addressing the comparison of different graphs. The implemented tool realized in Java, captures essential attributes and relationships within the system. It generates similarity matrices for pattern detection, allowing users to select desired patterns and automatically identify instances with minimal human intervention. The adaptability of the methodology to new patterns presents opportunities for its expansion into evolving software design practices. This research advances automated software engineering by providing a structured and efficient approach to detecting design patterns, enhancing software quality, and contributing to architectural comprehension and maintenance.

The methodology's effectiveness is validated through assessments of three open-source projects: JHotDraw 5.1 [39], JRefactory 2.6.24[22], and JUnit 3.7. Evaluation outcomes highlight the methodology's accuracy, even in the presence of pattern instance modifications. Significantly, the methodology exhibits impressive precision, achieving zero false positives across all patterns investigated. However, instances of false negatives attributed to specific pattern modifications or structural intricacies underline the approach's reliance on manual code validation. Analyses of efficiency reveal that pattern complexities and system sizes influence the time taken for pattern detection. However, this method does have limitations. It doesn't consider dynamic information from sequence diagrams, and it might use a lot of memory in bigger systems. Also, the similarity scoring algorithm mainly looks at how similar nodes are to each other, rather than comparing entire graphs.

The software tool discussed in Shi et al. [33], PINOT (Pattern INference and recOvery Tool), performs pattern detection based on structural characteristics of programs. It deploys static program analysis techniques such as data flow and control flow analysis to identify method collaborations. PINOT employs specific keywords to pinpoint design patterns. For example, in detecting the "template method" pattern, PINOT explicitly seeks final methods, leading lower likelihood of false positives in detection. The tool relies on the Java compiler (Jikes) for pattern search. However, adding or modifying patterns within PINOT is involved and requires substantial effort.

Manjari Gupta et al. [6] present an innovative strategy based on inexact graph matching to identify design patterns in software systems. Inexact graph matching specifies that it can handle the noise and distortions commonly found in objects effectively. The research's core insight involves applying graph-based techniques to automate pattern identification. This approach uses graph structures to represent software relationships, making the pattern detection process more adaptable and resilient. The proposed algorithm introduces a unique approach by breaking down the graph-matching process into multiple phases controlled by parameter K. This novel method enhances the exploration of potential matches, reducing the search space and enhancing the precision of pattern detection. Additionally, the algorithm's ability to adapt to various levels of noise and distortion enhances its relevance in real-world scenarios where precise matches might be challenging to achieve. In terms of implementation, the Gupta et al. [6] introduce a novel algorithm inspired by the A* method, focusing on intelligent exploration to find optimal matches. This algorithmic innovation has the potential to significantly improve the efficiency and accuracy of design pattern detection. Furthermore, the paper acknowledges the limitations of traditional exact graph matching when dealing with noisy data and distortions. The research contributes to a more comprehensive and flexible approach to design pattern detection by addressing this challenge through inexact graph matching.

7.4.5 Pattern Pre-detection

Washizaki et al. [40] introduced a novel approach to design pattern detection by analyzing a program's source code before its actual implementation. This method introduces both "forward" and "backward" techniques to identify potential design pattern candidates and their applications.

In the "forward" method, the researchers sequentially examine the source code to determine if it satisfies certain conditions known as "smell conditions." Smells are indicators or hints that a specific design pattern might be present. This approach is particularly valuable for detecting potential pattern applications that conventional methods might overlook.

Conversely, the "backward" method examines the source code in the reverse direction, seeking patterns that fit known design pattern specifications.

To validate their approach, the study leverages version control systems to track changes in the source code over different time periods. When specific smell conditions are satisfied, the research corroborates these findings through pattern specifications and manual assessment.

The study uses the State and Strategy design patterns in Java programs as examples to illustrate the effectiveness of their proposed techniques. In the case of the State pattern, conditions of smells (referred to as C1-C3) are successfully identified, indicating that the State pattern has been applied. However, for the Strategy pattern, conditions (C4-C5) are not met, suggesting that this particular pattern is not present in the code.

The research further evaluates its technique through experiments involving pairs of Java programs before and after applying the State pattern. Comparing proposed technique alongside conventional methods versus using only conventional methods, the research illustrates that this approach boosts the capacity to identify patterns and precisely detect their existence.

From a quality standpoint, the suggested technique can aid in evaluating the influence of implementing design patterns on the maintainability of a system. Measuring metrics like Lack of Cohesion in Methods (LCOM), Weighted Methods per Class (WMC), and McCabe Cyclomatic Complexity (MCC) before and after applying the State pattern reveals improvements in code quality. This suggests that the approach not only aids in pattern detection but also contributes positively to the maintainability of software.

7.4.6 *Dynamic Pattern Detection*

Pande et al. [36] proposed a novel approach in software engineering called DNIT (Depth-Node-Input Table) to identify design patterns in object-oriented software systems. Design patterns play a vital role in solving common design challenges and enhancing the flexibility and reusability of software. DNIT transforms the detection process by using rooted directed graphs and graph matching techniques, allowing for the automatic identification of design patterns at different levels of complexity.

The core concept of DNIT revolves around representing UML diagrams of both the target system and design patterns as relationship-directed graphs. These graphs capture the inherent structure of classes and their connections within the software.

The DNIT algorithm creates Depth-Node-Input Tables for both system designs and design patterns by assigning depths to nodes and labelling edges with reachability-based inputs. These tables act as templates for pattern detection. The research highlights the adaptability of DNIT in detecting design patterns through full-, partial-, and no-match scenarios. A full match occurs when all relationship-directed graph patterns for a design pattern correspond to those in the system design. A partial match signifies that some relationships match, but others do not due to differing nodes or paths. A no-match occurs when a design pattern's structure is absent from the system design. DNIT's capabilities are demonstrated through various design patterns, including Singleton, Facade, Strategy, and Composite. Singleton and Facade pattern detection relies on specific depth-node-input combinations, showcasing the algorithm's potential for simple and complex patterns. Strategy and Composite pattern detection involves matching multiple relationships, revealing DNIT's versatility in handling diverse patterns.

7.4.7 Ontology Based Detection

The research conducted by Thongrak et al. [41] introduces a fresh approach to identifying design patterns in UML class diagrams. This approach integrates ontology principles [42] and utilizes the Semantic Query-Enhanced Web Rule Language (SQWRL) [43]. Using these frameworks, the study aims to create a more organized and accurate way to find patterns, especially focusing on the Strategy pattern. One significant contribution of this research is using ontology to represent UML class diagrams. This means that class diagrams, which show how different parts of a program relate, are represented in a structured and formal way using specific rules. This structured representation helps us understand complex relationships among different parts of a class diagram, making it easier to find patterns. Another innovative aspect is using SQWRL, which stands for Semantic Query-Enhanced Web Rule Language. It's like asking detailed questions and making smart guesses when working with these structured diagrams. It's similar to using SQL, a language for working with databases, but it's adapted to work with these structured representations called ontologies [44].The study uses a collection of SQWRL rules to identify design patterns. The aim is to harness this querying capacity to unveil complex patterns that might pose difficulties for traditional graph-based techniques. However, the work also confronts a significant challenge in pattern detection—distinguishing patterns with similar structural characteristics. This issue is exemplified in scenarios where the Strategy and State patterns exhibit resemblances, potentially leading to misclassification. This situation highlights the complexity involved in distinguishing patterns and raises the question of whether there's a need to improve the detection rules for achieving higher precision. While the research is centred around the Strategy pattern, it implicitly suggests broader implications. The methodology proposed has the potential to extend to various pattern types, offering a generalizable framework for pattern detection. Moreover, the integration of tools like Protégé[6] exemplifies

[6] http://protegewiki.stanford.edu/wiki/Protege3DevDocs.

the practical implementation of the approach, bridging the gap between theoretical concepts and real-world application.

Ren et al. [45] proposed a novel approach for detecting the Observer design pattern in Java code, employing a multi-phase analysis methodology. This process initiates with the extraction of pattern rules from code analysis, followed by their conversion into the Semantic Web Rule Language (SWRL) [44] format. This transformation enables a formal representation of the rules and supports automated detection. Additionally, the incorporation of attributed abstract syntax trees (AST) [46] for modeling code structures establishes a seamless link between the code and ontology-based reasoning [47].

Moreover, their work introduces dynamic analysis as a complementary step in pattern detection. This dynamic analysis validates candidate instances identified during static analysis by assessing their real-world behavior against predefined pattern rules. This dynamic verification significantly enhances the accuracy and reliability of pattern detection. To further enrich their approach, the researchers integrated OWL ontology models [48], providing an extensive vocabulary to define software concepts and relationships. This enrichment not only aids in pattern detection but also extends the framework's utility to a broader spectrum of software analysis tasks.

To quantify the degree of conformance of candidate instances to the Observer pattern, the research introduced a matching confidence algorithm, offering a comprehensive evaluation of pattern detection. However, the manual translation of rules into SWRL and the reliance on dynamic analysis might present limitations. The process of manually translating rules could be labour-intensive and susceptible to errors, potentially impeding scalability. Furthermore, the precision of dynamic analysis relies significantly on the presence of execution traces and thorough test cases. Moreover, the approach's language specificity to Java and its fixed threshold for accepting or rejecting candidate instances raise questions about its adaptability to other programming languages and patterns and its robustness in various software contexts. Despite these limitations, the research forms a valuable foundation for advancing design pattern detection methodologies. Combining formal rule representation, ontology modeling, dynamic analysis, and a matching confidence algorithm, the approach delivers a comprehensive framework for advancing the precision and automation of pattern detection.

Wang et al. [49], extensively looked at how computer programs find design patterns. The examination of tools was carried out that use things like XML, Prolog, and ontologies to help with data organization and handling. These tools use different ways to analyze patterns, like checking how the code is structured, how it behaves, and what it means. Various methodologies was used for review process like searching, asking questions, using rules, and studying how the code runs.

The research emphasized the importance of tools that can find different versions of design patterns and sort them based on when they are found: before, during, or after writing the code. Different tools use different methods to find these different versions. For example, a tool called Sempatrec [39] uses ideas from the semantic web and rules from Semantic Web Rule Language (SWRL) [43] and reasoning from Web Ontology Language—Description Logic (OWL-DL) [44] to find design pattern versions more accurately. This tool uses a smarter way of representing knowledge to improve how it finds patterns in the code.

On the other hand, tools like ePAD [37], Visual Studio, Net's Add-In [50], and D-CUBED [51] take a different approach by utilizing query-based techniques to identify pattern variants during the pattern detection process. These tools employ queries or specific search patterns to identify instances of design patterns with variations in the source code, enhancing their ability to detect diverse pattern instances and deviations from standard patterns.

7.5 Discussion

Various methods for detecting design patterns are discussed below.

Table 7.1 provides a comprehensive overview of research papers addressing design pattern detection techniques. Each entry highlights the author(s), the type of detection method employed, specific techniques used, the paper's significance, and any associated limitations.

The results indicate that among many detection processes in software design pattern applications, the most used process to detect patterns are supervised learning-based, clustering-based, graph-based, ontology-based, and feature based. For instance, Graph algorithms can be used to find relationships and patterns in a set of texts. Pattern pre detection aids in reducing the search space and enhancing effectiveness. Different detection techniques results can also be combined to form a new detection process. Based on the table, the most commonly used type of detection is **graph-based** detection, which (e.g., papers [6, 22, 31, 34]). involves utilizing graph algorithms and structures to identify design patterns within software code.

The techniques most frequently used across the papers include **machine learning** algorithms (e.g., papers [4, 14, 16, 22]) and **clustering** methods (e.g., papers [16, 31, 34]). These techniques are applied in various combinations to detect design patterns effectively.

In contrast, **ontology-based** detection (e.g., papers [41]) is less commonly used across the papers, and it involves leveraging semantic knowledge representations and ontologies for pattern detection.

A common limitation observed across many of the works is the potential for **false positives and false negatives** in pattern detection. This means that some approaches may incorrectly identify patterns that are not present (false positives) or fail to identify patterns that are actually present (false negatives). These inaccuracies can impact the overall reliability and effectiveness of the design pattern detection process. Additionally, **scalability** is a recurring limitation, especially when dealing with larger and more complex software systems. Many of the approaches may face challenges in terms of performance and efficiency when applied to extensive codebases.

Moreover, **limited pattern coverage** is another common limitation. Some approaches may focus on specific types of design patterns or may not be comprehensive in detecting a wide range of patterns, which can restrict their applicability.

Table 7.1 Extracted data from primary studies

S. No.	Author(s)	Type of detection	Technique used	Significance	Limitations
1	Chaturvedi et al. [14]	Supervis learning	Machine learning, PCA, Regression, SVM	Presents a supervise learning methodology for detecting design patterns and showcases the superior effectiveness of Support Vector Machines (SVM) compared to alternative algorithms	Limited feature usage, limited number of ML algorithms, limited evaluation metrics, time-consuming feature extraction
2	Uchiyama et al. [16]	Clustering-based	Metrics, Machine learning, Clustering	Presents a clustering-based approach for detecting design patterns by roles, highlights challenges in role judgment and scalability	Limited to certain pattern types, potential false positives, accuracy impacted by project complexity and role judgment
3	Zheng et a [8]	Feature-based	Feature extraction, Word2Vec, ML classifier	Proposes a feature-based detection technique using semantic representation, demonstrates improvements in precision and recall	Limited precision for certain patterns, complexity for larger projects, potential false positives/negatives
4	Zanoni et al. [22]	Graph-based	Graph matching, Machine learning, Micro-structure	Introduces a graph-based detection approach using similarity scoring, presents novel method for instance identification	Limited handling of dynamic information, computational performance challenges for larger projects
5	Oruc et al. [31]	Graph-based	Graph mining, Isomorphic sub-graph mining	Proposes a graph-based approach with focus on microstructures, addresses challenges in role assignment and detection efficiency	Potential false positives/negatives, efficiency impact in larger systems

(continued)

Table 7.1 (continued)

S. No.	Author(s)	Type of detection	Technique used	Significance	Limitations
6	Gupta et al. [6]	Matrix representation	Matrix-based, NCC	Facilitates matrix-based visualization for design pattern detection, offers insights into impact on maintainability	Complexity and time-consuming link matrix calculation
7	Pradhan et al. [34]	Graph-based	Graph isomorphism, NCC, Candidate filtering	Introduces an innovative approach using graph techniques and NCC for recurring pattern detection, enhances accuracy	Complexity of larger codebases, potential false positives/negatives, scalability of techniques
8	Tsantalis et al. [20]	Automated detection	Similarity scoring, Graph matching	Presents a novel methodology for automated detection of design patterns, addresses modified versions and novel patterns	Lack of dynamic information from sequence diagrams, memory consumption in larger systems
9	Washizaki et al. [40]	Pattern pre-detection	Source code analysis, Smell conditions	Introduces a pre-detection approach based on smell conditions, offers potential for refining pattern detection accuracy	May miss certain patterns, requires manual validation, efficiency variations based on project size and complexity
10	Pandel et al. [52]	Dynamic Detection	Rooted directed graphs, Graph matching	Presents a dynamic pattern detection approach using depth-node-input tables, offers adaptablepattern detection at varying depths	Full, partial, and no match detection scenarios, demonstrated adaptability for various pattern types
11	Thongrak et al. [41]	Ontology-based	Ontology, SQWRL, Protege, XSLT	Introduces ontology-based approach for pattern detection, emphasizes semantics, demonstrates practical implementation	Focus on strategy pattern, potential to misclassify patterns with resemblances, practical implementation tools

(continued)

Table 7.1 (continued)

S. No.	Author(s)	Type of detection	Technique used	Significance	Limitations
12	Ren et al. [45]	Dynamic detection	Multi-phase analysis	Proposes a multi-phase dynamic detection approach for the observer pattern, offers a structured methodology	Focus on observer pattern detection, multi-phase approach
13	Wang et al. [49]	Variant detection	Intermediate representations (XML, Prolog, Ontologies)	Explores variant detection strategies using different representations, highlights the importance of representation tools	Focus on intermediate representations, enhancement of data specification/manipulation
14	Shi et al. [33]	Automated detection	Static analysis	Presents PINOT, a tool using structural characteristics and static program analysis for pattern detection	Relying on specific keywords may lead to false positives, complexity in adding/modifying patterns

Lastly, several works rely on **manual validation and labeling** of pattern instances, which can be time-consuming and subjective. This manual intervention can introduce biases and may not scale well to large software projects. These common limitations highlight the ongoing challenges in design pattern detection, and researchers are actively working to address these issues and improve the accuracy and applicability of detection techniques.

We can conclude that in general techniques improve the performance for solving design problem with a drawback that some of these techniques may miss certain patterns and may require large amount of dataset.

7.6 Conclusions

Design pattern detection using machine learning techniques is attracting the focus of researchers. These machine learning approaches have shown promising results, outperforming baseline classifiers and demonstrating improved accuracy in pattern detection. However, these approaches are constrained by factors such as dataset size, scarcity of pattern instances for certain patterns, the need for standardized benchmarks, subjectivity in pattern labelling, and the limited selection of algorithms.

Future research may focus on addressing these limitations by expanding datasets, enhancing collaboration among researchers, establishing agreed-upon benchmarks, and exploring a wider range of algorithms.

Declaration of Interest None.

References

1. P. Kuchana, *Software Architecture Design Patterns in Java* (CRC Press, 2004)
2. S. Hukerikar, C. Engelmann, Resilience design patterns-a structured approach to resilience at extreme scale (version 1.0). arXiv preprint arXiv:1611.02717 (2016)
3. F. Wedyan, S. Abufakher, Impact of design patterns on software quality: a systematic literature review. IET Software **14**(1), 1–17 (2020)
4. N. Nazar, A. Aleti, Y. Zheng, Feature-based software design pattern detection. J. Syst. Softw. **185**, 111179 (2022)
5. E. Gamma, R. Helm, R. Johnson, J. Vlissides, Design patterns: abstraction and reuse of object-oriented design, in *ECOOP'93-Object-Oriented Programming: 7th European Conference Kaiserslautern, Germany, July 26–30, 1993 Proceedings 7* (Springer, 1993), pp. 406–431
6. M. Gupta, A. Pande, R.S. Rao, A.K. Tripathi, Design pattern detection by normalized cross correlation, in *2010 International Conference on Methods and Models in Computer Science (ICM2CS-2010)* (IEEE, 2010), pp. 81–84
7. S. Alhusain, S. Coupland, R. John, M. Kavanagh, Towards machine learning based design pattern recognition, in *2013 13th UK Workshop on Computational Intelligence (UKCI)* (IEEE, 2013), pp. 244–251
8. R. Zheng, J. Li, H. Chen, Z. Huang, A framework for authorship identification of online messages: writing-style features and classification techniques. J. Am. Soc. Inf. Sci. Technol. **57**(3), 378–393 (2006)
9. S.Patel, S. Sihmar, A. Jatain, A study of hierarchical clustering algorithms, in *2015 2nd International Conference on Computing for Sustainable Global Development (INDIACom)* (IEEE, 2015), pp. 537–541
10. M. Needham, A.E. Hodler, *Graph Algorithms: Practical Examples in Apache Spark and Neo4j* (O'Reilly Media, 2019)
11. S. Hayashi, J. Katada, R. Sakamoto, T. Kobayashi, M. Saeki, Design pattern detection by using meta patterns. IEICE Trans. Inf. Syst. **91**(4), 933–944 (2008)
12. L. Wendehals, A. Orso, Recognizing behavioral patterns atruntime using finite automata, in *Proceedings of the 2006 International Workshop on Dynamic Systems Analysis* (2006), pp. 33–40
13. D. Kirasić, D. Basch, Ontology-based design pattern recognition, in *Knowledge-Based Intelligent Information and Engineering Systems: 12th International Conference, KES 2008, Zagreb, Croatia, September 3-5, 2008, Proceedings, Part I 12* (Springer, 2008), pp. 384–393
14. S. Chaturvedi, A. Chaturvedi, A. Tiwari, S. Agarwal, Design pattern detection using machine learning techniques, in *2018 7th International Conference on Reliability, Infocom Technologies and Optimization (Trends and Future Directions)(ICRITO)* (IEEE, 2018), pp. 1–6
15. J. Lin, W. Dongbo, Automatic extraction of domain terms using continuous bag-of-words model. Data Anal. Knowl. Discov. **32**(2), 9–15 (2016)
16. S. Uchiyama, A. Kubo, H. Washizaki, Y. Fukazawa et al., Detecting design patterns in object-oriented program source code by using metrics and machine learning. J. Softw. Eng. Appl. **7**(12), 983 (2014)
17. M. Lorenz, J. Kidd, *Object-Oriented Software Metrics: a Practical Guide* (Prentice-Hall, Inc., 1994)

18. Collective Intelligence. Programming collective intelligence (2008)
19. H. Hirano, Neural network implemented with c++ and java. *Personal Media* (2008)
20. N. Tsantalis, A. Chatzigeorgiou, G. Stephanides, S.T. Halkidis, Design pattern detection using similarity scoring. IEEE Trans. Softw. Eng. **32**(11), 896–909 (2006)
21. V.R. Basili, D.M. Weiss, A methodology for collecting valid software engineering data. Trans. Softw. Eng. (6), 728–738 (1984)
22. M. Zanoni, F.A. Fontana, F. Stella, On applying machine learning techniques for design pattern detection. J. Syst. Softw. **103**, 102–117 (2015)
23. S.A. Inamdar, S.M. Narangale, G.N. Shinde, Preprocessor agent approach to knowledge discovery using zero-r algorithm. Int. J. Adv. Comput. Sci. Appl. **2**(12) (2011)
24. R. Choudhary, H.K. Gianey, Comprehensive review on supervised machine learning algorithms, in *2017 International Conference on Machine Learning and Data Science (MLDS)* (IEEE, 2017), pp. 37–43
25. V. Parsania, N.N. Jani, N. Bhalodiya, Applying naïve bayes, bayesnet, part, jrip and oner algorithms on hypothyroid database for comparative analysis. Int. J. Darshan Inst. Eng. Res. Emerg. Technol **3**(1):1–6 (2014)
26. G. Biau, E. Scornet, A random forest guided tour. Test **25**, 197–227 (2016)
27. S. Ruggieri, Efficient c4. 5 [classification algorithm]. Trans. Knowl. Data Eng. **14**(2), 438–444 (2002)
28. T. Joachims, *Making Large-Scale SVM Learning Practical* (Technical Report, 1998)
29. T. Kanungo, D.M Mount, N.S. Netanyahu, C. Piatko, R. Silverman, A.Y. Wu, The analysis of a simple k-means clustering algorithm, in *Proceedings of the Sixteenth Annual Symposium on Computational Geometry* (2000), pp. 100–109
30. Y. Yang, X. Guan, J. You, Clope: a fast and effective clustering algorithm for transactional data, in *Proceedings of the Eighth ACM SIGKDD International Conference on Knowledge Discovery and Data Mining* (2002), pp. 682–687
31. M. Oruc, F. Akal, H. Sever, Detecting design patterns in object-oriented design models by using a graph mining approach, in *2016 4th International Conference in Software Engineering Research and Innovation (CONISOFT)* (IEEE, 2016), pp. 115–121
32. L. Tang, H. Liu, Graph mining applications to social network analysis. in *Managing and Mining Graph Data* (2010), pp. 487–513
33. N. Shi, R.A. Olsson, Reverse engineering of design patterns from java source code, in*21st IEEE/ACM International Conference on Automated Software Engineering (ASE'06)* (IEEE, 2006), pp. 123–134
34. P. Pradhan, A.K. Dwivedi, S.K. Rath, Detection of design pattern using graph isomorphism and normalized cross correlation, in *2015 Eighth International Conference on Contemporary Computing (IC3)* (IEEE, 2015), pp. 208–213
35. Y.M. Fouda, A.R. Khan, Normalize cross correlation algorithm in pattern matching based on 1-d information vector. Trends Appl. Sci. Res. **10**(4), 195 (2015)
36. A. Pande, M. Gupta, A.K. Tripathi, A new approach for detecting design patterns by graph decomposition and graph isomorphism, in *Contemporary Computing: Third International Conference, IC3 2010, Noida, India, August 9-11, 2010, Proceedings, Part II 3* (Springer, 2010), pp. 108–119
37. A. De Lucia, V. Deufemia, C. Gravino, M. Risi, An eclipse plug-in for the detection of design pattern instances through static and dynamic analysis, in *2010 IEEE International Conference on Software Maintenance* (IEEE, 2010), pp. 1–6
38. A. Borodin, G.O. Roberts, J.S. Rosenthal, P. Tsaparas, Link analysis ranking: algorithms, theory, and experiments. ACM Trans. Internet Technol. (TOIT) **5**(1), 231–297 (2005)
39. A. De Lucia, V. Deufemia, C. Gravino, M. Risi, Behavioral pattern identification through visual language parsing and code instrumentation, in *2009 13th European Conference on Software Maintenance and Reengineering* (IEEE, 2009), pp. 99–108
40. H. Washizaki, K. Fukaya, A. Kubo, Y. Fukazawa, Detecting design patterns using source code of before applying design patterns, in *2009 Eighth IEEE/ACIS International Conference on Computer and Information Science* (IEEE, 2009), pp. 933–938

41. M. Thongrak, W. Vatanawood, Detection of design pattern in class diagram using ontology, in *2014 International Computer Science and Engineering Conference (ICSEC)* (IEEE, 2014), pp. 97–102
42. K. Robles, A. Fraga, J. Morato, J. Llorens, Towards an ontology-based retrieval of UML class diagrams. Inf. Softw. Technol. 54(1), 72–86 (2012). ISSN 0950-5849. https://doi.org/10.1016/j.infsof.2011.07.003
43. M.J. O'Connor, A.K. Das, Sqwrl: a query language for owl, in *OWLED*, vol. 529 (2009), pp. 1–8
44. I. Horrocks, P.F. Patel-Schneider, H. Boley, S. Tabet, B. Grosof, M. Dean, et al., Swrl: a semantic web rule language combining owl and ruleml. W3C Member Submission **21**(79):1–31 (2004)
45. W. Ren, W. Zhao, An observer design-pattern detection technique, in *2012 IEEE International Conference on Computer Science and Automation Engineering (CSAE)* (IEEE, 2012), vol. 3, pp. 544–547
46. I. Neamtiu, J.S. Foster, M. Hicks, Understanding source code evolution using abstract syntax tree matching, in *Proceedings of the 2005 International Workshop on Mining Software Repositories* (2005), pp. 1–5
47. R. Möller, B. Neumann, Ontology-based reasoning techniques for multimedia interpretation and retrieval, in *Semantic Multimedia and Ontologies: Theory and Applications* (Springer, 2008), pp. 55–98
48. R. Falco, A. Gangemi, S. Peroni, D. Shotton, F. Vitali, Modelling owl ontologies with graffoo, in *The Semantic Web: ESWC 2014 Satellite Events: ESWC 2014 Satellite Events, Anissaras, Crete, Greece, May 25–29, 2014, Revised Selected Papers 11* (Springer, 2014), pp. 320–325
49. Y. Wang, C. Zhang, F. Wang, What do we know about the tools of detecting design patterns? in *2018 IEEE International Conference on Progress in Informatics and Computing (PIC)* (IEEE, 2018), pp. 379–387
50. G. Rasool, I. Philippow, P. Mäder, Design pattern recovery based on annotations. Adv. Eng. Softw. **41**(4), 519–526 (2010)
51. P. Wkegrzynowicz, K. Stencel, Relaxing queries to detect variants of design patterns, in *2013 Federated Conference on Computer Science and Information Systems* (IEEE, 2013), pp. 1571–1578
52. A. Pandel, M. Gupta, A.K. Tripathi, Dnit-a new approach for design pattern detection, in *2010 International Conference on Computer and Communication Technology (ICCCT)* (IEEE, 2010), pp. 545–550

Chapter 8
Machine Learning Techniques for the Measurement of Software Attributes

Somya R. Goyal

Abstract Measuring the attributes of software is a pivotal umbrella activity throughout the development process of software. The accuracy of measurements controls the successful completion of the project within the pre-decided timeframe and money constraints. With the changing needs of the current software industry, it is desirable to have advanced techniques apt for accurate measurements of software in alignment to the grown complicated software systems. Machine Learning (ML) techniques have been proven to be effective for accurate measurements in past years of research. Further, it needs enhancements to serve with more effective measurements of metrics associated with different tasks and activities of software development. This chapter aims to elaborate on the existing methodologies for software measurements using machine learning techniques with the intent to gain insight into the techniques being deployed in literature. Another major contribution is to elaborate on the tasks from the software measurement domain that can be modeled as machine learning tasks. It also highlights the most suitable machine learning model from the existing literature for software. This chapter will serve as an enlightenment for the academicians and researchers who are willing to work in the domain of applications of machine learning in software development.

Keywords Software measurements (SM) · Machine learning (ML) · Effort or cost estimation · Defect or fault prediction · Reliability estimation

8.1 Introduction

Measuring the software attributes implies the measurements for attributes of processes, products, or project [1]. Throughout the development, measurement and metrics are ongoing activity (see Fig. 8.1). At every phase, task or activity, one or

S. R. Goyal (✉)
Manipal University Jaipur, Jaipur, Rajasthan, India
e-mail: somya.goyal@jaipur.manipal.edu

T. R. Benala et al. (eds.), *Boosting Software Development Using Machine Learning*, Artificial Intelligence-Enhanced Software and Systems Engineering 7,
https://doi.org/10.1007/978-3-031-88188-6_8

more attributes are needed to be monitored and measured to control and get tracked. These may include size of module or the software, development cost, project effort, resource allocation, reliability of process or product, defect removal, reusability, software release timing, productivity, execution times, and testability of program modules and much more. Measurements are crucial for the timely delivery of good quality products within budget limits. De Marco says, "You cannot control what you cannot measure" [2].

In the present changing environment, software development and maintenance are a big challenge. Measuring the Software is one of the most necessitated but tedious tasks which consumes lots of effort. Any inaccuracy in software measurements may fail the software. In 1960, due to inaccurate estimations, the software industry had to go through the phase of software crisis [3].

Since the term software engineering was coined, huge research has been conducted to handle the shortcomings of classical estimation models. It has resulted into modern empirical models based on regression and classification. These models applying one or another machine learning model has raised the success rates of projects and are in total alignment with the current development approaches including agile and scrum. Machine learning models are apt and suitable because a wide range of development tasks in the software industry can be modelled as learning tasks.

The most focused research area in software development is estimation and prediction of a variety of attributes of either the process or tasks or the project itself. The agenda is to improve the overall quality of the product by fostering quality at each and every step of the development project. Application of machine learning based models in collaboration with the statistical and classical methods has done wonders and

Fig. 8.1 Measurements and SDLC

resulted in more effective resource management. No doubts, further improvements are always welcomed for effective attribute selection or extractions, hyperparameter tuning, selecting the models and hybridization of methods etc. [4].

The chapter discusses the formulation of software development tasks as learning problems; methodology to measure software attributes using machine learning; evaluation approaches, benchmarks, and case studies [5]. The study will be done in alignment of the following objectives which will be serving milestones for the research work:

1. To elaborate on the formulation of software development tasks as learning problems.
2. To elaborate on the existing prediction models based on ML techniques available for measuring software concerning their characteristics.
3. To analyze and compare the models from the state-of-the-practice in SE measurements using ML techniques based on the accuracy, dataset used, technique implemented, computational efficiency, and other parameters.
4. To improve the existing models using advanced ML techniques so that the measurements can be done more accurately.

The chapter is organized as follows: Sect. 8.2 brings light to the background concepts for this chapter. It includes the detailed discussion of the machine learning techniques available at our disposal. The elaboration on the software attributes whose measurement can be modelled as machine learning tasks. Section 8.3 summarizes the related literature work. It covers the survey of selected papers from the latest and relevant literature works. The contribution made by the researchers has been analyzed thoroughly to find the suitability of technique for a specific task. Section 8.4 draws the inferences from the literature survey conducted in Sect. 8.3 and detailed discussion is given for the results obtained. Section 8.5 concludes the study with a special reference to the future scope of the work.

8.2 Background Concepts

This section brings light to the background concepts for this chapter including software attributes whose measurement can be modelled as machine learning tasks along with the detailed discussion of the machine learning techniques available at our disposal.

Measuring the software refers to the all the measurements related to the development of the software since the inception of idea of the software. It covers all the development tasks ranging from requirement gathering through designing, coding to the delivery of the final product. All of the four P's of software development i.e., Project, Process, Product and People are in the inclusion. Here the author discussed few SDLC phases and mentions a few task that can be modelled as learning problems. Neither the list is exhaustive not it is complete. Author starts with the planning phase where the scheduling, budgeting, staffing and other activities are taken care of

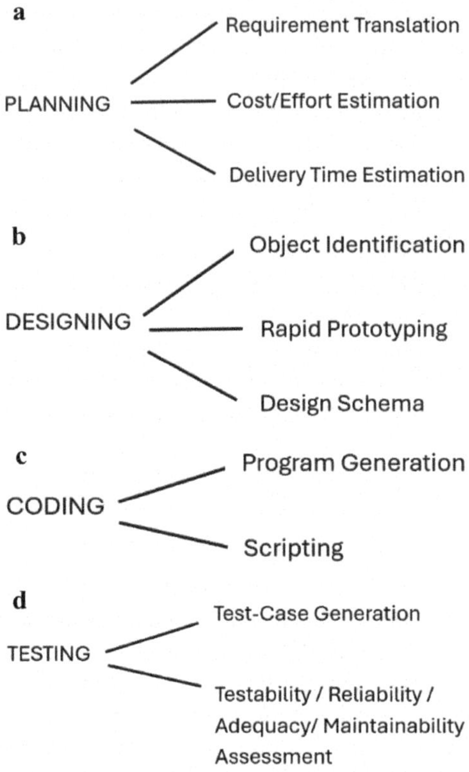

Fig. 8.2 **a** Planning phase of SDLC. **b** Designing phase of SDLC. **c** Coding phase of SDLC. **d** Testing phase of SDLC

(refer Fig. 8.2a). Here, the effort of the project, duration, time to deliver, overall cost can be predicted using regression models of ML.

Next, the designing phase (see Fig. 8.2b), where the solutions are carved as prototypes or blueprint as design documents, ML models can be used to identify the objects in the design, and compile the design schema. Here the size, time, effort to develop the design can be measured using ML models. ML helps to estimate the cost and reliability of the rapid prototype that can be assessed by customers for approval or further upgradations.

In coding, size of program, testability, reliability, and other metrics can be computed using ML (refer Fig. 8.2c).

In testing phase, effort in testing, correction cost can be measured (Fig. 8.2d). Test-case optimization vis-à-vis the coverage and effectiveness of the test cases can be done using ML models.

8.2.1 *Measuring Software Using Machine Learning*

Machine Learning (ML) is about writing the programs that learn through their experience and hence improves their own performance at a specific taskset. Significant application areas for machine learning techniques are the domains where the scenario is as follows:

a. Problem domains where scarce knowledge is available to develop effective solutions by human.
b. Problem cases where huge data with hidden patterns exists which has practical importance to solve the same; or
c. Problem domains that require constant change in the solutions as per the changing situations.

The field of software measurements sets a platform with huge applications of machine learning methods to complement the software development taskset [1]. ML is a subfield of AI as shown in Fig. 8.3.

The ML techniques can be classified into following categories as shown in Fig. 8.4.

(a) Decision tree learning (DT)—CHAID: chi-squared automatic interaction detection, CART: classification and regression trees, ADT: alternating decision tree, RF: random forest
(b) Neural networks (NN)—MLP: multilayer perceptron, PNN: probabilistic neural network, RBF: radial basis function,
(c) Support vector machines
(d) Bayesian learning (BL)—NB: Naïve Bayes, BN: Bayesian networks, ABN: augmented Bayesian networks, WNB: weighted Bayesian networks, TNB: transfer Bayesian networks
(e) Ensemble learning (EL)—LB: Logit boost, AB: AdaBoost
(f) Search-Based Techniques (SBT)—GP: genetic programming, ACO: ant colony optimization

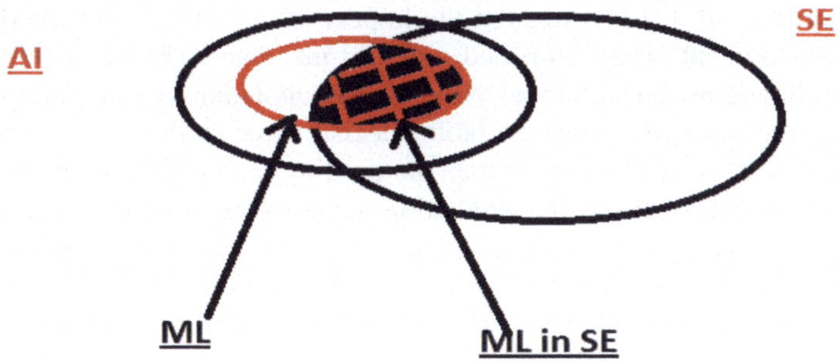

Fig. 8.3 ML approaches in SE

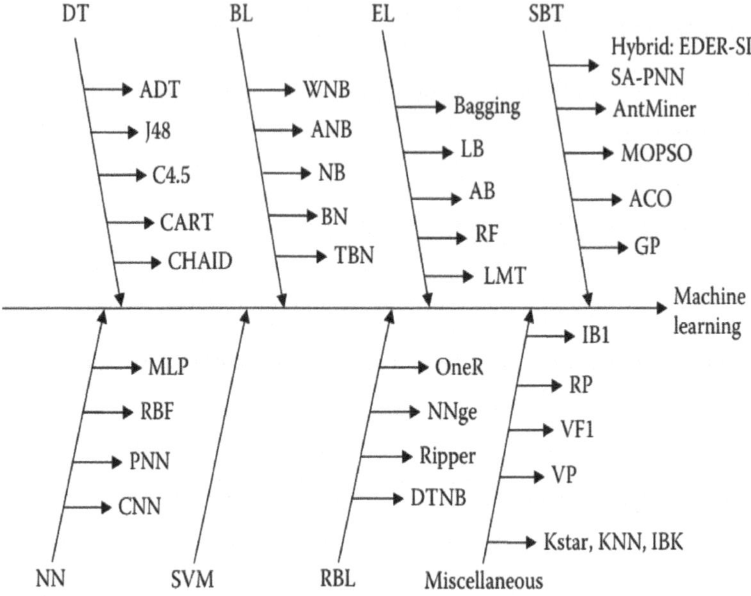

Fig. 8.4 ML techniques at our disposal for SM tasks

8.3 Review of Literature

This section brings review of the current state of the practice to identify the current trends and shifts in the paradigms for ML in Software Measurements.

Hameed et al. [6] proposed a Case-Based Reasoning (CBR) model for effort estimation. They addressed the issue of parameter setting for CBR models and introduced a genetic algorithm for assessing all possible parameter settings. Effort estimation is an important element to develop a project plan and accurate estimation raises the chances of successful completion of the project.

Chennappan et al. [7] investigated the impact of automated fault prediction in software development. They proposed a hybrid machine learning model for fault prediction. They deployed feature selection with the application of the genetic method and by using the selected feature set, built a decision tree model for automated fault prediction. Results were 50% better than the existing fault prediction models.

Khalid et al. [8] analyzed the prediction accuracy of support vector machine (SVM), Naïve Bayes (NB), Random Forest (RF) and ensembles to predict the software defects. Also, proposed the optimization methods for these ML techniques and later compared the performance of base models with the optimized models. The outcome was that the optimization by feature selection and k-means clustering improves the prediction accuracy of SVM, NB and RF.

Ali et al. [9] proposed a novel effort estimation model using the ensemble of expert judgement, use case points and neural networks. They advocated the ensembles as better than the estimation models based on one machine learning technique. The results show that the ensembles have better estimation accuracy and precision over the single models.

Rodríguez Sánchez et al. [10] estimated cost for agile based Scrum model using a machine learning bagging method over a story board analysis. They ensembled ADABoost, Random Forest, Decision tree as a bootstrap aggregation model. They concluded that ensembles handle regression problem effectively with high accuracy and low error.

Rashid et al. [11] presented a survey of 52 papers on software cost and effort estimation. They reported the results with the most popular techniques, namely multilayer perceptron and COCOMO. They came up with MMRE as the most used metric. They reviewed five digital repositories and followed the SLR guidelines to report the trends and practices in software effort estimation domain.

Butt et al. [12] proposed a novel user scenario and expert judgement based model as an enhancement to COCOMO for cost estimation in agile SDLC. It allows the work and re-work to be managed effectively in sync with the key feature of agile modelling i.e., flexibility. They contributed a more precise and accurate estimation model for scrum model.

Pan et al. [13] investigated test case selection and prioritization by studying 29 papers over the period of 14 years. They highlighted the most popular ML techniques and future prospects of reinforcement learning in the domain of software testing. They reflected upon the utility of ML for conduction of regression testing rigorously and efficiently.

Matloob et al. [14] reviewed systematically the ensemble methods for software defect prediction and reported the highlights of the defect prediction using ensemble learning. Random Forest, bagging, boosting are the popular methods with accyracy, AUC, MCC measures.

Ramessur et al. [15] developed a novel effort estimation model for sprints. They utilized multilayer perceptron which shows better results than linear regression, and SVR. Their contribution helps to accurately estimate the effort and cost along the sprints using ML.

Prenner et al. [16] worked on the unavailability of large datasets in software engineering which is a necessary condition for the application of machine learning techniques in automation of SE tasks. They built pre-trained models from 13 small and medium sized datasets and provided guidelines for conducting further research.

Mahmud et al. [17] conducted a SLR for risk prediction in software development. They analyzed 15 years of research and came up with the most popular models and the techniques in ML which are better than the statistical approaches. They reflected upon the evaluation metrics, size of datasets and performance of different models.

Li et al. [18] conducted a survey on the utilization of unsupervised techniques for software defect prediction over the period of 15 years. They reported concrete observations after experimental tests, that Fuzzy CMeans and Fuzzy SOMs are the most popular ones.

Tahir et al. [19] presented a model to automatically predict the software defect density in early phases of the development of software. They have benchmarked the datasets for public use. They deployed logged transformations along the multilayer perceptron model for the predictions.

Karimi et al. [20] developed fuzzy neural network with differential algorithm to estimate effort and compared the performance with other evolutionary models including genetic algorithm and reported an improvement of 7% in the performance of estimation models.

Alsolai et al. [21] investigated the effectiveness of ensemble techniques for maintenance change prediction in software development. They experimented with SVM, NB, kNN, and random forest with two feature selection methods and deployed sampling techniques. The results revealed that ensemble with sampling yield better results.

Alsolai et al. [22] contributed a review of the 56 studies over 1991–2018, showed that how the growth of ML techniques in the software maintenance change prediction is helping to have better prediction and planning.

From the review, it is observed that machine-learning techniques complement the existing approaches to software measurements. The review is summarized in Table 8.1. It is found that the ML approach is extensively being used to predict or estimate a variety of software metrics. ML allows for forecasting of the qualitative and quantitative attributes of the software project including the cost, effort, defect, defect density, reliability, maintenance change, etc.

8.4 Results and Discussions

This section provides insights from the review of existing studies related to prediction models based on ML techniques available for measuring software attributes. From review, it is evident that classification and regression techniques are heavily being utilized for software measurements. Two most popular tasks are effort estimation and defect prediction. Beyond these, risk prediction, change prediction, defect density prediction are also being carried out. Few unsupervised techniques namely clustering is also practiced for estimation and prediction activities. Here the main highlights of the current state of art includes the dimensionality reduction techniques, imbalance handling methods, ensemble of learning models, search based models and advanced deep architectures for learning.

8.4.1 Dimensionality Reduction Techniques

High dimensions of datasets can cause high computational resources and detriments the accuracy of the model due to some irrelevant or redundant features. Dimensionality reduction techniques can be of feature ranking type or feature selection type [23, 24]. The software data sets originally have not been created to be used for building the machine learning models. The dimensions or the attributes may or may

Table 8.1 Various ML approaches for software measurement tasks

Study	Measurement contributed	ML technique used
Hameed et. al. [6]	Effort estimation	Case-based reasoning (CBR)
Chennappan et al. [7]	Fault prediction	Decision tree
Khalid et al. [8]	Predict Software defects	SVM, NB and RF
Ali et al. [9]	Effort estimation model	Ensemble
Rodríguez Sánchez et al. [10]	Estimated cost for agile based Scrum	Bagging
Rashid et al. [11]	Software cost and effort estimation	Regression
Butt et al. [12]	Cost estimation in agile	Multilayer perceptron
Pan et al. [13]	Test case selection and prioritization	Reinforcement learning
Matloob et al. [14]	Software defect prediction	Ensemble methods
Ramessur et al. [15]	Effort estimation	Multilayer perceptron
Prenner et al. [16]	Automation of SE tasks	Pre-trained models
Mahmud et al. [17]	Risk prediction	Unsupervised techniques
Li et al. [18]	Software defect prediction	Fuzzy cmeans and Fuzzy SOMS
Tahir et al. [19]	Predict the software defect density	Multilayer perceptron
Karimi et al. [20]	Estimate effort	Fuzzy neural network
Alsolai et al. [21]	Maintenance change prediction	Ensemble techniques
Alsolai et al. [22]	Maintenance change prediction	Feature selection and sampling techniques

not be useful for building the estimation model and prediction model. Further, high dimensional datasets cause increased training time and low performance. Therefore, dimensionality reduction is necessary to be carried out for the dataset being utilized for building the prediction or estimation models. The attributes that are not significant or having high inter correlation with other attributes are required to be either eliminated or transformed to result into effective ML based estimation models.

Dimensionality reduction can be attribute selection including filter, embedded or wrapper techniques. It also includes PCA and LDA. Search based and evolutionary algorithms including GA, PSO etc. are highly popular for dimensionality reduction.

8.4.2 Imbalance Handling Techniques

Another set of ML techniques trending in the current scenario are class imbalance handling mechanisms. Class imbalance is a condition when the distribution of instances among the classes is non -uniform. This skewness causes biasness in the decisions of ML models. Hence, some class imbalance handling mechanisms are needed to be applied. It may be at data-level or algorithmic level. Data level solutions are applied as pre-processing steps before the building of the model [23, 24]. Algorithmic solutions are to be embedded with the ML model steps. Sampling and cost-sensitive algorithms are the most common examples of the class imbalance handling mechanisms. Combining the class imbalance handling techniques with dimensionality reduction methods results into better prediction or estimation models.

8.4.3 Ensemble of Learning Models

Ensemble of learning models refers to combining the two or more same or different learning algorithms to attain a better estimator or predictor [9, 10]. It may be any of the popular methods named bagging, boosting, random forest, adaboost, stacked ensembles. For measuring the software attributes, both homogenous and heterogenous ensembles are popular. Random forest is the most popular tree-based ensemble learner. Ensemble of different weak learners to build a strong learner is also gaining popularity. Stacking of models with two or three layers is another fashion with high influx of research. For estimation, SVR, and ANN are commonly being combined. Decision tree based homogenous ensembles are promising choices including bagging and boosting. For classification based tasks, SVM and ANN are combined to achieve high accuracy. Ensembles have built in capability to handle class imbalance issue. Further, ensembles in alliance with feature selection and class imbalance handling increased the prediction power of models in multiple cases.

8.4.4 Search Based Machine Learning

Search-based learning models are those techniques which models the learning problem as an optimization function. They apply some meta-heuristics to find an optimal solution in the search space of exponential range functions. There are huge feasible solutions, but the search based algorithm utilizes the principle of exploration and exploitation to find the optimal solution when the exhaustive search is not possible. Genetic algorithms, tabu search, particle swarm optimization, ant colony and many more algorithms are available in literature [25].

In the domain of software measurements, these algorithms find application for feature selection and hyperparameter tuning. Hybridization of these algorithms with other regression and classification models like PSO + SVM, PSO + ANN, GA + ANN yields promising results.

8.4.5 Advanced Deep Architectures for Learning

Advanced learning machines is about extending the neural networks to deep architectures with specific features suitable to the learning problem. Convolutional neural networks (CNNs), deep belief networks (DBNs), and staked denoising autoencoders (SDAEs) are most popular deep learning models for measuring software attributes [26]. Major advantage of utilizing deep models is that there is no need of feature extraction or feature engineering to develop these models as they are capable of extracting the features on their ownselves.

8.5 Conclusion

Measuring the Software is effectively is an essential task to the successful development of a good quality product. It is found that ML approach is being used to predict or estimate software attributes. This includes the measurement of software quality, software size, software development cost, project or software effort, maintenance task effort, defective module, testing effort. ML is a complement to the existing approaches to software measurements. Further, the software measurement models can be improved with advanced machine learning techniques and deep learning. This chapter has summarized the current trends in the domain of applications of machine learning in the software measurements and metrics. It reviewed latest research and review papers to assess the state of art models. Then, it uncovers the major category of ML techniques on the basis of the application in the software measurements. It is observed that dimensionality reduction, class imbalance handling, ensemble of models, search based models and advanced deep models are skyrocketing the software measurements applications. Effort estimation and fault predictions are the areas with high influx of research. In future, the hybridization of techniques and models has promising implications. Further, the data set generation is another important aspect to be taken care of. Industrial datasets with huge number of instances for training the models and transfer the learning among cross project metrics are desirable.

References

1. S. Goyal, Software measurements using machine learning techniques—a review. Recent. Adv. Comput. Sci. Commun. **16**(1), 38–55(18) (2023) (Bentham Science Publishers). https://doi.org/10.2174/2666255815666220407101922
2. S. Goyal, Software measurements from machine learning to deep learning, in *Computational Intelligence Applications for Software Engineering Problems* (Apple Academic Press, 2023), pp. 119–134. ISBN: 9781000575927, 1000575926
3. S. Goyal, Open challenges in software measurements using machine learning techniques, in *Computational Intelligence Applications for Software Engineering Problems* (Apple Academic Press, 2023), pp. 19–31. ISBN: 9781000575927, 1000575926

4. S. Goyal, Metaheuristics for empirical software measurements, in *Computational Intelligence in Software Modeling*, ed. by V. Jain, J.M. Chatterjee, A. Bansal, U. Kose, A. Jain (De Gruyter, Berlin, Boston, 2022), pp. 67–80. https://doi.org/10.1515/9783110709247-005
5. S. Goyal, A. Gupta, H. Jha, Current trends in methodology for software development process, in *Communication, Software and Networks. Lecture Notes in Networks and Systems*, vol. 493 (Springer, Singapore, 2023). https://doi.org/10.1007/978-981-19-4990-6_58
6. S. Hameed, Y. Elsheikh, M. Azzeh, An optimized case-based software project effort estimation using genetic algorithm. Inf. Softw. Technol. **153**, 107088 (2023)
7. R. Chennappan, An automated software failure prediction technique using hybrid machine learning algorithms. J. Eng. Res. **11**(1), 100002 (2023)
8. A. Khalid, G. Badshah, N. Ayub, M. Shiraz, M. Ghouse, Software defect prediction analysis using machine learning techniques. Sustainability **15**(6), 5517 (2023)
9. S.S. Ali, J. Ren, K. Zhang, J. Wu, C. Liu, Heterogeneous ensemble model to optimize software effort estimation accuracy. IEEE Access **11**, 27759–27792 (2023)
10. E. Rodríguez Sánchez, E.F. Vázquez Santacruz, H. Cervantes Maceda, Effort and cost estimation using decision tree techniques and story points in agile software development. Mathematics **11**(6), 1477 (2023)
11. C.H. Rashid, I. Shafi, J. Ahmad, E.B. Thompson, M.M. Vergara, I.D.L.T. Diez, I. Ashraf, Software cost and effort estimation: current approaches and future trends. IEEE Access (2023)
12. S.A. Butt, T. Ercan, M. Binsawad, P.P. Ariza-Colpas, J. Diaz-Martinez, G. Pineres-Espitia, E. De-La-Hoz-Franco, M.A.P. Melo, R.M. Ortega, J.D. De-La-Hoz-Hernández, Prediction based cost estimation technique in agile development. Adv. Eng. Softw. **175**, 103329 (2023)
13. R. Pan, M. Bagherzadeh, T.A. Ghaleb, L. Briand, Test case selection and prioritization using machine learning: a systematic literature review. Empir. Softw. Eng. **27**(2), 29 (2022)
14. F. Matloob, T.M. Ghazal, N. Taleb, S. Aftab, M. Ahmad, M.A. Khan, S. Abbas, T.R. Soomro, Software defect prediction using ensemble learning: a systematic literature review. IEEE Access **9**, 98754–98771 (2021)
15. M.A. Ramessur, S.D. Nagowah, A predictive model to estimate effort in a sprint using machine learning techniques. Int. j. inf. tecnol. **13**, 1101–1110 (2021). https://doi.org/10.1007/s41870-021-00669-z
16. J.A. Prenner, R. Robbes, Making the most of small software engineering datasets with modern machine learning. IEEE Trans. Software Eng. **48**(12), 5050–5067 (2021)
17. M.H. Mahmud, M.T.H. Nayan, D.M.N.A. Ashir, M.A. Kabir, Software risk prediction: systematic literature review on machine learning techniques. Appl. Sci. **12**(22), 11694 (2022)
18. N. Li, M. Shepperd, Y. Guo, A systematic review of unsupervised learning techniques for software defect prediction. Inf. Softw. Technol. **122**, 106287 (2020)
19. T. Tahir, Ç. Gencel, G. Rasool, U. Tariq, J. Rasheed, S.F. Yeo, T. Cevik, Early software defects density prediction: training the international software benchmarking cross projects data using supervised learning. IEEE Access (2023)
20. A. Karimi, T.J. Gandomani, Software development effort estimation modeling using a combination of fuzzy-neural network and differential evolution algorithm. Int. J. Electr. Comput. Eng. **11**(1), 707 (2021)
21. H. Alsolai, M. Roper, A systematic literature review of machine learning techniques for software maintainability prediction. Inf. Softw. Technol. **119**, 106214 (2020)
22. H. Alsolai, M. Roper, The impact of ensemble techniques on software maintenance change prediction: an empirical study. Appl. Sci. **12**(10), 5234 (2022)
23. M. Azzeh, Y. Elsheikh, A.B. Nassif, L. Angelis, Examining the performance of kernel methods for software defect prediction based on support vector machine. Sci. Comput. Program. **226**, 102916 (2023). https://doi.org/10.1016/j.scico.2022.102916
24. S.C. Rathi, S. Misra, R. Colomo-Palacios, R. Adarsh, L.B.M. Neti, L. Kumar, Empirical evaluation of the performance of data sampling and feature selection techniques for software fault prediction. Expert Syst. Appl. **223** (2023). https://doi.org/10.1016/J.ESWA.2023.119806
25. M. Gupta, K. Rajnish, V. Bhattacharjee, Software fault prediction with imbalanced datasets using SMOTE-Tomek sampling technique and genetic algorithm models. Multimed. Tools Appl. (2023). https://doi.org/10.1007/s11042-023-16788-7

26. A. Alqarni, H. Aljamaan, Leveraging ensemble learning with generative adversarial networks for imbalanced software defects prediction. Appl. Sci. **13**(24), 13319 (2023). https://doi.org/10.3390/app132413319

Chapter 9
An Effective Analysis of New Meta Heuristic Algorithms and Its Performance Comparison

Bijayalaxmi Panda, Chhabi Rani Panigrahi, Bibudhendu Pati, and Manaswinee Madhumita Panda

Abstract Metaheuristics represent a promising field of study that offers significant advancements in solving difficult optimization problems. Since the first metaheuristic was proposed, significant progress has been made, and new algorithms are still being proposed on a daily basis. Without a doubt, in the near future, research in this area will continue to advance. Nonetheless, there is a clear need to identify the top-performing metaheuristics that are predicted to last forever. In this chapter, we highlight a few of the exceptional and novel metaheuristics that have emerged in the last 20 years (2000–2020), in addition to the traditional ones like tabu search, genetic, and particle swarm. After the definite foundations of the new age group of metaheuristics are presented, the absence of theoretical foundations, hybrid metaheuristics, fresh research prospects, unresolved issues, and developments in parallel metaheuristics are discussed. Intractable problems are being successfully solved with a variety of metaheuristic algorithms in recent times. These algorithms are appealing because they produce the best/optimal answers in a short amount of time, even for very large problem sizes. Metaheuristic approaches have been drawn to a wide range of optimization problems, which can be single- or multi objective, continuous or discrete, constrained or unconstrained. Because of their complex behaviour, solving these problems is not an easy task. In this era of metaheuristic algorithm research, many new metaheuristics motivated by behavioural or evolutionary processes are introduced. For many unsolved benchmark problem sets, the best solutions are produced by this new wave of metaheuristic approaches. Because of the success and widespread appeal of metaheuristic studies, as well as the growing number of publications. Our focus has been on researching notable metaheuristic algorithms, which we refer to as "new generation" metaheuristic algorithms. We consider the quantity of citations received relative to the metaheuristic's introduction year when

B. Panda (✉) · M. M. Panda
Gita Autonomous College, Bhubaneswar, India
e-mail: pandabijayalaxmi1234@gmail.com

C. R. Panigrahi · B. Pati
Rama Devi Women's University, Bhubaneswar, India

T. R. Benala et al. (eds.), *Boosting Software Development Using Machine Learning*, Artificial Intelligence-Enhanced Software and Systems Engineering 7,
https://doi.org/10.1007/978-3-031-88188-6_9

choosing or determining the new generation metaheuristics. As a result, numerous scientists' experimental investigations confirm the efficiency of the metaheuristic.

Keywords Metaheuristics · Optimization problems · Tabu search · Genetic algorithms · Particle swarm optimization

9.1 Introduction

Several key reasons motivate the choice of metaheuristics to solve optimization problems, such as complexity of problems, [1] especially in engineering, logistics, and science, hard to be solved directly using methods like linear programming or calculus-based optimizations. Metaheuristics are strong tools to handle large and complex search spaces. On-Linearity and Non-Differentiability metaheuristics are apt at solving problems which are non-linear, non-convex, or non-differentiable, for which the traditional optimization techniques seem less efficient or even inapplicable. Metaheuristics can solve large-scale problems successfully.

While many traditional optimization methods tend to converge to local optima [2] quite easily, metaheuristics benefit from techniques like randomness and a trade-off between exploration and exploitation, allowing them to search for global solutions rather than fall into local optima. Throughout numerous real-world issues, it is desired to simultaneously optimize more than one objective. Metaheuristics like NSGA-II or SPEA2 are designed with multi-objective [3, 4] problems in mind. They can deal with uncertainty and randomness in uncertain or stochastic conditions, thus being more beneficial in dynamic or real-time optimization problems like scheduling or resource allocation problems. Metaheuristics are highly tuneable and hybridisable [5] in the sense that they may be hybridized with other optimization techniques when trying to improve some of the efficiency, robustness, or adaptability of the metaheuristic, or to fit characteristics of a problem.

9.2 Comprehensive Overview of Classical Metaheuristic Algorithms

In reality, most of the optimization techniques actually apply classical metaheuristic algorithms. The following presents technical application use cases in detail, along with accuracy and performance comparisons for each of the classical metaheuristic algorithms:

9.2.1 *Genetic Algorithm (GA)*

A wider category of evolutionary algorithms includes genetic algorithms, which are adaptive heuristic search algorithms. Natural selection theory and genetics are the foundation of genetic algorithms. With the aid of past data, which guides the search into the area of solution space where performance is higher, these represent an intelligent exploitation of random searches. Their application in producing superior solutions for search and optimization issues is widespread.

Because genetic algorithms replicate natural selection, organisms that are able to adapt to changes in their surroundings have a better chance of surviving and procreating to create offspring. Put differently, they address problems by mimicking the "survival of fittest" among individuals in successive generations. All generations are made up of people, and every person is a point in the search space and a potential answer. A string of characters, integers, floats, or bits is used to represent each individual. This string is regarded as the chromosome's equivalent [5].

9.2.1.1 Applications of GA

There's a great load of medical fields- radiology, radiotherapy, oncology, pediatrics, cardiology, endocrinology, surgical care, obstetrics, gynecology, pulmonology, infectious disease, orthopedics, rehabilitation medicine, neurology, pharmacotherapy, and health care management-that could definitely gain by the application of the genetic algorithm. This study highlights a few of these applications to provide the insights to physicians on way metaphysical thinking can be used to their benefit.

9.2.2 *Particle Swarm Optimization (PSO)*

Kennedy and Eberhart first introduced PSO-a swarm-based stochastic algorithm that actually utilizes the principles behind animal social movement, such as fish schooling and bird flocking. PSO perceives every possible solution to a problem as if it is a particle that travels around the problem space at a specific speed, just like to a flock of birds. Each particle then combines, with a little bit of randomness and disturbances, some characteristic of the record of its own historical best location and the current location with that of one or more swarm agents in order to compute its next movement across the search space. This iteration ends when each particle has travelled a sufficient distance. The whole swarm is likely to converge step by step towards the optimum of the objective function. The most significant advantage of PSO is that it requires less parameter tuning. PSO works on the interaction of particles to obtain the solution.

$$v(t+1) = v(t) + c_1 r_1 (pbest(\text{t}) - \text{X(t)} + c_2 r_2 (gbest(t) - x(t)), x(t+1) = x(t) + v(t+1), \quad (9.1)$$

where $v(t)$ stands for the particle's current velocity and $x(t)$ defines the current position of the particle, where acceleration co-efficient are c_1 and c_2 having the value equal to 2. $r_1(t)$ and $r_2(t)$ are two random variables defined in the range of [0,1]. $pbest(t) and gbest(t)$ represents the best previous position of each particle and best previous position of all particles respectively [6].

This phenomenon is because of both the potential variation of particle velocities, which restrict the consecutive range of trials inside a sub-plain of the whole search hyper-plain, and the phenomenon of local optima trap [7].

9.2.2.1 Applications of PSO in Different Area

(i) ***Energy storage optimization***: The goal of recent developments in the purpose of Particle Swarm Optimization (PSO) to energy storage optimization is to increase hybrid energy systems' efficiency. Weighted Particle Swarm Optimization (WPSO) is a noteworthy method for improving grid-connected hybrid system management that incorporates renewable energy sources like wind and solar power. By optimizing battery power utilization, this technique lowers total expenses and increases battery life. When precise energy forecasts and effective power distribution are essential, WPSO performs especially well [8].

(ii) ***Disease detection and classification***: The field of illness identification and classification has benefited greatly from the application of Particle Swarm Optimization (PSO), which has improved the precision and efficacy of machine learning models. PSO has been applied, for example, to improve the hyper parameters of CNNs for the purpose of identifying brain tumours and Alzheimer's disease. PSO and CNNs used in a hybrid technique has demonstrated notable gains in classification accuracy, with rates for Alzheimer's detection [9] reaching 98.83%.

(iii) ***Medical image segmentation***: Recent studies have shown very promising developments in using Particle Swarm Optimization in medical image segmentation, especially to improve the accuracy and reliability of such processes. Such a study that incorporated PSO with Histogram Equalization was proposed to improve the quality of segmentation in both lung CT scans and chest X-ray images. The method of PSO-HE was done with much improved clearness and precision of images, outperforming earlier techniques such as thresholding and K-means clustering.

9.2.3 ACO-Ant Colony Optimization

Ant colony optimization -ACO represents a metaheuristic optimization method that drew inspiration from the behaviour of ants. The early 1990s saw its development by [10]. Originally, the inspiration comes from observing how ants exploit food resources. Even though individually, ants possess only very limited cognitive capabilities, they can collectively find the shortest path from a food source to their nest.

In the foraging process, the ants begin with random explorations in the neighbourhood of their nest, searching for food. If an ant finds a source of food, it initiates the process of evaluating the source and returns to the nest with some food. On its return, it deposits a trail of pheromone along the path it has followed. These pheromones guide other ants toward the same food source. In order to determine the quickest routes between their nest and food, ants use a form of indirect communication known as stigmergy communication, which is constituted by peptide trails. Where a model parameter is the constant $Q > 0$.

$$\tau_i \leftarrow (1 - \rho)\tau_i i = 1, 2. \tag{9.2}$$

where the rate of pheromone evaporation is controlled by the model parameter $\rho \in (0,1)$. One way to think about the ACO metaheuristics is as a framework that needs to be appropriately tailored to the issue at hand.

The pheromone model, which is a set of pheromone values T that establishes the probabilistic search process, is what drives the optimization. Using assigned pheromone levels $\tau_i \in T$ as a weight, it creates candidate solutions from previously discovered candidates in a manner akin to pathways leading to food sources. The ACO framework is iterative and follows two general steps [11]:

(1) Solution generation on the basis of model of pheromone.
(2) After evaluating the potential solutions, the pheromone levels are adjusted to favorably bias subsequent sampling toward high-quality candidates.

9.2.3.1 Applications of ACO

(i) ***Reservoir optimization***: To get the most possible performance out of the reservoir system, optimization of the reservoir is essential [12]. When making decisions about reservoir release and storage over time, it is helpful to take fluctuations in demand and inflow into account. Many papers compare the use of GAs in place of classic ACO methods while addressing subsequent reservoir operations challenges.

ACO has been utilized for enhancing the operations of multi-purpose reservoir system, Hirakud Reservoir, in the state of Odisha in India. It concludes that ACO is far better than genetic algorithms with respect to global-optimum solution to be obtained for long-time horizon reservoir operation [13, 14]. The recently updated techniques of ACO can be used to solve challenging water resource issues since they offer certain advantages over the usual algorithm of ACO when it comes to handling multi-reservoir of large-scale operation difficulties [15].

(ii) ***Coastal Aquifer Management–Salt Water Intrusion:*** In the example of controlling saltwater intrusion, an algorithm maintained a coastal aquifer as efficiently as possible. Subsequently, the same method was used to protect wells from saltwater intrusion while simultaneously controlling drawdown restrictions and optimizing the overall water pumping rate. The researchers have not yet fully utilized the ACO variants, which is concerning given the current state of saltwater incursion.

(iii) ***Optimal Crop and Irrigation Water Allocation***: It is possible to dynamically adjust the choice of decision variables and incorporate problem knowledge. This formulation steers the search in favor of choosing crops with the highest net return and water allocations that, for a given total volume of water, produce the highest net return for the selected crop [16].

9.2.4 Differential Evolution (DE)

Another metaheuristic optimization technique that functions similarly to the GA algorithm's basic idea is differential evolution. It begins by producing a set of "agents," or potential solutions. These agents combine the positions of agents in the population using a straightforward mathematical algorithm to search the space of search. The new position will become a part of the population if it gets better; if not, it will be ignored. Storn and Price first suggested this technique in 1997 [17]. It has been improved recently to address the issue of path planning. The DE algorithm was used by Guo et al. on the Hongqi autonomous automobile, HQ430. After the controller has been designed using a model predictive control (MPC) scheme, the DE algorithm is used to find the best route. Nevertheless, no theoretical statistical analysis or comparative evaluation of alternative algorithms was conducted [18].

Storn and Price introduced DE, a population-based stochastic [17]. It solves optimization problems in a more direct, efficient, dependable, and powerful way. Performs well across a wide range of tasks [19–21]. The DE algorithm consists of the following steps [22]:

(1) Starts initializing a random population.
(2) Mutation process to produce a mutant vector.
(3) Crossover operation is applied to each pair of the target vector.

A candidate solution in mutation, V_i^t, is produced with the Eq. 9.1, where $x_{r_1}^t$, $x_{r_2}^t$, and $x_{r_3}^t$ are three dissimilar random results, and t represent the present iteration. Scaling factor F has a value of 0.5 for the version that has been chosen.

$$V_i^t = X_{r_1}^t + FX_{r_2}^t - X_{r_3}^t \tag{9.3}$$

The crossover operator is applied to the mutant vector V_i^t to produce a new vector $U_i^t = u_{i,1}, u_{i,2}, \ldots u_i$, dim, where the number of dimensions represented by *dim*. The solution of candidate X_i^t and the vector V_i^t are measured in the crossover method that uses a possibility criterion c_r as displayed in Eq. 9.2. In selection procedure, the new result vector U_i^t passes to the fresh iteration only if it is fitter that the candidate solution X_i^t

$$u_{i,j} = \begin{matrix} v_{i,j,}^t \; if\; rand\, 01 \leq C_r \\ x_{i,j}^t, \; otherwise \end{matrix} \tag{9.4}$$

Source Code Comparison: Tools like Diff, Meld, and WinMerge allow developers to visually compare code changes, highlighting differences in lines, blocks, or entire files. Binary Comparison: Tools like BinDiff and IDA Pro can compare binary files, identifying differences in instructions, data structures, and control flow. File System Comparison: Tools like Beyond Compare and WinMerge can compare file systems, identifying added, modified, or deleted files and directories.

Automated Tools for Diffing: Source Code Comparison: Developers can visually compare code changes by highlighting differences in individual lines, blocks, or entire files using tools like Diff, Meld, and Win Merge. Binary Comparison: By comparing binary files, programs such as BinDiff and IDA Pro can spot variations in control flow, data structures, and instructions. File System Comparison: By comparing file systems, programs such as WinMerge and Beyond Compare can detect newly added, edited, or removed files and directories.

Software development using Differential Evolution: In software development, differential evaluation is evaluating many approaches, strategies, or solutions to ascertain which works best for particular objectives. Because of better tools, procedures, and automation capabilities brought about by software development developments, the differential evaluation process has undergone significant change.

Tools for Performance Monitoring: Real-time performance evaluation of software under various configurations is made possible by tools such as JMeter, Datadog, and New Relic. Frameworks for automated testing: Selenium, Cypress, and JUnit are a few examples of frameworks that make functional and regression testing across versions easier.

Making Decisions Based on Data: Objective comparisons are made possible by the incorporation of analytics into software evaluation:

A/B testing: To evaluate several software revisions, platforms like as Optimizely and Google Optimize allow for real-time user feedback.

Telemetry and Logging: With the use of runtime data, contemporary logging solutions (such as Splunk and ELK Stack) enable differential analysis and offer insights into program behaviour. AI & ML in Evaluation: Using historical data, machine learning models identify trends and forecast the top-performing solutions [23].

9.2.5 *Tabu Search (TS)*

By avoiding "backwards" moves, the goal of tabu search is to prevent the search from being stuck in local minima. Usually, to accomplish this, a list of the most recent n variable-value assignments is created. Those on the list are tabu, or prohibited, when selecting the subsequent variable-value assignment. Several other nondeterministic processes can be applied to enhance the efficiency of local search algorithms. Tie-breaking rules, for instance, between equally good flips, or between two or more numbers that would imply the same greedy increase, could be influenced by past data.

Assume, for instance, that we choose the variable that has been flipped the least amount of times to break ties. Several recommendations also rely on the value diffusion principle: for example, in the event of a local minimum, applying value diffusion techniques like unit propagation or arc-consistency over unmet limitations, or applying value diffusion to improve the initial assignment on each attempt [24].

9.2.5.1 TS Applications

(i) ***Minimize power losses***: Reduce the amount of power outages losses change depending on the compensation level and network design. They are connected to the resistive components of HV/MV transformers and wires. Since the loads at MV/LV transformers are not changing with current, it is acceptable to assume a constant current model loss for them. It is also possible to ignore other losses, such as those brought on by capacitors and line insulation.

(ii) ***Maintaining voltage qualit*y**: If the voltage at the terminal nodes is as near to the rated value as feasible, a steady supply of loads is ensured.

(iii) ***Service reliability assurance***: The goal of network reconfiguration is to modify the status of the current sectionalizing switches and ties in order to lower power losses and increase power supply dependability. When it comes to running distribution systems, the operator defines service dependability as having the capacity to handle sudden spikes in load and release other feeds in the event of a fault.

In Software Development and Testing using Tabu Search: A metaheuristic optimization method known as Tabu search (TS) is utilized in a variety of software development and testing problems. It is very helpful for combinatorial optimization problems like software configuration optimization, test case generation, and test suite minimization.

Test Case Generation: Produces effective test cases that optimize fault detection or code coverage. Takes into account a number of limitations, including those related to time and resources. Minimizing the size of a test suite without sacrificing test coverage is known as test suite minimization. Finds test cases that are unnecessary or duplicate. Optimizing software configurations to increase dependability, performance, or other metrics is known as software configuration optimization finds the optimal design by examining a large solution space.

Here is a description of the software's advanced tabu search implementation. A combination of duties given to team members is known as task scheduling. Code refactoring is the process of converting classes into modules or micro services. Reduced time or expense of resources needed to finish. Increased the amount of code that was tested. Enhanced coherence of the code or decreased coupling while refactoring.

9.2.6 Greedy Randomized Adaptive Search Procedure (GRASP)

This is a constructive and consists of both a search of local phase and a construction phase [25]. For a predetermined number of times, the aforementioned process is repeated. The algorithm's output is finally determined by taking the best-obtained solution.

Construction phase: As long as there is at least one candidate element that can be added to the solution, this phase is repeated starting with an empty solution. Every time this phase is repeated, a blank Candidate List CL is created initially. Next, components that can be employed to enhance the existing partially developed solution in light of the problem's restrictions are identified and added to CL. Afterwards, a portion of the CL candidates take part in forming the RCL. Different approaches can be used to build the RCL. The following equation is used for construction of it:

$$\mathrm{RCL} = \{\mathrm{e} \in \mathrm{CL} | (\mathrm{benefit})\, \alpha \times \mathrm{benefit}\, (\mathrm{best})\} \tag{9.5}$$

Benefit is the value that each $e \in CL$ is linked to. The member with the highest value in CL is considered the best, and the parameter $\alpha \in [0, 1]$ needs to be adjusted based on the issue at hand [26].

To clarify, this phase involves investigating the area around the answer created in the first stage to see whether a better solution can be found [27].

A better outcome is achieved at the end of the first phase [28]. Because GRASP's local search part starts with a high-quality solution, there is a greater chance that a solution closer to the optimum will be produced, and this is typically appropriate for accelerating the rate of convergence. Hence, POP typically improves both the final solution's quality and the GRASP method's rate of convergence.

9.2.6.1 Applications of GRASP

(i) ***Software Testing and Debugging***: Software engineering finds an application of GRASP in test case generation, fault detection, and debugging, which can be modelled as an optimization problem, whereby test coverage is maximized with the use of a minimum number of test cases. GRASP algorithm applied to the automation software. The improved test coverage and reduced testing time of the algorithm based on prioritization of impactful test cases. Significant improvements to fault detection efficiency were evident from the results, especially for large software projects [29].

(ii) ***Machine Learning and Feature Selection***: GRASP has been employed in machine learning to perform the optimization of hyper parameters, feature selection, and model tuning to bring out the best among algorithms by finding the most optimal combination of either features or parameters. Recent study applied to GRASP for feature selection optimally in the classification task on big data analytics. It outperformed a few traditional methods of feature selection by improving model accuracy while reducing computational costs, becoming highly suitable for handling large datasets in machine learning pipelines [30].

(iii) ***Cloud Computing and Resource Allocation***: GRASP has been used in the optimization of resource allocation in cloud computing for a number of purposes, mainly around the optimal use of computation resources in an effort to minimize costs and maximize performance. A recent study discusses the problem of resource allocation in cloud data centres using GRASP to achieve two important objectives: minimizing energy consumption and preserving quality of service. The proposed algorithm was scalable and efficient in handling real-time resource demands within cloud operations [31].

Grasp Analysis using Software Development and Testing: A collection of object-oriented design guidelines known as GRASP (General Responsibility Assignment Software Patterns) assists programmers in choosing how to allocate tasks to classes. Developers can produce software systems that are more reusable, flexible, and maintainable by putting GRASP concepts to use. The GRASP analysis facilitates the creation, testing, and maintenance of software by ensuring that the duties are distributed across the system in an acceptable manner.

9.2.7 Artificial Immune Algorithm (AIA)

The AIA is a biologically inspired computational approach, whose working procedure depends basically on certain principles of the human immune system, particularly its mode of recognition and elimination of pathogens. Because of its properties of adaptability, diversity maintenance, and self-learning, AIA has been used to a broad variety of areas that includes optimization, machine learning, anomaly detection, and robotics [32, 33].

9.2.7.1 Applications of AIA

(i) ***Optimization Problems***: AIA has been widely applied to resolve a wide range of intractable problem of optimization: combinatorial optimization, multi-objective optimization, and dynamic optimization. IA-based algorithm solves a multi-objective optimization problem in supply chain management in the year 2023. It had shown very superior performance in balancing cost and service levels under the dynamic changing environment [34].
(ii) ***Machine Learning***: IA has been applied in feature selection, classification, and clustering in machine learning, especially in those problems that require robust and adaptive learning. Recent study, proposed an AIA-based feature selection method to enhance the accuracy of a classification model on medical diagnosis [35].
(iii) ***Clustering and Data Mining***: IA has been used in clustering issues especially for high-dimensional datasets so that meaningful groups can be discovered from the data. The new algorithm achieved better clustering quality and scalability compared with traditional clustering methods like k-means [36].

9.2.8 Chaos Optimization Method (COM)

The COM is characterized by a metaheuristic class of algorithms, which draws upon the chaotic behaviour of nonlinear dynamical systems for the purposes of scanning the search space in solving optimization problems. Chaos is actually a regime of deterministic yet unpredictably behaving systems, and these factors altogether contribute to ensure that the COM would not converge prematurely to local optima and therefore enhances its global search abilities.

9.2.8.1 COM Applications

(i) ***Control of Dynamic Systems***: COM has been used on nonlinear dynamic systems control, the challenge in this task being to maintain stability while optimizing performance. COM for finding optimized values of control parameters associated with the chaotic robot arm. The algorithm performed well regarding the nonlinearities in the system and hence depicted smooth and precise movement control of the arm.
(ii) ***Image Processing***: COM has found applications in image segmentation, feature extraction, and image enhancement and benefited from the fact that the algorithm is able to delve into a very large solution space. COM applied in the image segmentation of medical images. Chaotic search produced better segmentation accuracy than the other methods under noisy or complex image data.
(iii) ***Artificial Neural Networks (ANN) Training***: This has found its application in optimizing the weights and architecture of a neural network to improve

its performance for certain classification and regression tasks and COM used to train an ANN for stock price prediction. In fact, the chaotic approach outperforms the gradient-based methods.

ANN Is Utilizing Software Development: Predict continuous values using regression (e.g., sales prediction). Classification: Sort inputs into different categories (e.g., spam detection). Clustering: Assemble related data points (e.g., segmentation of customers). It is also utilized for Preparation: To ensure consistency, clean, standardize, and transform data. Splitting: Separate data into sets for testing, validation, and training. Feature Scaling: To guarantee effective training, normalize features.

9.2.9 Scatter Search (SS)

A population-based metaheuristic optimization technique that is intended to find optimal or near-optimal solutions for complex combinatorial and continuous optimization problems. First proposed by Fred Glover in the 1980s, the Scatter Search method searches for an optimal solution by systematically generating a set of solution candidates which are combined with the diversification of various search strategies [37].

9.2.9.1 Applications of Scatter Search

Applications of Scatter Search includes [38] Traveling Salesman Problem, Job Shop Scheduling, Vehicle Routing Problem and so on.

(i) ***Traveling Salesman Problem (TSP)***: SS has been successful in application to solve TSP, where the objective is finding the shortest possible route visiting a set of cities and returning to the start. TSP instances by applying Scatter Search. This study showed that SS yielded high-quality solutions and competitive results against other metaheuristic techniques.
(ii) ***Job Shop Scheduling***: SS applies to job shop scheduling with the objective of optimization and allocation of jobs to machines for minimum make span while meeting deadlines. Scatter Search methodology used to solve the job shop scheduling problem with complicated constraints and demonstrated improved results over other scheduling techniques.
(iii) ***Vehicle Routing Problem (VRP)***: SS has been applied to VRP to obtain an optimal routing for a fleet of vehicles in order to serve customers efficiently. SS applied to VRP with time windows and capacity constraints. The results proved that SS had better route planning with cost reduction compared to other heuristic algorithms.

9.2.10 Shuffled Frog-Leaping Algorithm (SFLA)

Itis a metaheuristic optimization algorithm that takes inspiration from the foraging behaviour of frogs and the idea of shuffling to enhance the efficiency of a search. It was an extension of the Frog-Leaping Algorithm by incorporation into it of mechanisms of shuffling with enhanced performance and diversity [39].

9.2.10.1 Applications of SFLA

(i) ***Function Optimization***: SFLA has been used to solve different function optimization problems, including multi-dimensional and multi-modal functions. SFLA applied to high-dimensional benchmark functions optimization. According to the study's results, SFLA demonstrated superior convergence speed and solution quality compared to some traditional algorithms such as Genetic Algorithms and Particle Swarm Optimization.
(ii) ***Job Shop Scheduling***: SFLA is put to work in job shop scheduling whereby the objective would be the optimal allocation of jobs to machines, also adhering to a number of other constraints. Employing SFLA for solving dynamic-constrained complex job shop scheduling problems, SFLA provided high-quality solutions and showed that SFLA could handle the problem dynamics with much efficiency [40].
(iii) ***Vehicle Routing Problem (VRP)***: SFLA was used in the vehicle routing problem with the aim of finding the optimal routes with a fleet of vehicles to service customers. SFLA used in solving the VRP with time windows. Their algorithm had improved route planning with a reduced operational cost compared to other heuristic methods.

9.2.11 Comparison Criteria of Classical Metaheuristic Algorithms

When comparing classical metaheuristic algorithms, the following criteria are typically considered:

Accuracy: The ability of the algorithm to find high-quality or optimal solutions.

Performance: The speed of convergence and computational efficiency.

Scalability: The algorithm's effectiveness and efficiency as the problem size increases.

Robustness: The capability of algorithm to handle different kind of optimization problems and avoid local optima (Table 9.1).

Table 9.1 Comparison of classical metaheuristic algorithms

Name of algorithm	Accuracy	Performance	Scalability	Robustness
Genetic algorithms	Good	Moderate convergence	Good for both continuous and combinatorial problems	Robust and expensive
Particle swarm optimization	High	Fast convergence	Good for continuous problems	Robust and sensitive to parameter settings
Ant colony optimization	High	Moderate convergence speed	Effective for discrete problems	Robust but can be slow
Genetic programming	High	Slow convergence	Effective for symbolic problems	Robust and expensive
Differential evolution	High	Fast convergence	Good for high-dimensional problems	Robust against local optima
Simulated annealing	Good	Slow convergence	Effective for discrete and combinatorial problems	Robust but sensitive to cooling schedule
Tabu search	High	Moderate to slow convergence	Effective for medium-sized problems	Robust but complex to implement
GRASP	Good	Moderate to slow convergence	Effective for various problem sizes	Robust but requires multiple iterations
Artificial immune algorithm	Good	Moderate convergence speed	Effective for specific problems	Effective in dynamic environments
Iterated local search	Good	Moderate convergence	Effective for various problem sizes	Robust but may require careful tuning
Chaos optimization method	Moderate	Variable convergence speed	Effective for certain problems	Less intuitive and less commonly used
Scatter search	Good	Slow convergence	Effective for various problem sizes	Effective but computationally intensive
SFLA	Good	Moderate convergence speed	Effective for medium-sized problems	Robust but less commonly used
Variable neighbourhood search	Good	Moderate convergence speed	Effective for combinatorial and continuous problems	Robust and adaptable

9.2.12 Classical Algorithms May no Longer Be Sufficient

Following are the parameters due to which classical algorithms are not implemented in current research problems.

(i) ***Increased Problem Complexity***: Modern optimization problems are often more complex and high-dimensional, which can exceed the capabilities of classical algorithms. For example, the need for real-time solutions and large-scale problems may demand more efficient algorithms.
(ii) ***Computational Demands***: Classical algorithms, especially those like GA and GP, can be computationally expensive. With the increasing size and complexity of problems, their computational demands can become prohibitive.
(iii) ***Lack of Theoretical Foundation***: Many classical algorithms lack solid theoretical underpinnings, which can limit their effectiveness and understanding in certain complex scenarios.
(iv) ***Emergence of New Algorithms***: The development of new metaheuristic algorithms and hybrid approaches has provided more efficient and effective solutions for many problems. Algorithms such as Artificial Bee Colony (ABC) and Whale Optimization Algorithm (WOA) offer better performance and adaptability.
(v) ***Adaptation to Modern Needs***: Newer algorithms are designed to address the specific needs of contemporary optimization problems, such as adaptability to dynamic environments and multi-objective optimization, which classical algorithms may struggle with.

Classical metaheuristic algorithms have provided significant contributions to optimization, their limitations in handling modern, complex, and large-scale problems have led to the development of more advanced algorithms that better meet contemporary needs. We are discussing the advanced algorithms in the next section of this chapter, which are in general we have defined as "new generation" metaheuristic algorithm.

9.2.13 Limitations of the Proposed Model

Several metaheuristic methods were employed in this paper different algorithms used metaheuristic algorithms. Some of the algorithms are computationally demanding. Traditional methods are unreliable. Security aspects not covered. Genetic algorithms (GAs) having high computational cost, Premature Convergence like.

Loss of Diversity: GAs can converge to suboptimal solutions if diversity in the population is not maintained, leading to local optima., Problem-Specific Design: Fitness Function Complexity: Designing an effective fitness function for software development tasks, such as bug prediction, test case optimization, or feature selection, can be difficult. Limited local search capabilities were discovered when using an artificial bee colony. Weak Exploitation: ABC may be less successful at fine-tuning solutions in intricate, multi-modal search spaces due to its dependence on random perturbations for local search. While using BFO algorithm in software development find displaying issues Encoding software artefacts like code, design models, or test cases into a format suitable for BFO can be challenging. While using cuckoo search algorithm Complex Constraints come out while using in Software development. It often involves hard constraints (e.g., resource limits, time constraints), which CSA may struggle to handle without significant customization.

9.3 New Generation Metaheuristic Algorithms

9.3.1 Artificial Bee Colony (ABC)

The ABC algorithm's main goal is to solve optimization problems by identifying a global, optimal, or nearly optimal solution. The way this is done is by mimicking the foraging behaviours seen in bee colonies. There are two different groups of bees in the population: working bees and observer bees. Bees that are employed go on missions to gather nectar from food sources while also gathering relevant positional data about these sources. In turn, unemployed bees share information with onlooker bees, which they use to further their own search for food [41, 42].

Phase initialization: ABC must produce a predetermined quantity of solutions to serve as the starting population. ZN initial solutions are created at random over the search space in the initialization phase using Eq. (9.7). Wi = $w_{i,1}$, $w_{i,2}$, $w_{i,D}$ is the population's ith answer, and D is the problem's dimensions. Suppose {1, 2..., ZN} = i, j = {1, 2..., D}, $wmin_{,j}$ and $w_{max,j}$ are the below and higher bounds dimensional explore range of jth, respectively, and rand(1) is a random amount uniformly circulated over [0,1].

$$w_{i,j} = w_{min,j} + rand(1) \cdot \left(w_{max,j} - w_{min,j}\right) \tag{9.6}$$

Phase of Employee Bee: Using the following equation, each employed bee searches the neighbours of the solution wi in this phase to find a potential solution y_i [43–45].

$$y_{i,j} = w_{i,j} + \varphi_{i,j} \cdot \left(w_{i,j} - w_{z,j}\right) \tag{9.7}$$

where w_i represents the present solution, w_z indicates a randomly chosen resolution from the entire population ($z^{\sim} = i$), j indicates a randomly chosen element from {1, 2..., A}. Using the greedy selection strategy, determine which of w_i and y_i is superior. This implies that in the event that y_i proves to be superior, w_i will be substituted with y_i and its answer will be reset to 0. If not, the present solution will be maintained with a plus-one counter.

Onlookerphase for bee: During the stage, bees investigate the surrounding areas of the food sources. However, in contrast to employed bees, spectator bees prefer to search for quality sources for food based on the knowledge that employed bees provide. Observer bees use the probability Q_i, which may be expressed as follows, to determine which food source they will look for further [46].

$$Qi = \frac{fit(w_i)}{\sum_{i=1}^{zn} fit(w_i)} \tag{9.8}$$

where $fit(w_i)$ signifies the fitness value of the present food source w_i, and it is considered by the follow equation:

$$fit(w_i) = \begin{cases} \frac{1}{1+f(w_i)}, & \text{if, } f(w_i) \geq 0 \\ 1 + |f(w_i)|, & f(wi) < 0 \end{cases} \tag{9.9}$$

value of the ith solution's objective function is symbolized as $f(w_i)$. According to Eq. (9.10), there is a greater likelihood of selecting the solution with a higher fitness value. Using the roulette selection procedure, each spectator bee selects the food source to hunt for. Equation (9.8) is then utilized to produce the candidate solution. The optimal solution selected by greedy selection method and counter updated same like in the employed bee phase [47].

Phase of scout Bee: Scout bees keep an eye on every food source throughout this phase, and they will choose the one with the highest counter. The relevant solution will be promptly rejected and its employed bee will be replaced by a scout bee to reinitialize a new food supply by Eq. (9.7) if its counter exceeds the predetermined value limit. Subsequently, the scout bee will become an employed bee once again when its counter is reset to 0.

A well-known meta-heuristic algorithm with excellent exploration capabilities is the artificial bee colony (ABC) [48] (Table 9.2).

Table 9.2 Comparison of ABC with GA

Attribute	GA	ABC
Global search	Strong exploration but requires careful tuning to prevent premature convergence	Balances exploration (scouts) and exploitation (onlookers and employed bees)
Convergence speed	Slower due to the genetic operations' overhead	Often faster due to direct search around promising solutions
Adaptability	Works well for diverse problem types (binary, discrete, continuous)	Effective for continuous and multi-modal optimization problems

9.3.1.1 Innovative Perspective Using ABC Algorithm

Implementation of API: For quicker response times and lower expenses, optimize the architecture and implementation of APIs.

Goal: Increase throughput while reducing delay.

- Execution: API setups, such as endpoint architectures and data payload volumes, serve as "food sources."
- The fitness function uses error rates, throughput, and latency to assess API performance.
- ABC maximizes the efficiency of API design.

Optimization of Software Configuration Parameters- Optimize program setup settings for optimal performance.

Goal: Increase software's dependability, speed, or resource consumption.

- Cache size and thread counts are examples of software setups that serve as "food sources."
- Performance metrics such as error rates, throughput, and response time are measured by the fitness function.
- For optimal outcomes, ABC optimizes the configuration options.

9.3.2 Bacterial Foraging Optimization

An AI method called BFOA can be used to approximate solutions to very challenging or unsolvable numerical maximization or minimization problems. BFO is a probabilistic process for solving numerical problems of optimization in situations without a workable deterministic solution. It replicates the food-seeking and behaviour of reproductive bacteria of general, such as E. coli [49].

Original BFO Parameters

Parameters	Description
W	Bacterial number = 1, 2..., W
X	Search of dimension
W_{ed}	Steps counted for elimination-dispersal d = 1, 2, ...W_{ed}
W_{re}	Reproduction steps c = 1, 2...W_{re}
W_c	Number of Chemotix steps b = 1, 2...Wc
W_s	Swimming steps in number S = 1, 2, ...W
C(a)	Step size of chemotix
B(a)	Direction of a random [− 1, 1]
X_{ed}	Dispersal elimination

(continued)

(continued)

Parameters	Description
B(a,b,c.d)	ath bacterium fitness value at chemotaxis bth, reproduction on cth, elimination on dth, ith fitness value of bacterium at chemotaxisjth, reproductionkth, lth elimination-dispersal

It is a nature inspired optimization algorithm, modelled after the foraging behaviour of *Escherichia coli* (E. coli) bacteria. The algorithm mimics the way these bacteria search for nutrients in a highly dynamic environment, especially those that are nonlinear, multimodal, or noisy.

9.3.2.1 Applications of BFO

Optimization of Engineering problems
Parameter tuning of machine learning
Processing of images
Optimization of Control systems (Table 9.3).

9.3.2.2 BFO Used in Software

Optimize test case selection, prioritization, and execution order in software development by utilizing BFOA. Automate the localization of bugs in intricate codebases, dynamically distribute resources for cloud-based apps, Assign and schedule work to developers and optimize software system setup parameters.

9.3.3 Firefly Algorithm (FA)

Based on the way fireflies flash, the Firefly Algorithm (FA) is an additional optimization method inspired by nature. Xin-She Yang created it in 2008, and it is frequently used to resolve multimodal, complicated, and nonlinear optimization issues. Fireflies

Table 9.3 Comparison of ACO with BFO

Attribute	ACO	BFO
Global Search	Effective for combinatorial optimization problems	Robust in handling noisy or dynamic optimization landscapes
Convergence Speed	Can be faster in discrete spaces, especially for routing problems	Handles complex fitness landscapes but may converge slowly
Scalability	Performs well in moderate-sized problem spaces	Suitable for high-dimensional and dynamic optimization problems

communicate by flashing, either to entice possible mates or to signal to others. To direct fireflies toward the best answers in a search space, this attraction is modelled within the algorithm.

9.3.3.1 Applications of FA

Design issues in engineering
Neural networks and machine learning
image manipulation
Choose features
Optimization of power systems.
Scheduling issues

The Firefly Algorithm is a well-liked option in many fields due to its versatility, resilience, and ease of use in solving multimodal optimization issues [50].

Firefly Algorithm
Random initialization the population R fireflies W_j, j = 1, 2…R
Compute the fitness value of each firefly W_j, j = 1, 2…R
FEs = R;
While FEs ≤ MAX_FEs do
For j = 1 to R do
For k = 1 to R do
If $f(W_k) < f(W_j)$ then
Move W_j toward W_k
Compute the fitness value of new W_j
FEs++;
End if
End for;
End for;
End while;

9.3.4 BAT Algorithm

Xin-She Yang created the BAT algorithm, a bio-inspired optimization technique, in 2010. It uses bat echolocation as a model to tackle challenging optimization problems. Bats employ this technique to locate prey, navigate, and steer clear of obstructions [51, 52]. To identify the optimal answer in a search space, the BAT method combines this idea with optimization techniques [53].

9.3.4.1 Application of BAT Algorithm

Design issues in engineering
Optimization of functions
Processing Images
Data mining and machine learning
Optimization Networks.

Combining aspects of controlled exploration with randomization, the BAT algorithm is easy to use, adaptable, and strong for a wide range of optimization tasks [54].

BAT Algorithm
1. The fitness function need to be specified
2. The population of bats is created at random
3. Xi, w, B must be initialized
4. Iterations must be carried out till $n < 30$
5. Updated solution created by applying the below formula Eq. (9.2) and Eq. (9.3)
6. **if** (rand > wi)
7. Solution of local find the best solution using Eq. (9.4)
8. **end if**
9. **if** (rand < Bi and f(wi) < f(w))
10. The stored solution is the best solution
11. The values of W and A are updated using Eq. (9.5) and Eq. (9.6) [Ref. 13]
12. **end if**
13. The best answer available right now is found by ranking the bats
14. Ended the algorithm
15. Display the optimum solution

Real-Time Software Applications Using Bat Algorithm: The Bat Algorithm (BA) is a meta-heuristic algorithm, which is motivated by the echolocation behaviour of bats. It has earned a big reputation due to its capability and efficiency in solving optimization problems. Though it is not relevant to make real-time applications just like classical algorithms such as PID controllers or Kalman filters; realizing the outcome of the bat algorithm can be used in the offline training or optimization phase to improve its performance in online systems. Here are some examples of possible real-time applications that can be used indirectly by BA.

A Real-Time Optimization of Control Systems

The basic control parameters of Model Predictive Control (MPC) can be tuned using BA in order to optimize the parts, thereby rendering the system faster and more economic than using the existing techniques.

Adaptive Control: In this area, one might find applications of BA in real-time tuning of parameters for adaptive controllers, which include gain scheduling and self-tuning controllers.

Real-Time Signal Processing

Adaptive Filtering: In this form of application, BA will use adaptive filter coefficient optimization for maintenance of noise or echo cancellation.

Feature Extraction: BA may optimize algorithms of feature extraction, such as PCA (principal component analysis) or ICA (independent component analysis), for applications in real time.

Hitting the Buffoonery with Real-Time Applications

Computational Efficiency: For BA, computational efficiency is what matters in real time. Parallel processing and GPU acceleration would prove to be of great help in performance enhancements.

Convergence Speed: In order to assure that optimum solutions are reached within the time constraints; BA needs to converge quickly.

Robustness: Noise and uncertainty in the real-time environment should be accounted for by BA.

Hybridization: Combination of BA with other optimization algorithms and machine learning techniques can further augment the performance and applicability of BA in real-time scenarios.

BA may not be finally available in real-time systems, but it makes a huge difference when it comes to optimally designing and parameterizing such systems. These would be effective improvements on application performances and efficiencies. With these specifications, BA would stand as one of the greatest tools in the development of advanced applications of software for real time.

9.3.5 GSA-Gravitational Search Algorithm

Applications of Gravitational Search Algorithm (GSA) can be found in Artificial Intelligence (AI), especially in learning, optimization, and pattern recognition domains. Finding the best answers to complicated problems can be difficult, and AI frequently solves these kinds of difficulties [55]. The GSA improves the functionality of AI systems in the following ways by exploring and exploiting search spaces.

9.3.5.1 GSA Applications

(i) ***Optimization problems***: Many tasks in artificial intelligence (AI) include optimization, including selection of feature, selection of model, and hyper parameter tuning.
(ii) ***Selection of Feature***: Feature selection is the process of choosing the most pertinent characteristics from huge datasets in order to reduce computational load and enhance machine learning model performance. Through its exploration and exploitation process, GSA assists in determining the optimal combination of traits.
(iii) ***Tuning of hyper parameter***: To achieve optimal performance, machine learning models frequently depend on hyper parameters that require fine tuning. The best hyper parameters for algorithms such as Support Vector Machines (SVM), Neural Networks, and Decision Trees can be found using GSA.

9.3.5.2 Comparing GSA with Some Other Algorithms

(i) ***Genetic Algorithms (GA) versus GSA***: GSA bases its search strategy on gravitational interactions, whereas GAs rely on evolutionary concepts like crossover and mutation. Although GSA can converge more quickly, it occasionally becomes trapped in local optima [56].
(ii) ***Particle Swarm Optimization (PSO) against GSA***: PSO modifies particle locations and velocities in response to local and global optimal placements. GSA, in contrast, makes advantage of gravitational force, which in some circumstances might provide better exploration.

9.3.6 Biogeography-Based Optimization (BBO)

Artificial intelligence (AI) professionals are using Biogeography-Based Optimization (BBO) more frequently to address a range of optimization issues [57]. Finding the best or almost best answers in challenging AI scenarios is made possible by its innate inspiration from the migration of species and habitat suitability. Using BBO in AI is explained in full below.

9.3.6.1 Selection of Feature Using Machine Learning

Feature selection aims to retrieve the most relevant features in any dataset, which can strengthen any machine learning model by enhancing its predictive capabilities. This is because irrelevant and redundant features have a tendency to increase computation time while decreasing model performance [58].

9.3.6.2 BBO Applications

BBO generates an improved feature set continuously through migration and mutation in search of better performance of the learning model with lesser complexity.

BBO optimizes the weights and biases of neural networks by representing every possible configuration as a habitat. Each configuration's performance is measured on a training dataset by mean squared error or cross-entropy loss. The migration mechanism of BBO lets it study a number of configurations to come up with the best set of weights leading to better training accuracy and generalization [59].

By considering every group of centres as a habitat, BBO in clustering maximizes the location of cluster centres [60].

9.3.7 Grey Wolf Algorithm (GWO)

A metaheuristic optimization method inspired by nature, the Grey Wolf Optimization (GWO) algorithm is based on the social structure and hunting techniques of grey wolves in the wild. Since its introduction in 2014 by Seyedali Mirjalili et al., it has grown in popularity because of its adaptability, simplicity, and capacity to resolve challenging optimization issues [61]. An extensive examination of GWO and its uses in AI is provided below:

(i) ***Initialization***: Within the search space, a population of potential solutions, or "wolves," is randomly initialized. Every wolf stands for a possible fix for the optimization issue.
(ii) ***Assessment***: An objective function that establishes the calibre of the solution is used to assess each wolf's fitness.
(iii) ***Update of Hierarchy***: The population is categorized according to fitness; the top three answers are called δ, ω, and α wolf, while the remaining answers are called β wolves.
(iv) ***Prey Encircling***- Wolves modify their locations according to the best options available at the moment (δ, ω, and α wolves). This encircling behaviour can be represented mathematically as follows:

$$Z = A\left|W^P(t) - W(t)\right| \\ W(t+1) = W^p(t) - XZ \tag{9.10}$$

$W^P(t)$—It is the position of prey (δ, ω, and α).
W (t)—It is the present position of wolf.
The coefficient vectors are X and A calculated as:

$$X = 2b.I1 - b$$
$$W = 2.I2$$

b linearly decreases from 2 to 0.

I1 and I2 are random vectors in the range [0,1].

(v) ***Hunting***: Wolves simulate hunting by updating their positions based on the locations of δ, ω, and α wolves. These top three answers have an impact on a wolf's new position.

(vi) ***Attacking Prey***: When the algorithm runs, the value of *a*a drops. This lowers the exploration capacity and makes it easier for wolves to take advantage of the best solutions, which causes convergence.

(vii) ***Termination***: Until a halting requirement is satisfied, the process is repeated (e.g., maximum iterations or a suitable fitness level).

9.3.7.1 Applications of GWO

A subset of features is represented by each wolf. The aim was to optimize selected characteristics while optimizing the model's performance (such as accuracy). GWO imitates the hunting behaviour to iteratively enhance the feature subset.

Each wolf in a neural network symbolizes a set of neural network parameters (weights and biases) in training. Using a training dataset, GWO modifies these settings to minimize error.

GWO can optimize the location of cluster centroids in tasks involving clustering. In order to reduce intra-cluster distances, the algorithm modifies the positions of each wolf, which represents a potential set of centroids.

To improve control performance, GWO adjusts the parameters of fuzzy logic controllers [62] (Table 9.4).

Table 9.4 Comparison of PSO with GWO

Attribute	PSO	GWO
Individuality	Social behaviour of birds and fish	Bubble-net hunting strategy of humpback whales
Search mechanism	Handles exploration and exploitation through update of velocity	Adaptive mechanism based on locality of leader
Parameter dependency	Relatively simple with social components like inertia, cognitive etc	Parameter dependency is lower for simplification of tuning
Diversity maintenance	Less in some part of iterations	Because of adaptive weights diversity is better
Exploration/exploitation trade-off	Better parameter tuning is required to handle	From exploration to exploitation transition is dynamic
Convergence speed	Speed is faster but decline at local optima	Speed is slower but global search strategy is better
Applications	Used across several fields	Used in different fields of engineering

New metaheuristic algorithms such as GWO are more suitable for complex problems and problems with constraint and high dimensionality but in case of PSO resources are limited for computation [63–65].

9.3.8 Krill Herd Algorithm (KH)

Inspired by the aggregate behaviour of individual krill in nature, the Krill Herd (KH) algorithm is an optimization method that draws inspiration from nature [66]. To increase their chances of surviving, krill, which are tiny crustaceans, behave like a swarm, frequently arranging themselves to best take advantage of opportunities for breeding, protection, and food gathering [67]. In order to solve optimization challenges in AI and other domains, Amir H. Gandomi and Amir Hossein Alavi proposed the Krill Herd method in 2012.

9.3.8.1 Applications of KH

KH has been used in a number of sectors, such as:

(i) ***Social Spider Optimization***

Based on the social behaviour of spiders, Social Spider Optimization (SSO) is a met heuristic optimization method inspired by nature. In particular, it solves challenging optimization issues by simulating the cooperative behaviours—such as sharing information, cooperation, and communication—that are seen in social spider colonies. When searching over expansive, multidimensional search spaces for global optima, the algorithm is especially helpful.

(ii) ***Selection of Features***

Feature selection is important because it includes choosing the most pertinent features from a dataset to enhance model performance and shorten computation times. In this situation, SSO can be used to identify the best subset of features by considering each spider as a possible solution for a feature subset [68].

Usually, a combination of feature count and classification accuracy determines the fitness function. Optimizing accuracy while reducing feature count is the aim [69].

SSO employs vibrational communication to drive its feature space exploration toward subsets that exhibit superior performance, ultimately striking a balance between redundancy and feature significance.

9.4 Overview of the Selected Methods Used for Software Development

By using the Java decoding approach, the author presented the component-based test suite sequence method along with a Genetic Algorithm implementation. For test cases used at the component-based software module level, this prioritization method was introduced. With the help of this technique, we were able to make considerably more accurate predictions about the frequency of severe defects or problems [70].

(i) ***Generating Test Cases***: It might be difficult to manually create a complete set of test cases that cover all the paths, conditions, and edge situations in many software systems. Test cases can be automatically generated using meta-heuristics based on specific criteria, like: Coverage of paths, Coverage of codes, Conditions at the boundaries, Satisfaction of constraints by experimenting with various input combinations, can assist in producing optimal test cases [71].

(ii) ***Minimizing the Test Suite***: The abundance of redundant test cases, which can be ineffective and time-consuming, is a significant obstacle in software testing. Test suite minimization can be accomplished with metaheuristics. The goal is to minimize the quantity of test cases while ensuring adequate software coverage [72]. By choosing test cases from an initial pool that offer the most coverage while removing redundancy, genetic algorithms can develop a reduced collection of test cases.

(iii) ***Generation of Test Data***: Creating test data that satisfies specific input criteria or domain constraints is frequently necessary for software testing. Test data that pushes the limits of the software system, such as edge situations, incorrect inputs, and combinations that could cause unexpected behaviour, can be produced via metaheuristics. By changing populations of test inputs over many generations, evolutionary algorithms, for example, can produce a variety of test data [73].

(iv) ***Optimizing Tests***: Prioritizing test cases or test scenarios according to criteria like risk, likelihood of failure, or relevance is a common optimization challenge in software testing. By giving test cases weights and giving priority to the most important ones for execution, metaheuristics can aid in test optimization [74].

The selection of test cases can be optimized to improve efficacy while minimizing resources (e.g., time, budget, or personnel) by using different algorithms.

(v) ***Selection of Regression Tests***: Running previously completed test cases is known as regression testing, and it is used to make sure that software updates don't bring new flaws. Regression tests that are more likely to identify errors based on system modifications can be chosen using metaheuristics [75]. Regression test subsets that concentrate on the most pertinent or dangerous sections of the code can be found using genetic algorithms or Ant Colony Optimization (ACO) [76].

9.5 Common Type of Metaheuristic Used in Software Testing

(i) ***Genetic Algorithm***

One of the most often used metaheuristics for creating and refining tests is GA. It imitates natural selection by using crossover, mutation, and selection to evolve populations of possible solutions (such as test cases) [72]. In test data generation, test suite minimization, and test case generation, GAs are especially helpful.

(ii) ***Simulated Annealing (SA)***

SA is a probabilistic method that functions similarly to metallurgical annealing. Finding an approximate solution to an optimization issue is its purpose, especially when the solution space is vast and intricate [77]. SA is used in software testing to prioritize test cases and minimize test suites, without compromising coverage.

(iii) ***Optimization of Particle Swarms (PSO)***

The social behaviour of fish schools or flocks of birds serves as the model for PSO. The swarm as a whole investigates the solution space, with each "particle" representing a potential solution (such as a collection of test cases) [78].

(iv) ***Ant Colony Optimization (ACO)***

ACO is used for pathfinding and optimization problems and was inspired by the foraging activity of ants. ACO can be used in software testing to optimize test suites for cost-effectiveness and coverage, or to choose test cases that optimize code coverage [74].

(v) ***Tabu Search***

An optimization approach called Tabu Search avoids local optima by combining memory structures with local search strategies. Examining several test setups and choosing a subset that optimizes the testing objectives is a common practice in test case optimization [79].

From test suite simplification and prioritization to test case and data development, metaheuristics provide a robust toolkit for optimizing software testing. They can greatly increase efficiency and efficacy in vast and complex testing scenarios, but they cannot always ensure that they will find the optimum option. They are very useful in contemporary software testing, particularly in agile or continuous delivery scenarios, due to their versatility, agility, and capacity to manage wide search spaces.

9.6 A Comparative Analysis of Different Metaheuristic Approach with Respect to Strength, Weakness and Use Case

See Tables 9.5, 9.6, 9.7, 9.8 and 9.9.

Table 9.5 Comparative Analysis of PSO

Algorithm	Strength	Weakness	Use case
PSO	PSO algorithms are applied towards solving a wide variety of optimization problems, such as subset-sum problems, by finding the optimal solution among a set of candidate solutions. Overall, the strengths of PSO algorithms make them a favourite selection in solving a variety of problems that are connected to optimization problems such as subset-sums and were widely used in many fields, including engineering, finance, and biology among many others	PSO may converge slowly, particularly when addressing complex optimization problems. The rate of convergence also depends on the parameter of PSO, including the learning factor and the velocity clamping range. PSO is computationally expensive, especially for large-scale optimization problems, mainly because each particle should be evaluated at every iteration. In addition, PSO is sensitive to parameter values, making it hard to optimize the algorithm for different applications	It optimizes and looks to solve complex problems ranging from energy storage, logistics, machine learning, finance, and more. It is tasked to tune hyperparameters, produce an optimal route, design structures efficiently, and manage power grids. Fast convergence and adaptability of PSO improve it to become a suitable choice for dynamic, nonlinear optimization problems in real-time and large-scale applications [80]

9.7 Application of Metaheuristic for Test Case Generation in Different Fields

Metaheuristic algorithms such as Genetic Algorithm (GA), Particle Swarm Optimization (PSO), Ant Colony Optimization (ACO), and Grey Wolf Optimizer (GWO), have been practically employed in automated software testing. Here are some case studies which illustrate how effective they can be used.

(i) ***GA in Telecommunication***: In telecommunications GA is being deployed to automate the generation of test cases especially for testing protocols and network setup.

Case study: Use cases and other software engineering paradigms have confirmed that software testing is the most paramount and critical stage for any SDLC where even real-time coding projects are involved. Roughly speaking, testing in the development of software projects generally accounts for around 40–60% of the total time and resources spent on a project. The generation of test cases is the most crucial activity in the process of software testing. The system has been designed to solve automatically the problem of generating test case sets by using the minimal number of test cases necessary to achieve the desired level of adequacy and thus decreasing the software testing effort and costs. In the structural testing of a system under development, test cases that focus on specific paths within the design are difficult to generate efficiently thus the process is often treated as an optimization problem. Search based

Table 9.6 Comparative Analysis of GA

Algorithm	Strength	Weakness	Use case
GA	Some of the major benefits of using genetic algorithms are:	As with any technique, genetic algorithm also has its own disadvantages:	The optimization and search issues have been addressed through simulation of natural selection, the genetic algorithms. These perform extremely well in complicated tasks such as scheduling, selection of features in machine learning and engineering design, and provide sub-optimal solutions over wide and dynamic areas of search [81]
	1. Parallelism	1. The dilemma of defining the fitness function	
	2. Ramp rate	2. The dilemma of defining the encoding for the problem	
	3. Broader exploration of the solution space	3. There is a tendency towards premature convergence	
	4. Fitness landscape is non-trivial	4. The issues of deciding other parameters such as population size, mutation rate, crossover rate, selection method and its pressure	
	5. Smooth global optimization is simple	5. Gradients are not applicable	

techniques such as genetic algorithm GA and swarm optimizations have been looked at as solutions to this problem since it is regarded as an optimization problem. The importance of the research is aimed to the automatic test case generation optimization problem in search based software engineering coverage [85].

Telecom automation testing encompasses the processes of verification and validation of usability, performance and effective functionality of these services and systems. Its fundamental targeted aim is to keep the focus on the customers, the regulatory bodies with set quality and reliability limits. The coverage of telecom testing is vast as it includes functional, performance, security, API, compatibility and interoperability testing, and so on [86].

Outcome: As a result of employing a genetic algorithm approach, telecommunication providers have been able to derive a variety of efficient test cases for unique network environments thus increasing the efficiency of fault localization and decreasing the volume of manual testing.

Table 9.7 Comparative Analysis of ACO

Algorithm	Strength	Weakness	Use case
ACO	1. Similar standard deviation	1. Runtime on small and medium scale graph	Based on the behavior of ants while foraging for food, Ant Colony Optimization (ACO) is a method developed for solving different optimization problems. It is applied to solving problems such as those associated with finding the shortest routes, designing networks, and scheduling operations by looking for the best paths in extensive searching areas [82]
	2. Ensuring Convergence	2. Improve the algorithm to improve quality of solution	
	3. High Compatibility	3. Not suitable for complex problems	
	4. Decreasing heuristic algorithm	4. Improve the algorithm to improve quality of solution	
	5. High efficiency	5. Not improving the cost efficiency	
	6. Development of upper bound	6. Type of operation not clear not flexible	

(ii) ***Autonomous Robot Navigation***: Both an obstacle-detecting sensor and machine intelligence are required for path planning in the mobile robot to navigate in an environment. The mobile robot is equipped with the measurement device and records and either transmits or returns the data to the operator. The robot must have sensors to sense obstacles in its environment, while machine intelligence is applied to plan paths around those obstacles. The application of genetic algorithms is an example of application of machine intelligence to modern robot navigation. Genetic algorithms represent heuristic optimization techniques that have mechanisms analogous to biological evolution.

Outcome: It implies autonomy that the robot should decide on the traveling paths in the given environment. Basic information exists for the robot about the navigation area boundary, but there are unknown obstacles in the navigation area. This is called an uncertain environment; the robot must be able to detect and maneuver around these obstacles in order to reach its target point. "World space" refers to the navigation environment wherein the robot and the obstacles in reality share the same space.

(iii) ***Application of Particle Swarm Optimization (PSO) in Regression Testing of E-commerce Websites***: In selected e-commerce websites PSO has been implemented successfully for automating the regression test case selection process with an aim of maximizing the test case coverage and minimizing the test cases.

Table 9.8 Comparative analysis of ABC

Algorithm	Strength	Weakness	Use case
ABC	1. Easy to use, flexible and powerful	1. The utilization of secondary information is hardly limited	The ABC (Artificial Bee Colony) algorithm has found its applications in various optimization problems in many branches including engineering, image processing, and bioinformatics. It imitates the foraging routine of honey bees for locating their food source optimally [83]
	2. Local search capability	2. New fitness tests are needed for the parameters of the new algorithms	
	3. Objective cost measurement	3. A larger number of objective function evaluations are required	
	4. Capable of being put into practice easily	4. Has a cumbersome nature when in sequential computation mode	

Case Study: In a mobile application for an online shopping site, developers made slight adjustments to the overall design of the product screen. The testers did not think this improvement would cause problems with other functions like search, catalogue, cart, or billing. They checked the performance of the product page performing correct operations with different goods across different browsers and devices and approved it for publishing.

But there is always a flaw in the system. Because the developer reworked the obscure a function, which reduces the size of the input string, to now reduce the number of input characters from 80 to 40. The product screen was functioning perfectly fine. However, since the same function was used in the shopping cart screen, it failed to show some important information due to this modification. There was a 25% decline in checkouts and the company suffered a loss in sales.

Table 9.9 Comparative analysis of CSA

Algorithm	Strength	Weakness	Use case
CSA	1. Implementation is Straightforward—Thanks to the uncomplicated nature of the algorithm, it is quite possible to implement and comprehend	1. Capability of Pre-Convergence Effect: Similar to several optimization techniques, there exists the danger of falling into a trap of suboptimal solutions	The Cuckoo Search Algorithm (CSA), which is based on the brood parasitism of cuckoo birds, is employed in optimization problems in engineering, data mining, machine learning, etc. It performs remarkably in feature selection, tasks scheduling, routing optimization and so forth due to its exploration and exploitation capabilities, which is useful in high dimensional and complex problems [84]
	2. Proven Successful in Finding Global Optima—It has proved successful in locating global optima in different types of complicated terrains	2. Tuning Adjustments Might Be Needed: In some scenarios, the effectiveness of the Cuckoo Search Algorithm may be dependent on adjusting parameters like the discovery rate to fit a particular problem domain or class of issues	

Regression testing has been conducted, the screen for the shopping cart would have been included, and this incurred cost to the website would have been avoided with ease [87, 88].

Outcome: By automating test case selection, the platform significantly reduced testing time and resource usage, achieving quicker deployment cycles with high test coverage.

(iv) ***Ant Colony Optimization (ACO) in Performing GUI Testing of Mobile Aids***

Case Study: Mobile app developers employed ACO to help in the synthesis of test sequences for GUI testing of mobile applications, especially Android and I phone applications. With new applications comes an increased complexity. It is therefore expected that the GUIs are becoming more advanced and it is proving more difficult to create test cases for them by hand. This has created a challenge in application testing where good quality GUIs test case generation has to be done automatically to create what is referred to as an event sequence graph [89].

Outcome: ACO bettered user interface exploration which revealed unanticipated navigation path bugs that were impossible to pinpoint through manual exploration which ultimately improved the user experience.

(v) ***Application of Grey Wolf Optimization Techniques to Software with a Case Study***

Case Study: Fault localization in complex systems was optimized by applying Grey Wolf Optimization techniques in a software development company. One of the most challenging steps in software debugging is the process of fault location. This can be done by rating the program statements from the most to the least suspicious of containing an error based on the test case execution traces. [90].

Outcome: The algorithm has successfully recognized the modules that are more prone to having faults which makes it easy for the developers to cut down on most of the debug parts and concentrate on the important aspects increasing the quality of the software [91–93].

9.8 Conclusion

According to this study, meta-heuristic algorithms—which draw inspiration from nature processes and phenomena—have become essential for resolving challenging software development optimization issues. New meta-heuristic algorithms, including Ant Colony Optimization (ACO), Particle Swarm Optimization (PSO), Genetic Algorithms (GA), Grey Wolf Optimizer (GWO), and others, have been developed as a result of recent developments in this field. When tackling software engineering problems like requirement prioritization, testing, project scheduling, and resource allocation, these techniques seek to improve efficiency, scalability, and resilience. It has been shown that new meta-heuristic algorithms can outperform conventional methods in solving high-dimensional, multi-objective, and non-linear optimization problems. They do well in circumstances that are dynamic and unpredictable, such as software development. Numerous research demonstrates the promise of hybrid meta-heuristic algorithms, which combine the advantages of several approaches to get around their drawbacks.

References

1. M. Abdel-Basset, L. Abdel-Fatah, A.K. Sangaiah, Metaheuristic algorithms: a comprehensive review, in *Computational Intelligence for Multimedia Big Data on the Cloud with Engineering Applications* (2018), pp. 185–231
2. D. Steinley, Local optima in K-means clustering: what you don't know may hurt you. Psychol. Methods **8**(3), 294 (2003)
3. E. Zitzler, M. Laumanns, S. Bleuler, A tutorial on evolutionary multiobjective optimization, in *Metaheuristics for Multiobjective Optimisation* (2004), pp. 3–37
4. C.R. Yu, X. Liu, Q.C. Wang, D. Yang, Solving the comfort-retrofit conundrum through post-occupancy evaluation and multi-objective optimisation. Build. Serv. Eng. Res. Technol. **44**(4), 381–403 (2023)
5. R.C. Eberhart, Y. Shi, J. Kennedy, *Swarm Intelligence* (Elsevier, 2001)

6. https://www.geeksforgeeks.org/genetic-algorithms/. Accessed on 20 Oct 2024
7. https://www.sciencedirect.com/topics/physics-and-astronomy/particle-swarm-optimization. Accessed on 25 July 2024
8. U. Ramanathan, S. Rajendran, Weighted particle swarm optimization algorithms and power management strategies for grid hybrid energy systems. Eng. Proc. **59**(1), 123 (2023)
9. A. Mosavi, P. Ozturk, K.W. Chau, Flood prediction using machine learning models: literature review. Water **10**(11), 1536 (2018)
10. G. Battineni, M.A. Hossain, N. Chintalapudi, E. Traini, V.R. Dhulipalla, M. Ramasamy, F. Amenta, Improved Alzheimer's disease detection by MRI using multimodal machine learning algorithms. Diagnostics **11**(11), 2103 (2021)
11. A. Hemmati-Sarapardeh, A. Larestani, N.A. Menad, S. Hajirezaie, *Applications of Artificial Intelligence Techniques in the Petroleum Industry* (Gulf Professional Publishing, 2020)
12. R. Bhavya, L. Elango, Ant-inspired metaheuristic algorithms for combinatorial optimization problems in water resources management. Water **15**(9), 1712 (2023)
13. I. Fister, X.S. Yang, J. Brest, D. Fister, A brief review of nature-inspired algorithms for optimization. Elektrotech. Vestnik Electrotech. Rev. **80**, 116–122 (2013)
14. A. Ostfeld, Ant colony optimization for water resources systems analysis–review and challenges, in *Ant Colony Optimization-Methods and Applications* (2011)
15. R. Soto, B. Crawford, R. Olivares, C. Taramasco, I. Figueroa, Á. Gómez, C. Castro, F. Paredes, Adaptive black hole algorithm for solving the set covering problem. Math. Probl. Eng. **2018**(1), 2183214 (2018)
16. C. Blum, Beam-ACO—hybridizing ant colony optimization with beam search: an application to open shop scheduling. Comput. Oper. Res. **32**(6), 1565–1591 (2005)
17. R. Storn, K. Price, Differential evolution–a simple and efficient heuristic for global optimization over continuous spaces. J. Global Optim. **11**, 341–359 (1997)
18. H. Guo, D. Cao, H. Chen, Z. Sun, Y. Hu, Model predictive path following control for autonomous cars considering a measurable disturbance: implementation, testing, and verification. Mech. Syst. Signal Process. **118**, 41–60 (2019)
19. R.S. Rampriya, R. Suganya, RSNet: Rail semantic segmentation network for extracting aerial railroad images. J. Intell. Fuzzy Syst. **41**(2), 4051–4068 (2021)
20. H. Zhao, J. Shi, X. Qi, X. Wang, J. Jia, Pyramid scene parsing network, in *Proceedings of the IEEE Conference on Computer Vision and Pattern Recognition* (2017), pp. 2881–2890
21. L.C. Chen, Y. Zhu, G. Papandreou, F. Schroff, H. Adam, Encoder-decoder with atrous separable convolution for semantic image segmentation, in *Proceedings of the European Conference on Computer Vision (ECCV)* (2018), pp. 801–818
22. H. Noh, S. Hong, B. Han, Learning deconvolution network for semantic segmentation, in *Proceedings of the IEEE International Conference on Computer Vision* (2015), pp. 1520–1528
23. https://www.sciencedirect.com/topics/materials-science/simulated-annealing. Accessed on 10 Oct 2024
24. https://www.sciencedirect.com/topics/computer-science/tabu-search. Accessed on 20 Oct 2024
25. F. Glover, J.P. Kelly, M. Laguna, New advances for wedding optimization and simulation, in *Proceedings of the 31st Conference on Winter simulation: Simulation—A Bridge to the Future-Volume 1* (1999), pp. 255–260
26. R.C.R. Andrade, M.R. Resende, Aspectos legais do estágio na formação de professores: uma retrospectiva histórica. Educação em perspectiva **1**(2) (2010)
27. T.A. Feo, M.G. Resende, Greedy randomized adaptive search procedures. J. Global Optim. **6**, 109–133 (1995)
28. C. Fleurent, F. Glover, Improved constructive multistart strategies for the quadratic assignment problem using adaptive memory. INFORMS J. Comput. **11**(2), 198–204 (1999)
29. T.D. Hendrix, J.H. Cross, L.A. Barowski, K.S. Mathias, Providing enhanced visual support for software development and maintenance, in *Proceedings of the 36th Annual Southeast Regional Conference* (1998), pp. 23–28

30. V.B. Semwal, Y.K. Prajapat, R. Jain, Training a multi-task model for classification and grasp detection of surgical tools using transfer learning. SN Comput. Science **4**(5), 582 (2023)
31. J. Jeyaraman, S.V. Bayani, J.N.A. Malaiyappan, Optimizing resource allocation in cloud computing using machine learning. Eur. J. Technol. **8**(3), 12–22 (2024)
32. A. Phu-Ang, An improve artificial immune algorithm for solving the travelling salesman problem, in *2021 Joint International Conference on Digital Arts, Media and Technology with ECTI Northern Section Conference on Electrical, Electronics, Computer and Telecommunication Engineering* (IEEE, 2021), pp. 261–264
33. D. Dasgupta, (Ed.) *Artificial Immune Systems and Their Applications* (Springer Science & Business Media, 2012)
34. J.Y. Wu, Solving constrained global optimization problems by using hybrid evolutionary computing and artificial life approaches. Math. Probl. Eng. **2012**(1), 841410 (2012)
35. E. Mabrouk, Y. Raslan, A.R. Hedar, Immune system programming: a machine learning approach based on artificial immune systems enhanced by local search. Electronics **11**(7), 982 (2022)
36. R.J. Kuo, N.J. Chiang, Z.Y. Chen, Integration of artificial immune system and K-means algorithm for customer clustering. Appl. Artif. Intell. **28**(6), 577–596 (2014)
37. M. Kalra, S. Tyagi, V. Kumar, M. Kaur, W.K. Mashwani, H. Shah, K. Shah, A comprehensive review on scatter search: techniques, applications, and challenges. Math. Probl. Eng. **2021**(1), 5588486 (2021)
38. G.C. Onwubolu, B.V. Babu, F. Glover, M. Laguna, R. Martí, New ideas and applications of scatter search and path relinking, in *New Optimization Techniques in Engineering* (2004), pp. 367–383
39. M. Eusuff, K. Lansey, F. Pasha, Shuffled frog-leaping algorithm: a memetic meta-heuristic for discrete optimization. Eng. Optim. **38**(2), 129–154 (2006)
40. https://towardsdatascience.com/a-survey-on-shuffled-frog-leaping-algorithm-d309d0cf7503. Accessed on 02 Sep 2024
41. M.M. Eusuff, K.E. Lansey, Optimization of water distribution network design using the shuffled frog leaping algorithm. J. Water Resour. Plan. Manag. **129**(3), 210–225 (2003)
42. N. Sluijk, A.M. Florio, J. Kinable, N. Dellaert, T. Van Woensel, Two-echelon vehicle routing problems: a literature review. Eur. J. Oper. Res. **304**(3), 865–886 (2023)
43. I. Ahmad, F. Qayum, S.U. Rahman, G. Srivastava, Using improved hybrid grey wolf algorithm based on artificial bee colony algorithm onlooker and scout bee operators for solving optimization problems. Int. J. Comput. Intell. Syst. **17**(1), 111 (2024)
44. E. Forcael, I. Carriel, A. Opazo-Vega, F. Moreno, F. Orozco, R. Romo, D. Agdas, Artificial bee colony algorithm to optimize the safety distance of workers in construction projects. Mathematics **12**(13), 2087 (2024)
45. Y. Zhang, B. Pang, Y. Song, Q. Xu, X. Yuan, Artificial bee colony algorithm based on dimensional memory mechanism and adaptive elite population for training artificial neural networks. IEEE Access (2023)
46. X. Li, S. Zhang, P. Shao, Discrete artificial bee colony algorithm with fixed neighbourhood search for traveling salesman problem. Eng. Appl. Artif. Intell. **131**, 107816 (2024)
47. X. Ni, W. Hu, Q. Fan, Y. Cui, C. Qi, A Q-learning based multi-strategy integrated artificial bee colony algorithm with application in unmanned vehicle path planning. Expert Syst. Appl. **236**, 121303 (2024)
48. S. Phoemphon, Grouping and reflection of the artificial bee colony algorithm for high-dimensional numerical optimization problems. IEEE Access (2024)
49. X. Zhou, G. Tan, H. Wang, Y. Ma, S. Wu, Artificial bee colony algorithm based on multi-neighbor guidance. Expert. Syst. Appl. 125283 (2024)
50. Ş Öztürk, R. Ahmad, N. Akhtar, Variants of artificial bee colony algorithm and its applications in medical image processing. Appl. Soft Comput. **97**, 106799 (2020)
51. S. Mishra, I. Swain, Exploring the applications of the bacterial foraging optimization algorithm (BFOA) for the graph coloring problem: a review. Int. J. Comput. Sci. Manag. Stud. **13**(4).

52. H. Al-Khazraji, W. Guo, A.J. Humaidi, Improved cuckoo search optimization for production inventory control systems. Serb. J. Electr. Eng. **21**(2), 187–200 (2024)
53. S.Y. Yang, Y.H. Xiang, D.W. Kang, K.Q. Zhou, An improved cuckoo search algorithm for maximizing the coverage range of wireless sensor networks. Baghdad Sci. J. **21**(2 (SI)), 0568 (2024)
54. S.M. Hosseini, M.H. Shirvani, H. Motameni, Multi-objective discrete Cuckoo search algorithm for optimization of bag-of-tasks scheduling in fog computing environment. Comput. Electr. Eng. **119**, 109480 (2024)
55. R.M. Naji, H. Dulaimi, H. Al-Khazraji, An optimized PID controller using enhanced bat algorithm in drilling processes. Journal Européen des Systèmes Automatisés **57**(3), 767 (2024)
56. K. Prajapati, S. Jani, M. Singh, R. Brajpuriya, Contribution of AI and deep learning in revolutionizing gravitational wave detection. Astron. Comput. 100856 (2024)
57. Z. Qian, Y. Xie, S. Xie, MAR-GSA: mixed attraction and repulsion based gravitational search algorithm. Inf. Sci. **662**, 120250 (2024)
58. S. Suraya, S.M. Irshad, G.P. Ramesh, P. Sujatha, Biogeography based optimization algorithm and neural network to optimize place and size of distributed generating system in electrical distribution. J. Electr. Eng. Technol. **17**(3), 1593–1603 (2022)
59. H. Ma, D. Simon, P. Siarry, Z. Yang, M. Fei, Biogeography-based optimization: a 10-year review. IEEE Trans. Emerg. Top. Comput. Intell. **1**(5), 391–407 (2017)
60. S. Gupta, N. Singh, K. Joshi, Biogeography based novel AI optimization with SSSC for optimal power flow. Majlesi J. Electr. Eng. **12**(2), 39–45 (2018)
61. A.K. Bansal, V. Garg, Biogeography-based optimization (BBO) trained neural networks for wind speed forecasting, in *Proceedings of International Conference on Trends in Computational and Cognitive Engineering: TCCE 2019* (Springer Singapore, 2021), pp. 79–94
62. P. Palai, D.P. Mishra, S.R. Salkuti, Biogeography in optimization algorithms: a closer look. IAES Int. J. Artif. Intell. **10**(4), 982 (2021)
63. Q. Zhu, A. Shankar, C. Maple, Grey wolf optimizer based deep learning mechanism for music composition with data analysis. Appl. Soft Comput. **153**, 111294 (2024)
64. S. Saleem, I. Ahmad, S.H. Ahmed, A. Rehman, Artificial intelligence based robust nonlinear controllers optimized by improved gray wolf optimization algorithm for plug-in hybrid electric vehicles in grid to vehicle applications. J. Energy Storage **75**, 109332 (2024)
65. M. Melgarejo, M. Medina, J. Lopez, A. Rodriguez, Optimization test function synthesis with generative adversarial networks and adaptive neuro-fuzzy systems. Inf. Sci. 121371 (2024)
66. E. Korkmaz, A.P. Akgüngör, A hybrid traffic controller system based on flower pollination algorithm and type-2 fuzzy logic optimized with crow search algorithm for signalized intersections. Soft Comput. 1–23 (2024)
67. R. Karmakar, S. Chatterjee, D. Datta, D. Chakraborty, Application of harmony search algorithm in optimizing autoregressive integrated moving average: a study on a data set of Coronavirus Disease 2019. Syst. Soft Comput. **6**, 200067 (2024)
68. P. Kaliraj, B. Subramani, Intrusion detection using krill herd optimization based weighted extreme learning machine. J. Adv. Inf. Technol. **15**(1), 147–154 (2024)
69. M. Forghani, M. Soltanaghaei, F.Z. Boroujeni, Dynamic optimization scheme for load balancing and energy efficiency in software-defined networks utilizing the krill herd meta-heuristic algorithm. Comput. Electr. Eng. **114**, 109057 (2024)
70. M. Hosseinllu, M. Abbasi, O. Safarzadeh, F. Dehghani, Multi-objective loading pattern optimization of a soluble boron free core using social spider algorithm. Nucl. Eng. Des. **420**, 113034 (2024)
71. R. Chatterjee, M.A.K. Akhtar, D.K. Pradhan, F. Chakraborty, M. Kumar, S. Verma, R.A. Khurma, M. García-Arenas, FNN for diabetic prediction using oppositional whale optimization algorithm. IEEE Access (2024)
72. N. Khoshniat, A. Jamarani, A. Ahmadzadeh, M. Haghi Kashani, E. Mahdipour, Nature-inspired metaheuristic methods in software testing. Soft Comput. **28**(2), 1503–1544 (2024)
73. M. Kovačević, M. Madić, M. Radovanović, Software prototype for validation of machining optimization solutions obtained with meta-heuristic algorithms. Expert Syst. Appl. **40**(17), 6985–6996 (2013)

74. F.G. de Freitas, C.L.B. Maia, G.A.L. de Campos, J.T. de Souza, Optimization in software testing using metaheuristics. Revista de Sistemas de Informação da FSMA **5**(5), 3–13 (2010)
75. P.R. Srivastava, V. Ramachandran, M. Kumar, G. Talukder, V. Tiwari, P. Sharma, Generation of test data using meta heuristic approach, in *TENCON 2008–2008 IEEE Region 10 Conference* (IEEE, 2008), pp. 1–6
76. A.H. Halim, I. Ismail, S. Das, Performance assessment of the metaheuristic optimization algorithms: an exhaustive review. Artif. Intell. Rev. **54**(3), 2323–2409 (2021)
77. E. Ekinci, Z. Garip, K. Serbest, Meta-heuristic optimization algorithms-based feature selection for joint moment prediction of sit-to-stand movement using machine learning algorithms. Comput. Biol. Med. **178**, 108812 (2024)
78. B. Alhijawi, A. Awajan, Genetic algorithms: theory, genetic operators, solutions, and applications. Evol. Intel. **17**(3), 1245–1256 (2024)
79. N. Ahuja, P.K. Bhatia, Test suite minimization through PSO, in *2024 11th International Conference on Reliability, Infocom Technologies and Optimization (Trends and Future Directions) (ICRITO)* (IEEE, 2024), pp. 1–5
80. D. Shanthi, N. Swapna, A. Kiran, S. Anoosha, Ensemble approach of GP, ACOT, PSO, and SNN for predicting software reliability. Int. J. Eng. Syst. Model. Simul. **15**(2), 68–75 (2024)
81. E. Díaz, J. Tuya, R. Blanco, Automated software testing using a metaheuristic technique based on tabu search, in *18th IEEE International Conference on Automated Software Engineering, 2003. Proceedings* (IEEE, 2003), pp. 310–313
82. T.M. Shami, A.A. El-Saleh, M. Alswaitti, Q. Al-Tashi, M.A. Summakieh, S. Mirjalili, Particle swarm optimization: a comprehensive survey. IEEE Access **10**, 10031–10061 (2022)
83. https://www.bartleby.com/essay/Advantages-And-Limitations-Of-Genetic-Algorithm-PCXCTU8LAKU
84. https://www.sciencedirect.com/topics/computer-science/ant-colony-optimization. Accessed on 20 Oct 2024
85. D. Karaboga, B. Basturk, A powerful and efficient algorithm for numerical function optimization: artificial bee colony (ABC) algorithm. J. Global Optim. **39**, 459–471 (2007)
86. H. Faris, I. Aljarah, M.A. Al-Betar, S. Mirjalili, Grey wolf optimizer: a review of recent variants and applications. Neural Comput. Appl. **30**, 413–4 (2018)
87. https://www.linkedin.com/pulse/unveiling-wings-optimization-dive-cuckoo-search-algorithm-yeshwanth-n-xmoac
88. M. Rajagopal, R. Sivasakthivel, K. Loganathan, L.E. Sarris, An automated path-focused test case generation with dynamic parameterization using adaptive genetic algorithm (AGA) for structural program testing. Information **14**(3), 166 (2023)
89. https://www.linkedin.com/pulse/telecom-automation-testing-tools-test-cases-best-practices-white-i48ze Accessed on 20 Oct 2024
90. https://tuskr.app/learn/regression-testing Accessed on 20 Oct 2024
91. A. Bajaj, A. Abraham, S. Ratnoo, L.A. Gabralla, Test case prioritization, selection, and reduction using improved quantum-behaved particle swarm optimization. Sensors **22**(12), 4374 (2022)
92. J. Rodriguez, G.D. Rodriguez, Automatic generation of GUI test cases using ant colony optimization and greedy algorithm, in *CIbSE* (2015), p. 209
93. B. Yang, X. Ma, H. Guo, Y. He, F. Xu, An approach of improving the efficiency of software fault localization based on feedback ranking information. Appl. Sci. **13**(18), 10351 (2023)

Chapter 10
Empowering Software Security: Leveraging Machine Learning for Anomaly Detection and Threat Prevention

Lalit Kumar Vashishtha, Kakali Chattejee, Pranav Pant, Santosh Kumar Sahu, and Durga Prasad Mohapatra

Abstract The chapter investigates the amalgamation of defined security approaches and modern machine-learning techniques inside software development. It aligns with the book's core theme of utilizing machine learning to transform the development process. The chapter begins by outlining fundamental security concepts and the significance of identifying anomalies in software systems, setting the stage for a proactive approach to threat detection. It delves into various machine learning techniques, such as neural networks and unsupervised learning, that drive innovation in anomaly detection. A detailed analysis highlights the strengths and challenges of each method. The chapter addresses the challenges of integrating these methods into current systems while offering valuable insights into applying machine learning in security protocols by using cases from the real world. It also looks at ethical issues, like algorithmic bias, and how they affect the fairness and reliability of security measures. This chapter concludes by providing comprehensive guidance for project managers and software developers on successfully incorporating machine learning into security plans and guaranteeing safer and more effective software solutions.

L. K. Vashishtha · K. Chattejee
Department of Computer Science and Engineering, NIT Patna, Patna, India
e-mail: lalitv.phd20.cs@nitp.ac.in

K. Chattejee
e-mail: kakali@nitp.ac.in

P. Pant
School of Computing and Augmented Intelligence, Ira A. Fulton School of Engineering, Arizona State University, Tempe, AZ, USA

S. K. Sahu (✉)
Geodata Processing and Interpretation Centre(GEOPIC), ONGC, Dehradun, India
e-mail: santoshsahu@hotmail.co.in

D. P. Mohapatra
Department of Computer Science and Engineering, NIT Rourkela, Rourkela, India
e-mail: durga@nitrkl.ac.in

T. R. Benala et al. (eds.), *Boosting Software Development Using Machine Learning*, Artificial Intelligence-Enhanced Software and Systems Engineering 7,
https://doi.org/10.1007/978-3-031-88188-6_10

Keywords Anomaly detection · Large language model (LLM) · Retrieval-augmented generation (RAG) · Machine learning · Deep learning · Software security

10.1 Introduction

Software security encompasses the strategies, methodologies, and procedures to safeguard software applications against unauthorized access, data breaches, malicious attacks, and other potential threats. It involves designing and developing software with robust defences to mitigate vulnerabilities that attackers could exploit, thereby ensuring the confidentiality, integrity, and availability of the software and its data.

10.1.1 Overview of Software Security: Importance and Current State

Software security has become one of the most important issues for developers, companies, and end users in the ever-changing technological landscape. The stakes have never been higher as software applications increasingly support critical national security, healthcare, and financial operations. Software security is essential for preserving the dependability, availability, and integrity of software products that millions of people use daily in addition to protecting data and systems from unauthorized access [21].

Challenges in Traditional Security Methods Software security has historically been viewed as a secondary issue that is usually handled too late in the software development process or, worse, after a security breach. A never-ending firefighting cycle has resulted from this reactive strategy, in which vulnerabilities are corrected only after they have been exploited. Instead of preventing accidents, organizations are forced to respond to them, which is expensive and exposes systems to potentially disastrous outcomes [16]. For many years, the foundation of software protection has been established by conventional security techniques including intrusion detection systems (IDS), firewalls, and encryption. Nevertheless, these techniques primarily target recognized dangers and mainly depend on preset criteria or signatures [12]. This approach has significant limitations:

- Inability to Detect Novel Threats: Since these systems are rule-based, they struggle to identify new or previously unknown attack vectors, making them vulnerable to zero-day exploits.
- High False Positive Rates: The high number of false positives generated by conventional methods can lead to alarm fatigue and potentially the neglect of important concerns.

- Scalability Issues: As software systems grow in complexity, the ability to effectively monitor and secure every component using traditional methods becomes increasingly challenging.

These drawbacks highlight the need for more sophisticated, more flexible security solutions that can change to meet new threats.

10.1.2 Introduction to Machine Learning in Security

Transition from Traditional Methods to Machine Learning-Driven Approaches In response to the challenges posed by traditional security methods, the field of software security is experiencing a significant shift towards integrating machine learning (ML) techniques [3]. In contrast to traditional security systems that depend on static rules, machine learning models have the capacity to process vast volumes of data, identify trends, and adapt to new information over time. They are especially well-suited for security applications where threats are ever-evolving because of this capability [28].

The transition to machine learning-driven security approaches represents a move from reactive to proactive security postures:

- **Proactive Threat Detection**: Anomalies-differences from established patterns of typical behavior that can point to a security breach-can be identified using machine learning algorithms. This makes it possible to detect possible dangers before they have a chance to do serious harm [10].
- **Continuous Learning and Adaptation**:With time, machine learning systems can become more proficient at spotting subtle or new risks as they gain knowledge from new data [31].
- **Reducing False Positives**: Security teams may concentrate on real threats by lowering the number of false positives by using machine learning models that can distinguish between malicious and benign behavior [5].

This integration of machine learning into security processes is not just a technological enhancement but a paradigm shift in how we approach software security. It allows for a more dynamic, responsive, and ultimately more secure software environment, where threats can be anticipated and mitigated rather than merely reacted to.

As we continue in this chapter, we will explore how machine learning techniques are applied in various aspects of software security, particularly in anomaly detection and threat prevention, and discuss the practical implications of this technology in enhancing the security of software systems.

10.2 Fundamentals of Anomaly Detection in Software Development

Definition and Significance in the Context of Software Security Anomalies, in the context of software security, refer to deviations from the expected or normal behavior of a system. These deviations can manifest as unusual patterns of activity, unexpected changes in system performance, or any behavior that does not conform to the established norms of operation [18]. It is essential to comprehend abnormalities since they frequently indicate the existence of security risks like illegal access, data breaches, or the exploitation of system flaws.

In simpler terms, anomalies are like red flags in the software environment, indicating that something unusual or potentially harmful is occurring. For instance, a sudden spike in network traffic, a series of failed login attempts, or the unexpected modification of system files could all be considered anomalies that warrant further investigation.

Figure 10.1 shows the system architecture for network anomaly detection. The significance of detecting these anomalies cannot be overstated. In many cases, anomalies are the first indicators of a security breach. Early detection helps security teams respond quickly and reduce the impact of an attack before it causes major damage. Moreover, understanding the nature of anomalies helps in identifying the underlying causes, whether they are the result of malicious activity, software bugs, or configuration errors [32]. Additionally, Kishore et al. [19] used a variational mislabeled dataset from Virustotal to provide a two-stage process to identify mislabeled malware and defend the host against Black-Box attacks.

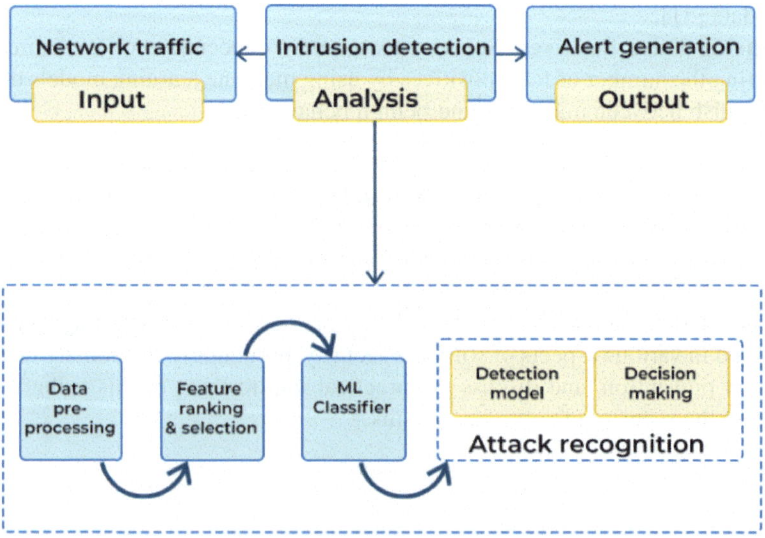

Fig. 10.1 System architecture of network anomaly detection. *Source* [6]

10.2.1 *Key Concepts in Security: Overview of Foundational Security Concepts Relevant to Anomaly Detection*

To effectively detect and respond to anomalies, it is essential to have a solid grasp of foundational security concepts. These concepts form the basis of any security strategy and are particularly relevant when developing and deploying anomaly detection systems. Some of the key concepts include:

- **CIA Triad (Confidentiality, Integrity, and Availability)**: The three fundamental ideas of information security are represented by this trio. Only those who are permitted can access sensitive information thanks to confidentiality. Information accuracy and unalteredness are ensured by integrity. Availability guarantees the availability of systems and data when needed. Finding anomalies is essential to preserving the overall security posture since they frequently indicate a breach of one or more of these principles.
- **Attack Vectors**: These are the techniques or routes that hackers take to enter a system without authorization. It's critical to comprehend popular attack vectors like phishing, malware, and SQL injection because their exploitation frequently results in anomalies [7].
- **Threat Modeling**: This is how possible security risks to a system are found and ranked. Systems for anomaly detection can be improved to identify pertinent suspicious activity by knowing which threats are most likely to target a system [27].
- **False Positives and False Negatives**: A false negative happens when a real threat is missed in anomaly detection, whereas a false positive happens when a typical behavior is mistakenly identified as an anomaly. Since too many false positives can cause warning fatigue and false negatives might expose a system to vulnerability, balancing the rate of false positives and negatives is a crucial difficulty in anomaly detection.
- **Behavioral Baselines**: These are recognized patterns of typical system functioning that serve as a guide for identifying abnormalities. These baselines are frequently used by machine learning models to spot deviations that might point to a security risk [33].

These foundational concepts are integral to understanding how anomaly detection systems function and how they can be effectively implemented in a security strategy.

10.2.2 *Importance of Anomaly Detection: Why Detecting Anomalies Is Crucial for Preemptive Threat Prevention*

To identify and stop security problems early on, anomaly detection is essential. Anomaly detection aims to find odd patterns or behaviors that are not typical, as opposed to conventional security techniques that utilize predefined signatures or criteria to identify known risks [34]. This proactive strategy is crucial for several reasons:

- Detection of Unknown Threats: The capacity of anomaly detection to detect novel or unidentified threats is among its most important benefits. Conventional security solutions frequently only identify threats that have already been detected and categorized [22]. Anomaly detection, however, can identify suspicious behavior that has not been seen before, providing a crucial layer of defense against emerging threats and zero-day exploits.
- Minimizing Damage from Attacks: Early detection of anomalies allows for a faster response to potential security incidents. By identifying and addressing a threat before it fully manifests, organizations can minimize the damage and reduce the impact of an attack. This can be particularly important in mitigating the effects of ransomware, data breaches, and other high-impact security incidents [26].
- Supporting Incident Response: Anomaly detection systems provide valuable insights that can inform the incident response process. By analyzing the nature and context of detected anomalies, security teams can better understand the scope and severity of an incident, allowing them to respond more effectively [29].
- Continuous Monitoring: Anomaly detection systems typically operate continuously, monitoring the system in real-time for any signs of suspicious activity. This constant vigilance is essential in a threat landscape where attacks can occur at any time, often without warning [15].
- Reducing Operational Disruptions: By catching anomalies early, organizations can prevent or mitigate the effects of security incidents that could disrupt operations. This is especially important in industries where downtime can lead to serious financial or reputational harm [14].

In conclusion, anomaly detection is a fundamental component of a modern security strategy. By focusing on identifying deviations from normal behavior, it provides a proactive and dynamic approach to threat prevention that complements traditional security methods. As we continue to explore the integration of machine learning into software security, understanding and implementing effective anomaly detection systems will be key to staying ahead of evolving cyber threats.

10.3 Machine Learning Techniques for Anomaly Detection

In order to identify odd patterns or behaviors that can indicate possible security risks, system failures, or fraudulent activity, machine learning approaches for anomaly detection are essential. Depending on the particular data types, anomaly incidence rates, and system requirements, several machine learning techniques are used to find anomalies in a variety of fields.

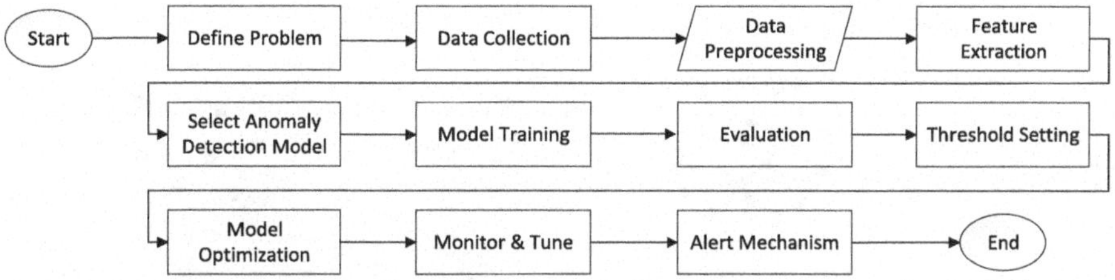

Fig. 10.2 Generic workflow of anomaly detection system

The general workflow diagram of the Anomaly detection system is shown in Fig. 10.2. This is a generic approach for designing and implementing the anomaly detection system.

The different types of learning approaches are described below.

10.3.1 Supervised Versus Unsupervised Learning: Differences and Applications in Security

In general, there are two types of machine learning techniques: supervised learning and unsupervised learning. Though they are both employed in anomaly detection, their functions and suitability for various data and security issues differ.

1. Supervised Learning involves training a machine learning model on a labeled dataset as shown in Fig. 10.3, where the input data is paired with the correct output (e.g., normal behavior versus anomalous behavior). The model learns to map inputs to the correct outputs, and once trained, it can classify new, unseen data based on the patterns it has learned [24].

   - **Applications in Security**: Supervised learning is particularly effective when there is a substantial amount of labeled data available. For example, in a network security context, if a dataset contains known examples of benign traffic and various types of attacks, a supervised model can be trained to distinguish between normal and malicious activity. This approach is effective for identifying specific types of threats that the model has been trained on, such as phishing attacks or malware [25].
   - **Challenges**: The primary challenge with supervised learning is the need for labeled data, which can be time-consuming and expensive to obtain. Additionally, supervised models may struggle to detect new or evolving threats that do not closely resemble the patterns in the training data [36].

   Neural Network, an important supervised machine learning algorithms is described below:

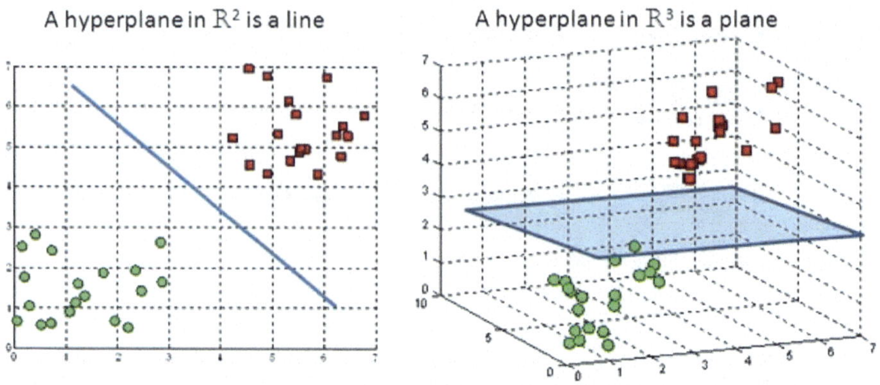

Fig. 10.3 An illustration of supervised learning method. *Source* [20]

- Neural Networks: The major role of Nural Network is given below:
  - **Feature Extraction**: Neural networks excel at automatically extracting relevant features from raw data, which is crucial for identifying complex patterns associated with anomalies. For example, in network traffic analysis, a neural network might learn to recognize subtle variations in packet sizes or timing that indicate a potential attack [17].
  - **Pattern Recognition**: Once trained, neural networks can recognize patterns in data that correspond to normal behavior and distinguish them from abnormal patterns. This ability to learn from vast amounts of data makes neural networks particularly effective for detecting sophisticated threats that may not follow obvious patterns [4].
  - **Deep Learning**: Deep neural networks, which consist of many layers, are capable of learning highly complex representations of data. This makes them suitable for tasks such as detecting advanced persistent threats (APTs) or identifying subtle anomalies in large datasets where simpler models might fail [11].
  - **Time Series Analysis**: Recurrent Neural Networks (RNNs), a type of neural network designed to handle sequential data, are particularly effective for anomaly detection in time series data, such as monitoring the behavior of a system over time. This is useful for detecting slow-moving threats or identifying patterns that evolve gradually [35].

Applications of Neural Networks in Security are described below:

- Intrusion Detection Systems (IDS): Neural networks can be integrated into IDS to identify unusual patterns of behavior that suggest an intrusion attempt. By continuously learning from network traffic data, these systems can detect both known and unknown threats.

- Fraud Detection: In financial systems, neural networks are widely used to detect fraudulent transactions by identifying anomalies in user behavior, transaction amounts, or spending patterns.
- Malware Detection: Neural networks can be trained to recognize the characteristics of malicious software by analyzing code, behavior, or communication patterns, enabling the detection of novel malware strains.

Support Vector Machines (SVM) a type of supervised machine learning algorithm is described below:

- Support Vector Machine: SVMs are supervised learning models that find the optimal boundary between different classes of data. In anomaly detection, one-class SVMs are often used, where the model learns the boundary of normal behavior and identifies anything outside this boundary as an anomaly.
- Applications in Security: SVMs are particularly useful in scenarios where the data is not linearly separable, making them effective for detecting complex anomalies in various types of security data as shown in Fig. 10.7.

2. Unsupervised Learning, does not rely on labelled data as shown in Fig. 10.4. Instead, the model identifies patterns and structures within the data on its own. Unsupervised learning methods usually provide a baseline of expected behaviour in the context of anomaly identification and mark any departures from this standard as possible abnormalities [37].

   - **Applications in Security**: When there is little or no tagged data available, supervised learning works quite well. For instance, an unsupervised model can be trained to identify any anomalous activity that would point to a security breach in a situation where typical system behaviour is well established but possible risks are unclear or unpredictable [8]. Techniques such as clustering, where the model bundles similar data points, can help identify outliers representing anomalies.
   - **Challenges**: The possibility of a high percentage of false positives is one of the primary issues with unsupervised learning. The model may identify harmless abnormalities as dangers because it was not trained on specific instances of harmful activity, necessitating additional research by security personnel [2].

The major steps typically involved in anomaly detection software design are shown in Fig. 10.2. These steps can be translated into a clear and structured flowchart, which you can use to develop or visualize your system as per the problem at hand. As per the requirements, the steps may be inserted or removed from the workflow.

In summary, the choice between supervised and unsupervised learning depends on the accessibility of labelled data and the specific security use case. Supervised learning is powerful for detecting known threats, while unsupervised learning is better suited for discovering new or unforeseen threats.

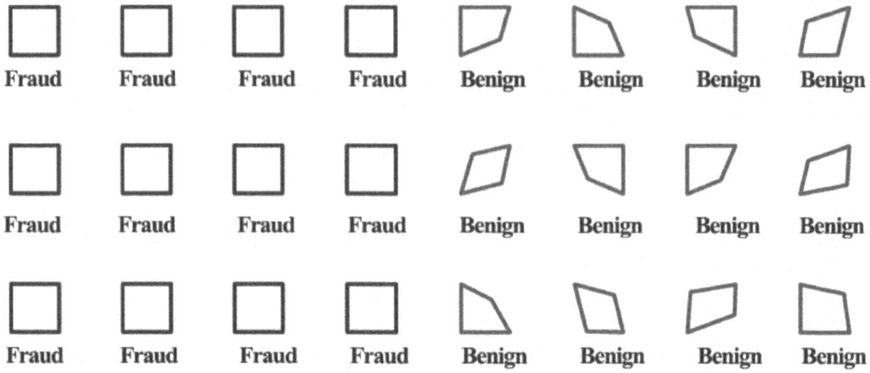

Fig. 10.4 An illustration of supervised learning method. *Source* [20]

K-Means, DBSCAN (Density-Based Spatial Clustering of Applications with Noise) and Autoencoders types of unsupervised algorithms are described below:

- K-Means Clustering: As seen in Fig. 10.5, this unsupervised learning method clusters data points according to how similar they are. Points that don't fit neatly into any cluster are called anomalies. Although K-Means is straightforward and efficient for some kinds of data, Sahu and Jena [23] highlight how it may not work well with complicated or high-dimensional datasets. The mathematical concepts used in K-means are centroid, Euclidean Distance and Inertia which makes K-Means simple, understandable and one of the efficient mechanisms for data clustering. Elbow and Silhouette Methods are used to select the optimal value of K.
- DBSCAN (Density-Based Spatial Clustering of Applications with Noise): DBSCAN is another clustering algorithm that identifies dense regions of data points as clusters and treats points in sparse regions as anomalies. It is particularly useful for detecting outliers in data with varying densities [9] as visualized in Fig. 10.6.
- Autoencoders: Neural networks called autoencoders are made to learn effective data representations, usually by compressing and then rebuilding input in a lower-dimensional space (Fig. 10.7). Anomalies are found using the reconstruction error, which is the difference between the original input and the reconstruction. The components of an Autoencoder are shown in Fig. 10.8.
- Applications in Security: Autoencoders are widely used for detecting anomalies in high-dimensional data, such as log files or network traffic. They are particularly effective when normal behaviour can be well-represented by a compact model, and deviations from this model indicate potential threats [1].

Other algorithms used for malware detection are described below.

Large Language Models (LLMs) and Generative AI for Designing Anomaly Detection Utilizing Large Language Models (LLMs) and Generative AI in the design

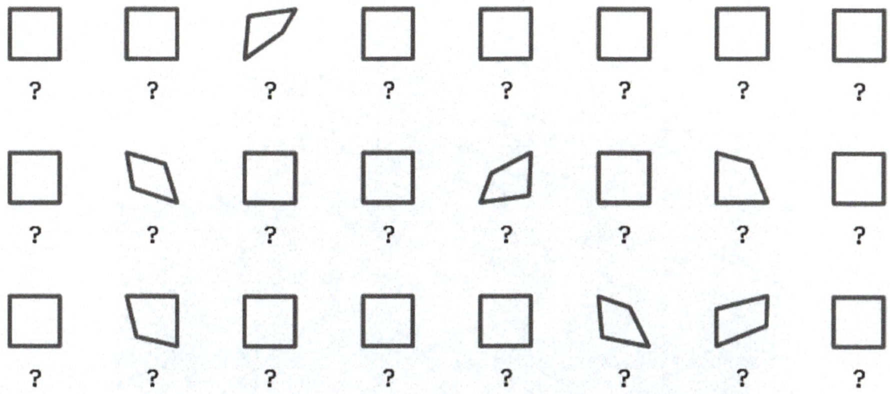

Fig. 10.5 Visualization of K-Means before and after clustering. *Source* Youtube

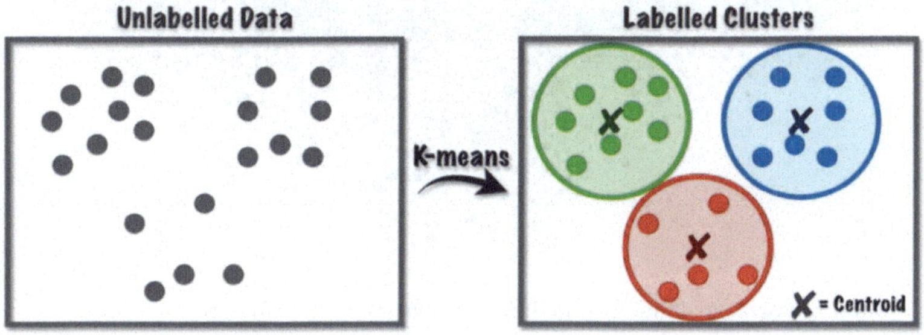

Fig. 10.6 DBSCAN formulation. *Source* [13]

of anomaly detection systems can significantly improve the capability to identify, forecast, and comprehend anomalies across diverse data types, including logs, transactions, sensor data, and software behaviour. Figure 10.9 shows the workflow diagram that use LLM for Common vulnerability exploitability using the Morpheus LLM engine supporting model-generated RAG tasks and multiple loops. Security analysts can easily assess whether a particular software payload has vulnerable and exploitable components or not.

It also investigates the individual CVEs, four times faster, and identifies the weakness of the software with high precision.

Another example is to generation of Synthetic data which provides 100% detection of spear phishing e-mails as shown in Fig. 10.10. Most of the malicious activities are carried out to Spear phishing emails.

The primary difference between a spear phishing email and a normal mail lies in the sender's intent. Due to the scarcity of training samples, its make difficult to identify the phishing spear mail using a supervised approach. Generating the synthetic data samples improves spear phishing email detection.

Here's how LLMs and Generative AI can be integrated into anomaly detection system design:

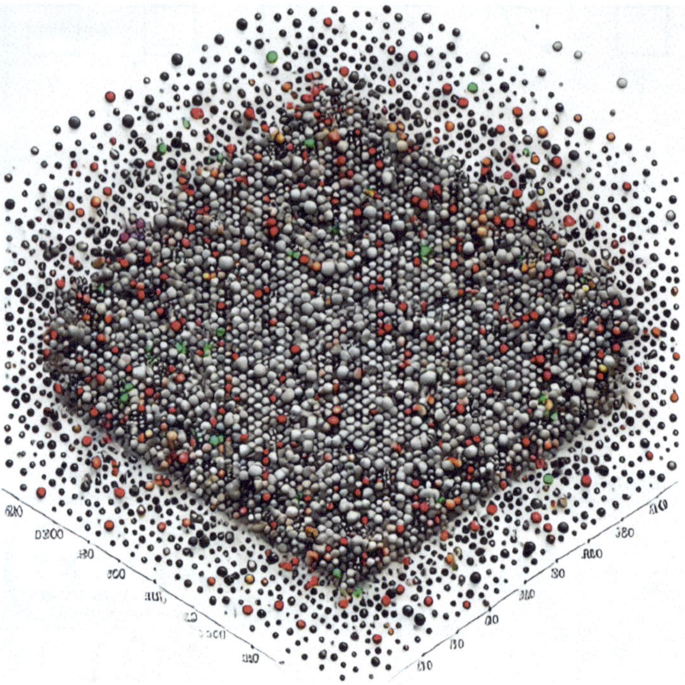

Fig. 10.7 Support vector machine and their hyperplane. *Source* [13]

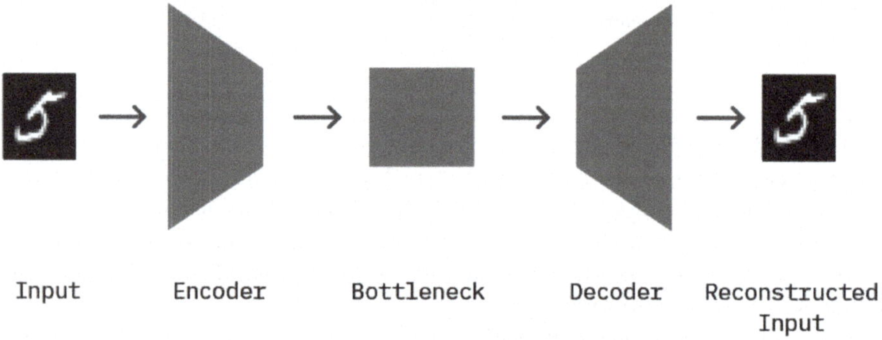

Fig. 10.8 Component-wise workflow of an autoencoder. *Source* [20]

1. **Log and Data Analysis with LLMs**
   - **Use Case**: Software systems frequently produce extensive logs that include both typical and anomalous behaviors. LLMs can be employed to parse and analyze these logs for anomaly detection.
   - **How**: Pre-trained LLMs (such as GPT or BERT) can be fine-tuned using a dataset of log files to recognize patterns of normal versus abnormal behavior.

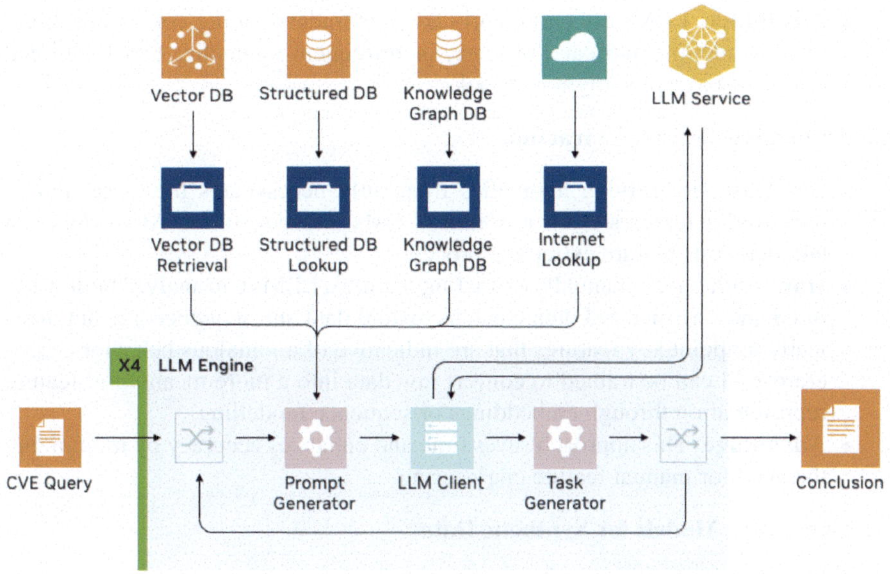

Fig. 10.9 Common vulnerability detection using Morpheus LLM engine and RAG. *Source* [30]

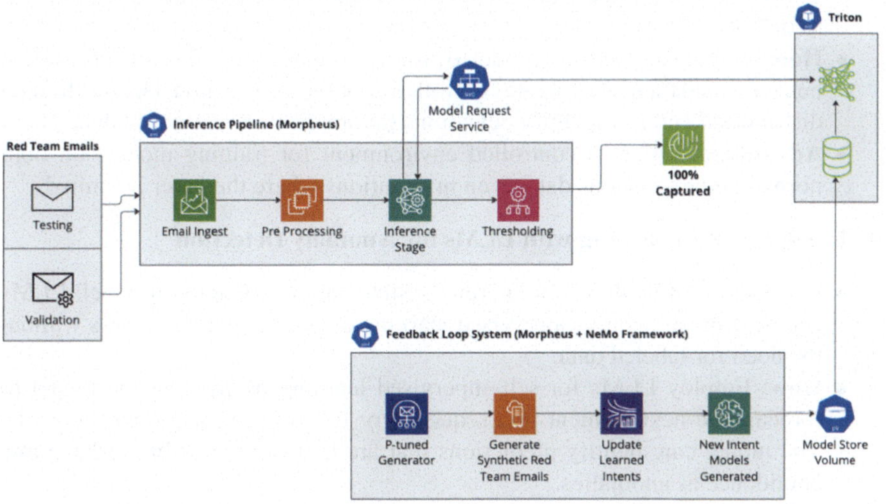

Fig. 10.10 Spear phishing detection pipeline constructed using synthetically generated spear phishing emails that align with specific behavioral intents. *Source* [30]

By training the model on labelled logs, you can utilize the LLM to classify new log entries as either normal or anomalous based on patterns and contextual information.

- **Advantage**: LLMs are particularly adept at understanding context, enabling them to identify intricate, obscure, or infrequent anomalies that traditional rule-based systems might overlook.

2. **Automated Feature Extraction**

- **Use Case**: Identifying anomalies frequently necessitates the extraction of meaningful features from raw data. LLMs and generative AI models can automate this feature extraction process.
- **How**: Rather than manually extracting features, LLMs can analyze both structured and unstructured data (such as textual data and sequences) to automatically pinpoint key features that are indicative of anomalous behaviors. Generative AI can be trained to convert raw data into a more meaningful feature representation through embeddings or sequence modelling.
- **Advantage**: This approach saves time and enhances accuracy by minimizing the need for manual feature engineering.

3. **Generative Models for Synthetic Data**

- **Use Case**: Generative AI, like GANs (Generative Adversarial Networks), can generate synthetic datasets to model both typical and abnormal behaviours. This is especially beneficial when real-world anomalies are scarce and difficult to obtain.
- **How**: Employ a generative model to produce synthetic data that mirrors normal behavior, and then incorporate controlled anomalies. The model learns the data distribution and can identify deviations (anomalies) in real-world data.
- **Advantage:** Offers a controlled environment for training models on both normal and anomalous data, even in situations where the latter is limited.

4. **Unsupervised Learning with LLMs for Anomaly Detection**

- **Use Case**: Anomalies in numerous systems are challenging to label. LLMs can be utilized in an unsupervised learning context to detect outliers without the need for labeled data.
- **How**: Employ LLMs for self-supervised learning by training the model to forecast the next element in a sequence or to fill in missing data segments. The model can identify predictions that are highly improbable or have low confidence as anomalies.
- **Advantage**: Effective in scenarios where labelling data is difficult or costly (e.g., fraud detection, software behavior monitoring).

5. **Time Series Anomaly Detection with Transformers**

- **Use Case**: Numerous anomalies occur in time series data (e.g., sensor readings, financial transactions, network traffic). LLM architectures such as Transformers are well-suited for understanding sequences and can be applied to time series data.

- **How**: Utilize Transformer-based models to learn the patterns in time series data. These models can predict the next expected value in the series and identify significant deviations as anomalies.
- **Advantage**: Transformers are particularly adept at managing long-range dependencies, which are prevalent in time series data, leading to more accurate anomaly detection.

6. **Interactive Anomaly Detection Using LLMs**

- **Use Case**: LLMs can aid domain experts in anomaly detection by offering explanations or interactive insights into detected anomalies.
- **How**: By incorporating LLMs into dashboards or monitoring systems, users can engage with the model to inquire about potential anomalies. For instance, they can ask, "Why was this anomaly flagged?" or "What patterns led to this detection?"
- **Advantage**: This fosters human-AI collaboration, where the system elucidates its findings, enhancing trust and facilitating validation.

7. **Root Cause Analysis with Generative AI**

- **Use Case**: After detecting an anomaly, understanding its root cause is essential. LLMs and generative models can assist in identifying why the anomaly occurred.
- **How**: Fine-tune an LLM using datasets that include historical anomalies and their root causes. When a new anomaly is detected, the LLM can propose likely causes based on patterns observed in past cases.
- **Advantage**: This expedites root cause analysis by automatically highlighting pertinent insights from the data, thereby reducing the time spent on troubleshooting.

8. **Contextual Anomaly Detection with Multi-Modal Data**

- **Use Case**: In software development, anomalies can encompass various types of data (e.g., code, logs, user interaction data). LLMs and generative models can efficiently manage this multi-modal data.
- **How**: Train an LLM to process and correlate anomalies across different data types (e.g., unstructured text logs, structured database records, code sequences). This enables more comprehensive anomaly detection across various layers of the software stack.
- **Advantage**: Offers a comprehensive view of anomalies, enhancing accuracy in detecting and diagnosing issues.

By incorporating LLMs and Generative AI, one can develop a highly flexible anomaly detection system that can effectively manage complex and dynamic data patterns.

10.3.2 Comparative Analysis: Benefits and Drawbacks of Different Machine Learning Approaches

The particular needs of the security application will determine which machine learning approach is best for anomaly detection, as each has advantages and disadvantages.

Supervised Learning

- Benefits: High accuracy when labelled data is available, effective for detecting known threats.
- Drawbacks: Requires large amounts of labeled data, and may not detect new or unknown threats.

Unsupervised Learning

- Benefits: Effective for detecting unknown threats, no need for labelled data.
- Drawbacks: Higher false positive rate, may require more manual tuning and interpretation.

In conclusion, the selection of machine learning techniques for anomaly detection in software security should be guided by the specific context, the nature of the data, and the type of anomalies that need to be detected. Often, a combination of techniques may be employed to balance their respective strengths and mitigate their weaknesses, leading to a more robust and comprehensive security solution.

10.4 Practical Applications and Case Studies

The implementation and practical applications across various sectors are outlined as follows:

10.4.1 Real-World Implementation: Examples of Machine Learning Applied to Software Security

Machine learning in software security is no longer a theoretical concept-many organizations have successfully implemented machine learning techniques to protect their systems from various types of cyber threats. These implementations span a wide range of security domains, from network security and malware detection to fraud prevention and user behavior analytics. Here are a few notable examples:

- **Network Intrusion Detection Systems (NIDS)**: In order to keep an eye on network traffic for indications of malicious behaviour, several organisations include machine learning models into their NIDS. To detect anomalous network traffic patterns that may point to an intrusion, for example, Cisco and Palo Alto Networks have included machine learning algorithms into their security solutions. By spotting abnormalities in network behaviour that depart from predetermined baselines, these systems analyse enormous volumes of data in real-time and spot possible dangers.

- **Malware Detection**: The detection and mitigation of malware has been transformed by machine learning. Conventional antivirus programs use signature-based detection, which is limited to detecting known threats. Companies like as Symantec and McAfee, on the other hand, have implemented machine learning models that examine program behaviour in order to identify malware based on actions rather than code signatures. This method greatly improves the efficacy of malware detection systems by enabling the identification of novel and unidentified malware strains.
- **Fraud Detection in Financial Systems**: To identify fraudulent transactions, financial institutions have implemented machine learning algorithms. Credit card issuers, for instance, employ machine learning algorithms to examine transaction patterns in real time and highlight any irregularities that could point to fraud. These algorithms are highly accurate in detecting and preventing fraud because they are trained on past transaction data to identify acceptable behaviour.
- **User Behavior Analytics (UBA)**: Machine learning is used by UBA systems to track user activity inside an application or network. By creating a baseline of typical user behaviour, these systems are able to identify deviations that could point to internal threats, compromised accounts, or illegal access. Businesses like IBM and Splunk have created UBA systems that employ machine learning to offer ongoing monitoring and notifications in response to departures from typical patterns of behaviour.

10.4.2 Case Studies: Detailed Examination of Specific Instances Where Machine Learning Improved Security

To better understand the impact of machine learning on software security, let's delve into a few detailed case studies where these techniques have been successfully applied:

Case Study 1: Enhancing Security in a Large Financial Institution A major international bank implemented a machine learning-based fraud detection system to safeguard its online banking platform. The system was designed to analyze transaction data in real-time, identifying fraudulent activities by detecting anomalies in transaction amounts, locations, and times.

- Implementation: The bank used a combination of supervised learning models, trained on historical transaction data, and unsupervised learning models to identify outliers in new transaction data. The models were integrated into the bank's transaction processing system, allowing for real-time fraud detection.
- Results: The machine learning system significantly reduced the number of false positives compared to the bank's previous rule-based system. It also detected several new types of fraud that had not been seen before, leading to a substantial decrease in fraud-related losses.
- Lessons Learned: One key insight from this implementation was the importance of continuously updating the machine learning models with new data to maintain their effectiveness. The bank also learned that a hybrid approach, combining supervised and unsupervised learning, provided the best results in detecting a wide range of fraudulent activities.

Case Study 2: Detecting Advanced Persistent Threats (APTs) in a Government Network A government agency faced the challenge of protecting its network from APTs-sophisticated, long-term cyber-attacks that are difficult to detect using traditional methods. To address this challenge, the agency deployed a machine learning-based anomaly detection system.

- **Implementation**: The agency used a deep learning neural network, specifically a Long Short-Term Memory (LSTM) model, to analyze network traffic and detect patterns indicative of an APT. The model was trained on historical network data to understand normal traffic patterns and was deployed to monitor real-time traffic for anomalies.
- **Results**: The system successfully identified several instances of APTs that had previously gone undetected by traditional security measures. The machine learning model was able to detect subtle anomalies that indicated the presence of an ongoing attack, allowing the agency to respond before significant damage was done.
- **Lessons Learned**: The success of this implementation highlighted the value of deep learning techniques in detecting sophisticated threats. However, the agency also noted the importance of maintaining a balance between detection sensitivity and the rate of false positives, as highly sensitive models can sometimes overwhelm security teams with alerts.

Case Study 3: Improving Email Security with Machine Learning A large multinational corporation sought to enhance its email security by implementing a machine learning-based phishing detection system. The goal was to reduce the number of successful phishing attacks, which had been a major source of security breaches.

- **Implementation**: The corporation deployed a supervised learning model trained on a large dataset of phishing and legitimate emails. The model analyzed various features of incoming emails, such as the sender's address, the content of the email, and embedded links, to determine the likelihood that an email was a phishing attempt.

- **Results**: The machine learning-based system significantly improved the detection rate of phishing emails, reducing the number of successful attacks by over 80%. The system was particularly effective at identifying sophisticated phishing attempts that mimicked legitimate emails.
- **Lessons Learned**: The corporation found that continuously retraining the model on new data was crucial to keeping up with evolving phishing techniques. Additionally, the integration of the machine learning model with user education efforts helped further reduce the impact of phishing by encouraging employees to report suspicious emails.

10.4.3 Lessons Learned: Insights from the Application of Machine Learning Techniques in Real-World Scenarios

The aforementioned case studies and real-world implementations offer insightful information about machine learning's applicability in software security. The following are some important takeaways from these implementations:

- Continuous Learning and Adaptation: The necessity of constant learning and adaptation is among the most crucial lessons. For machine learning models to remain successful against changing threats, fresh data must be added on a regular basis. This continuous procedure guarantees that models continue to be applicable and able to identify novel forms of assaults.
- Combining Multiple Techniques: The greatest outcomes are frequently obtained by using a hybrid strategy that blends many machine learning approaches, such as supervised and unsupervised learning or combining deep learning with conventional techniques. This makes it possible for security systems to minimise the drawbacks of each strategy while utilising its advantages.
- Balancing Sensitivity and Specificity: A model's sensitivity, or its capacity to identify threats, and specificity, or its capacity to prevent false positives, must be properly balanced. While overly detailed models could overlook real dangers, overly sensitive models might overload security professionals with false warnings. For security to work, models must be tuned to strike this equilibrium.
- Scalability and Performance: Deep learning-based machine learning models in particular can be computationally demanding. It is crucial to make sure that these models can scale efficiently to manage massive amounts of data in real-time, particularly in settings with high data flow.
- User Education and Collaboration: Machine learning-based security systems are most effective when combined with user education and collaboration. For example, educating users on recognizing phishing attempts can complement a machine learning model's efforts, leading to a more robust security posture.

- Ethical and Privacy Considerations: Lastly, it's critical to think about the privacy and ethical ramifications of using machine learning in security. Maintaining trust and compliance requires that models don't unintentionally introduce biases or break privacy laws.

In conclusion, while machine learning has proven to be a powerful tool in enhancing software security, its success depends on thoughtful implementation, continuous improvement, and the integration of human expertise. As cyber threats continue to evolve, the lessons learned from these real-world applications will be invaluable in guiding future developments in security technology.

10.5 Ethical Considerations in Machine Learning for Security

Ethical considerations in machine learning for security involve addressing the moral, social, and legal implications of using machine learning technologies in security applications. These considerations ensure that machine learning models protect users and data without infringing on rights or introducing biases. Key ethical aspects include:

10.5.1 Algorithmic Biases: Potential Biases in Machine Learning Models and Their Implications

Algorithmic bias has become a significant concern as machine learning models are incorporated into security systems more and more. When a machine learning model generates consistently unfair results because of biases in the training data, the model's architecture, or its implementation, this is known as algorithmic bias. These prejudices may show themselves in a number of ways, which might result in unfair treatment, discriminating actions, or even compromised security.

Sources of Algorithmic Bias

- **Biased Training Data**: If the data used to train a machine learning model reflects existing biases or lacks diversity, the model is likely to inherit these biases. For example, if a security system is trained primarily on data from a specific demographic or geographical region, it may perform poorly or unfairly when applied to a different context.
- **Model Design Choices**: The choices made during model development, such as the selection of features or the weighting of different factors, can introduce bias. For instance, if certain security features are overemphasized, the model might unfairly target specific groups or types of behavior as suspicious.

- **Deployment Context**: Even a well-trained model can produce biased outcomes if it is deployed in a context different from the one it was trained. For example, a model trained on corporate network data may not perform as effectively in a cloud-based environment, leading to inaccurate or biased threat detection.

Implications of Algorithmic Bias

- **Discrimination**: In the context of security, biased models can lead to discriminatory practices, such as disproportionately flagging individuals from certain demographic groups as security risks. This not only undermines the fairness of the system but can also erode trust in the security measures.
- **Ineffective Security**: Biases in machine learning models can result in false positives or false negatives, where legitimate activities are incorrectly flagged as threats, or real threats go undetected. This reduces the overall effectiveness of security measures, potentially leaving systems vulnerable to attack.
- **Legal and Reputational Risks**: Organizations that deploy biased machine learning models in their security systems may face legal challenges or damage to their reputation if their systems are found to discriminate unfairly or fail to protect users adequately.

Addressing algorithmic bias requires a proactive approach, including the use of diverse and representative training data, regular audits of model performance, and transparency in how models make decisions.

10.5.2 Security Fairness and Dependability: Ensuring That Machine Learning Methods Are Fair and Reliable

Ensuring fairness and dependability in machine learning-based security systems is essential for maintaining both the effectiveness of the security measures and the trust of the individuals and organizations that rely on them. Fairness refers to the principle that security measures should treat all individuals and entities equitably, without bias or discrimination. Dependability refers to the reliability and consistency of the security measures in effectively detecting and responding to threats.

Ensuring Fairness

- Diverse Training Data: To promote fairness, it is crucial to use diverse and representative training data that captures a wide range of behaviors, environments, and demographics. This helps ensure that the model performs equitably across different contexts and does not disproportionately target or overlook specific groups.
- Fairness Audits: Regular audits of machine learning models can help identify and mitigate biases. These audits should assess how the model performs across different groups and scenarios, ensuring that no group is unfairly disadvantaged by the security measures.

- Transparency and Explainability: Providing transparency in how machine learning models make decisions is key to ensuring fairness. By making the decision-making process understandable and explainable, organizations can build trust and allow for accountability when biases are detected.

Ensuring Dependability

- Robustness to Adversarial Attacks: Machine learning models must be designed to withstand adversarial attacks, where attackers manipulate inputs to cause the model to make incorrect decisions. Ensuring that models are robust to such attacks is critical for their dependability.
- Continuous Monitoring and Updating: Security threats evolve over time, and so should the machine learning models used to detect them. Continuous monitoring of model performance and regular updates based on new data are essential for maintaining the dependability of the security system.
- Fail-Safe Mechanisms: Implementing fail-safe mechanisms ensures that, in the event of an anomaly or unexpected behavior from the machine learning model, the system can fall back on traditional security measures or trigger human oversight to prevent security breaches.

By focusing on fairness and dependability, organizations can ensure that their machine learning-based security systems are both effective and just, protecting all users and environments equitably.

10.5.3 Ethical Challenges: Addressing Ethical Dilemmas in the Deployment of Machine Learning in Security

The deployment of machine learning in security systems raises several ethical challenges that must be carefully considered to avoid unintended consequences and maintain the integrity of security practices.

Balancing Privacy and Security

- **Data Privacy Concerns**: Machine learning models often require large amounts of data to function effectively, which can raise concerns about data privacy. Security systems must strike a balance between collecting enough data to ensure robust threat detection and respecting the privacy of individuals and organizations.
- **Surveillance Risks**: The use of machine learning for security can lead to increased surveillance, where the activities of users are closely monitored to detect anomalies. While this can enhance security, it also raises ethical concerns about the extent of surveillance and the potential for abuse of power.

Accountability and Transparency

- **Decision-Making Accountability**: When machine learning models make security decisions, it can be challenging to assign accountability, especially if the model's

decision-making process is not fully transparent. Organizations must establish clear lines of accountability for the actions taken by machine learning models, including mechanisms for addressing mistakes or biases.
- **Transparency Versus Security**: There is often a tension between the need for transparency in how machine learning models make decisions and the need to protect the security of the model and its decision-making process. Striking a balance between these competing priorities is an ongoing ethical challenge.

Impact on Employment and Workforce

- **Job Displacement**: As machine learning models take on more responsibilities in security, there is a risk of displacing human workers, particularly in roles related to monitoring and threat detection. Organizations must consider the impact on their workforce and explore ways to reskill employees or integrate human oversight into machine learning-driven security systems.
- **Human-Machine Collaboration**: The ethical deployment of machine learning in security should emphasize collaboration between humans and machines, rather than outright replacement. By combining the strengths of both, organizations can enhance security while preserving the role of human expertise.

Bias and Discrimination

- Addressing Discrimination: Machine learning models must be carefully designed and audited to prevent discrimination based on race, gender, age, or other protected characteristics. This includes ensuring that the data used for training is free from bias and that the models are regularly tested for fairness across different demographic groups.
- Inclusive Security Practices: Ethical machine learning deployment in security should consider the impact on all users, including marginalized or vulnerable groups. Security measures should be inclusive, ensuring that everyone is protected equally and that no group is unfairly targeted or excluded.

In conclusion, the ethical deployment of machine learning in security requires a holistic approach that considers the potential biases, fairness, dependability, and broader societal impacts of these technologies. By addressing these ethical challenges head-on, organizations can build security systems that are not only effective but also aligned with values of justice, transparency, and respect for individual rights.

10.6 Integration of Machine Learning Into Existing Security Systems

Integrating machine learning into existing security systems involves embedding ML models into traditional security frameworks to enhance their ability to detect, analyze, and respond to threats more effectively. The various challenges for machine learning integration into existing security systems are discussed below.

10.6.1 Challenges in Integration: Technical and Organizational Challenges in Merging Machine Learning with Traditional Security

Integrating machine learning into existing security systems presents several technical and organizational challenges that must be addressed to ensure a seamless and effective deployment.

Technical Challenges

- **Data Compatibility and Quality**: One of the primary technical challenges is ensuring that the data used for training machine learning models is compatible with existing systems and of high quality. Many organizations have legacy systems with data stored in various formats, making it difficult to gather and preprocess data for machine learning purposes. Additionally, the quality of the data is critical; noisy, incomplete, or biased data can lead to inaccurate models and poor security outcomes.
- **Scalability and Performance**: Machine learning models, particularly those involving deep learning, can be computationally intensive. Ensuring that these models can scale to handle large volumes of data in real-time, without causing delays or system slowdowns, is a significant challenge. Organizations need to consider the computational resources required and the potential impact on system performance when integrating machine learning into their security operations.
- **Integration with Existing Tools**: A variety of security solutions, including firewalls, intrusion detection systems (IDS), and antivirus software, are already in place in many organisations. It can be difficult to integrate machine learning models with these current technologies; it is important to carefully analyse how the models will work with and enhance conventional security measures. Compatibility issues, data exchange protocols, and the potential need for custom APIs or middleware solutions must be addressed.
- **Handling False Positives/Negatives**: Machine learning models, particularly in the early stages of deployment, may produce false positives (flagging benign activity as a threat) or false negatives (failing to detect actual threats). Managing these errors is crucial to prevent alert fatigue among security teams and to maintain trust in the system's effectiveness.

Organizational Challenges

- **Cultural Resistance**: Employees used to current procedures may object to the introduction of machine learning into traditional security settings. Concerns might exist around the apparent complexity of the new technology, the possibility of job displacement, or the dependability of machine learning models. Effective change management techniques, transparent explanations of the advantages of machine learning, and the participation of important stakeholders at every stage of the integration process are necessary to overcome this reluctance.

- **Skill Gaps**: Many organizations lack the in-house expertise required to develop, deploy, and maintain machine learning models for security purposes. This skills gap can hinder the successful integration of machine learning into existing security systems. Organizations must address this challenge through targeted hiring, training programs, or partnerships with external experts.
- **Compliance and Regulation**: Security-related machine learning models must adhere to industry standards and any data protection laws. It can be difficult to make sure the integration process complies with these rules, especially in highly regulated sectors like healthcare and finance, and it could call for more supervision and resources.

10.6.2 Best Practices for Integration

Strategies for Successful Integration To overcome the challenges of integrating machine learning into existing security systems, organizations can adopt several best practices.

Start with a Pilot Project: Pilot Implementation: Before rolling out machine learning across the entire security infrastructure, organizations should start with a pilot project. This allows them to test the model in a controlled environment, assess its performance, and identify potential issues. The insights gained from the pilot can inform the broader integration strategy, reducing the risk of disruptions and improving the likelihood of success.

Adopt a Hybrid Approach: Complementary Integration: Instead of replacing traditional security tools outright, organizations should consider a hybrid approach where machine learning models complement existing measures. For example, machine learning can be used to enhance the detection capabilities of an existing IDS by providing additional layers of analysis and anomaly detection. This approach allows organizations to leverage the strengths of both traditional and machine learning-based methods.

Focus on Data Management

- Data Preprocessing and Cleansing: Effective machine learning models require high-quality data. To guarantee that the data utilised for training is correct, comprehensive, and pertinent, organisations should make significant investments in strong data preparation and cleansing procedures. This entails resolving problems like missing values, normalising the data, and eliminating outliers that can distort the model's output.
- Data Governance: Maintaining the integrity and quality of the data used in machine learning requires the establishment of robust data governance procedures. This entails establishing data ownership, putting data access rules in place, and making sure data is utilised in accordance with applicable laws and guidelines.

Ensure Continuous Monitoring and Feedback

- **Model Monitoring**: Once machine learning models are integrated into the security system, continuous monitoring is essential to assess their performance and effectiveness. This includes tracking key metrics such as detection rates, false positives, and system response times. Organizations should implement feedback loops that allow for the ongoing refinement and tuning of the models based on real-world data and outcomes.
- **Human Oversight**: While machine learning models can automate many aspects of security, human oversight remains critical. Security teams should be involved in reviewing and interpreting the outputs of the models, particularly in cases where the models flag unusual or high-risk activity. This collaboration ensures that the insights provided by machine learning are actionable and aligned with the organization's broader security strategy.

Develop a Clear Integration Roadmap

- **Phased Implementation**: A clear and well-defined roadmap is essential for the successful integration of machine learning into security systems. This roadmap should outline the stages of integration, key milestones, resource requirements, and timelines. A phased approach allows for gradual adoption, minimizing disruptions and providing opportunities to address challenges as they arise.
- Engaging stakeholders from across the organization-including IT, security, compliance, and business units-is critical to the success of the integration process. Regular communication and collaboration ensure that all parties understand the goals, benefits, and implications of the integration and can provide input and support throughout the process.

10.6.3 Training and Skill Development: Essential Skills and Training Required for Professionals

Integrating machine learning into security systems requires a combination of technical and domain-specific skills. Organizations must invest in training and skill development to ensure that their teams are equipped to manage, maintain, and optimize these advanced technologies.

Core Skills for Security Professionals

- **Machine Learning Fundamentals**: Security professionals should have a solid understanding of machine learning principles, including how models are trained, validated, and deployed. This includes knowledge of key concepts such as supervised and unsupervised learning, neural networks, and clustering algorithms.
- **Data Science and Analytics**: Proficiency in data science is essential for working with machine learning models. This includes skills in data preprocessing, feature

engineering, and statistical analysis. Security professionals should also be familiar with data visualization techniques to effectively communicate the results of machine learning models.
- **Programming and Scripting**: Knowledge of programming languages such as Python, R, and Java is important for developing and implementing machine learning models. Security professionals should also be proficient in scripting languages like Bash or PowerShell, which are often used in security operations.
- **Cybersecurity Domain Expertise**: While technical skills are crucial, an understanding of cybersecurity concepts and practices is equally important. Security professionals must be familiar with threat landscapes, attack vectors, and security protocols to effectively apply machine learning in a security context.

Training and Development Programs

- **Continuous Learning**: Given the rapid evolution of both machine learning and cybersecurity, continuous learning is essential. Organizations should provide opportunities for ongoing education and skill development, such as online courses, certifications, and workshops. This helps security professionals stay up-to-date with the latest advancements and best practices in both fields.
- **Cross-Disciplinary Training**: Security teams should be encouraged to develop cross-disciplinary skills that bridge the gap between machine learning and cybersecurity. This could involve collaborations with data scientists, participation in interdisciplinary projects, or training programs that focus on the intersection of these fields.
- **Real-World Practice**: Hands-on experience is invaluable for building expertise in machine learning and security. Organizations should provide access to sandbox environments, simulated attack scenarios, and real-world case studies that allow security professionals to apply their skills in a practical context.

Collaboration and Knowledge Sharing

- **Internal Knowledge Sharing**: Creating a culture of collaboration and knowledge sharing within the organization can accelerate skill development and improve the effectiveness of machine learning integration. Regular meetings, workshops, and internal knowledge-sharing sessions can help disseminate best practices and lessons learned across teams.
- **External Collaboration**: Engaging with the broader cybersecurity and machine learning communities through conferences, forums, and online platforms can also provide valuable insights and opportunities for professional development. Collaboration with academic institutions, industry groups, and technology vendors can further enhance the organization's capabilities.

In conclusion, the successful integration of machine learning into existing security systems requires careful planning, a phased approach, and a commitment to ongoing skill development. By addressing the technical and organizational challenges, adopting best practices, and investing in training, organizations can harness the power of machine learning to enhance their security posture and better protect against emerging threats.

10.7 Future Trends in Machine Learning and Security

Future trends in machine learning and security are expected to shape the ways in which organizations protect against cyber threats, improve response times, and handle emerging challenges. As machine learning (ML) technologies continue to evolve, they will bring advancements and address complexities in security across various domains. Key trends are discussed below:

10.7.1 Emerging Technologies: Overview of New Advancements and Their Potential Impact on Security

The intersection of machine learning and security is rapidly evolving, driven by emerging technologies that promise to transform the way we approach cybersecurity. These advancements are poised to enhance threat detection, streamline security operations, and create more resilient systems. Here are some of the key emerging technologies and their potential impact on security:

1. **Federated Learning**

    - **Overview**: Federated learning is a decentralised method of machine learning in which models are developed without sharing the actual data among several servers or devices that store local data samples. By preserving sensitive data on local devices, this approach improves security and privacy while preserving the advantages of collaborative learning.
    - Impact on Security: Federated learning can be particularly useful in environments where data privacy is paramount, such as healthcare or finance. It enables the development of robust security models without compromising user privacy, and it can improve the detection of distributed threats by aggregating insights from multiple sources.

2. **Explainable AI (XAI)**

    - **Overview**: Explainable AI refers to machine learning models that provide transparent and interpretable results, allowing users to understand how decisions are made. This contrasts with "black-box" models, which offer little insight into their decision-making processes.

- **Impact on Security**: XAI is crucial in security contexts where understanding the rationale behind a model's decision is as important as the decision itself. This transparency can help security professionals trust and validate the actions taken by machine learning models, and it can also be critical in regulatory environments that require accountability and auditability.

3. **AI-Powered Threat Intelligence**

   - **Overview**: Threat intelligence solutions powered by AI examine enormous volumes of data from many sources in order to detect and forecast new risks. These platforms identify patterns and anomalies that can point to emerging attack methods or trends in cyberthreats by using machine learning.
   - **Impact on Security**: AI-powered threat intelligence can assist organisations in staying ahead of cyber risks by delivering actionable, real-time intelligence. Additionally, it may automate the correlation of data from many sources, saving time and effort in identifying and addressing problems.

4. **Autonomous Security Systems**

   - **Overview**: Autonomous security systems leverage AI and machine learning to automate the detection, analysis, and mitigation of cyber threats. These systems can operate with minimal human intervention, responding to threats in real-time and adapting to evolving attack techniques.
   - **Impact on Security**: The development of autonomous security systems has the potential to significantly reduce the time between threat detection and response, minimizing the damage caused by cyberattacks. As these systems become more advanced, they could eventually manage entire security infrastructures, freeing up human resources for more strategic tasks.

5. **Quantum Machine Learning**

   - **Overview**: Quantum machine learning solves complicated problems at previously unheard-of speeds by fusing regular machine learning methods with quantum computing. Although quantum machine learning is still in its infancy, it has the potential to completely transform a number of industries, including cybersecurity.
   - **Impact on Security**: New encryption techniques that are impervious to quantum attacks may be made possible by advances in cryptography brought about by quantum machine learning. Additionally, it might improve security algorithms' speed and effectiveness, enabling real-time threat analysis and response-even in extremely complex contexts.

10.7.2 Predicted Developments: Future Directions in the Intersection of Machine Learning and Software Security

As machine learning continues to evolve, several key trends are likely to shape the future of software security. These predicted developments reflect the ongoing convergence of AI, machine learning, and cybersecurity, leading to more sophisticated and effective security solutions.

(a) **Increased Integration of AI Across Security Domains**

- **Prediction**: AI and machine learning will become integral to every aspect of cybersecurity, from threat detection and prevention to incident response and recovery. In order to create unified systems that capitalise on the advantages of both automated analysis and human knowledge, AI-driven technologies will be progressively included into current security infrastructures as they develop.
- **Implications**: More proactive and flexible security measures that can foresee and eliminate threats before they have a chance to do serious damage will be the outcome of this integration. Additionally, it will simplify security operations, lowering the need for manual intervention and freeing up security staff to concentrate on more difficult problems.

(b) **Rise of AI-Driven Offensive Security**

- **Prediction**: While AI is primarily used for defensive purposes today, its potential for offensive security operations is likely to grow. Organizations may begin using AI to simulate sophisticated attacks, identify vulnerabilities, and test the resilience of their systems in a controlled environment.
- **Implications**: The rise of AI-driven offensive security could lead to the development of more robust defenses, as organizations gain a deeper understanding of how attackers might exploit their systems. However, it also raises ethical and legal questions about the use of AI in cyber warfare and the potential for these tools to be misused by malicious actors.

(c) **Evolution of AI in Threat Hunting and Incident Response**

- **Prediction**: AI will play an increasingly prominent role in threat hunting and incident response, automating the identification of threats and the orchestration of responses. Machine learning algorithms will become more adept at recognizing complex attack patterns, even in the early stages of an incident.
- **Implications**: The automation of threat hunting and incident response will enable faster and more effective mitigation of cyberattacks, reducing the time between detection and resolution. It will also empower security teams to manage a larger volume of incidents, improving overall security resilience.

(d) **AI-Augmented Human Decision-Making**

- **Prediction**: AI will enhance human skills by offering real-time insights, suggestions, and predictive analytics, rather than taking the place of human decision-makers. This collaborative approach will enhance the decision-making process, allowing security professionals to make more informed and strategic choices.
- **Implications**: AI-augmented decision-making will lead to more effective and efficient security strategies, as humans and machines work together to address complex threats. It will also help bridge the skills gap in cybersecurity by empowering less experienced professionals with AI-driven tools.

(e) **Advancements in Privacy-Preserving Machine Learning**

- **Prediction**: There will be notable developments in privacy-preserving machine learning methods as worries about data privacy continue to rise. Organizations will be able to take use of machine learning without jeopardizing sensitive data thanks to techniques like homomorphic encryption and differential privacy.
- **Implications**: Privacy-preserving machine learning will become a key component of secure data processing, allowing organizations to comply with stringent data protection regulations while still benefiting from AI-driven insights. This will be particularly important in industries such as healthcare and finance, where data privacy is critical.

10.7.3 Long-Term Implications: How the Evolving Role of Machine Learning Will Shape Software Security in the Future

Looking to the future, the evolving role of machine learning in security is likely to have profound and far-reaching implications. As AI and machine learning become more sophisticated and pervasive, they will not only change the way we approach security but also reshape the broader landscape of software development and IT infrastructure.

1. **The Shift Towards Proactive Security**

   - **Implication**: The transition from reactive to proactive security measures will be accelerated by the incorporation of machine learning into security systems. Rather than waiting for threats to materialize, organizations will use predictive analytics and AI-driven insights to predict and counter risks/attacks before they occur. This proactive approach will lead to more resilient and secure systems, capable of withstanding the evolving threat landscape.

2. **Democratization of Security**

   - **Implication**: As machine learning tools become more accessible and user-friendly, security will become democratized, enabling organizations of all sizes to benefit from advanced threat detection and prevention capabilities. This democratization will reduce the barriers to entry for effective cybersecurity, making it possible for smaller organizations to protect themselves against sophisticated threats.

3. **Ethical and Regulatory Evolution**

   - **Implication**: The widespread adoption of machine learning in security will drive the evolution of ethical standards and regulatory frameworks. As AI-driven tools become more powerful, there will be increased scrutiny of their fairness, transparency, and accountability. Governments and industry bodies will likely introduce new regulations to ensure that machine learning models are used responsibly and do not inadvertently cause harm.

4. **Human-Machine Collaboration as the Norm**

   - **Implication**: Collaboration between humans and machines will eventually become commonplace in security operations. Routine chores and data analysis will be handled by AI and machine learning, while complicated problem-solving and strategic decision-making will be handled by human professionals. This mutually beneficial partnership will improve security measures' efficacy and enable organisations to react to new threats faster and more strategically.

5. **Continuous Adaptation to Emerging Threats**

   - **Implication**: Machine learning models will need to constantly adjust to new attack methods and vectors as cyber threats continue to change. Effective security will depend on the models' capacity to be quickly updated and retrained in response to new threats. Businesses will be better equipped to ward off the upcoming wave of cyberattacks if they invest in adaptive machine learning systems.

In conclusion, there are a lot of interesting prospects and big problems for machine learning in security in the future. These technologies will become more and more important in safeguarding digital infrastructure and systems as they develop. By staying ahead of the curve and embracing these innovations, organizations can build more secure, resilient, and adaptive security frameworks that are equipped to meet the demands of the future.

10.8 Conclusion

In conclusion, there are advantages and disadvantages to the continuous development of software security in the machine-learning era. By embracing these advancements and addressing the associated ethical and technical issues, security professionals can lead the way in building a safer, more resilient digital world. The time to act is now-those who adopt and adapt to these new technologies will be better equipped to protect their systems, their data, and ultimately, their users in the years to come. The chapter also included actual-world scenarios and cases that illustrated how machine learning affects security in the actual world. From network intrusion detection systems to fraud prevention and malware detection, we saw how machine learning has been successfully implemented to enhance security measures. The lessons learned from these implementations emphasized the importance of continuous learning, hybrid approaches, and human-machine collaboration.

References

1. M.K. Almansoori, M. Telek, Anomaly detection using combination of autoencoder and isolation forest. Research Gate (2023)
2. O. Alshaikh, S. Parkinson, S. Khan, Exploring perceptions of decision-makers and specialists in defensive machine learning cybersecurity applications: the need for a standardised approach. Comput. Secur. **139**, 103694 (2024)
3. F. Alwahedi, A. Aldhaheri, M.A. Ferrag, A. Battah, N. Tihanyi, Machine learning techniques for IoT security: current research and future vision with generative AI and large language models. Internet Things Cyber Phys. Syst. (2024)
4. L. Alzubaidi, J. Zhang, A.J. Humaidi, A. Al-Dujaili, Y. Duan, O. Al-Shamma, J. Santamaría, M.A. Fadhel, M. Al-Amidie, L. Farhan, Review of deep learning: concepts, CNN architectures, challenges, applications, future directions. J. Big Data **8**, 1–74 (2021)
5. G. Apruzzese, P. Laskov, E. Montes de Oca, W. Mallouli, L. Brdalo Rapa, A.V. Grammatopoulos, F. Di Franco, The role of machine learning in cybersecurity. Digit. Threat. Res. Pract. **4**(1), 1–38 (2023)
6. H. Ashtari, What is network behavior anomaly detection? Definition, importance, and best practices for 2022 (2022)
7. Ö. Aslan, S.S. Aktuğ, M. Ozkan-Okay, A.A. Yilmaz, E. Akin, A comprehensive review of cyber security vulnerabilities, threats, attacks, and solutions. Electronics **12**(6), 1333 (2023)
8. Z. Azam, M.M. Islam, M.N. Huda, Comparative analysis of intrusion detection systems and machine learning based model analysis through decision tree. IEEE Access (2023)
9. D. Deng, Dbscan clustering algorithm based on density, in: *2020 7th International Forum on Electrical Engineering and Automation (IFEEA)* (IEEE, 2020), pp. 949–953
10. A. Diro, S. Kaisar, A.V. Vasilakos, A. Anwar, A. Nasirian, G. Olani, Anomaly detection for space information networks: a survey of challenges, techniques, and future directions. Comput. Secur. **139**, 103705 (2024)
11. V. Dutta, M. Choraś, M. Pawlicki, R. Kozik, A deep learning ensemble for network anomaly and cyber-attack detection. Sensors **20**(16), 4583 (2020)

12. O. Ebong, A. Edet, A. Uwah, N. Udoetor, Comprehensive impact assessment of intrusion detection and mitigation strategies using support vector machine classification. Res. J. Pure Sci. Technol. **7**(2), 50–69 (2024)
13. R. Gandhi, Support vector machine-introduction to machine learning algorithms. Towar. Data Sci. **7**(06) (2018)
14. A.S. George, T. Baskar, P.B. Srikaanth, Cyber threats to critical infrastructure: assessing vulnerabilities across key sectors. Partn. Univers. Int. Innov. J. **2**(1), 51–75 (2024)
15. N. Ghadge, Enhancing threat detection in identity and access management (iam) systems. Int. J. Sci. Res. Arch. **11**(2), 2050–2057 (2024)
16. A. Giannaros, A. Karras, L. Theodorakopoulos, C. Karras, P. Kranias, N. Schizas, G. Kalogeratos, D. Tsolis, Autonomous vehicles: sophisticated attacks, safety issues, challenges, open topics, blockchain, and future directions. J. Cybersecur. Priv. **3**(3), 493–543 (2023)
17. F. Hu, S. Zhang, X. Lin, L. Wu, N. Liao, Y. Song, Network traffic classification model based on attention mechanism and spatiotemporal features. EURASIP J. Inf. Secur. **2023**(1), 6 (2023)
18. K. Hwang, P. Dave, S. Tanachaiwiwat, Netshield: protocol anomaly detection with datamining against ddos attacks, in *Proceedings of the 6th International Symposium on Recent Advances in Intrusion Detection, Pittsburgh, PA* (2003), pp. 8–10
19. P. Kishore, S.K. Barisal, D.P. Mohapatra, R. Mall, An efficient two-stage pipeline model with filtering algorithm for mislabeled malware detection. Comput. Secur. **135**, 103499 (2023)
20. C.F.F. Labs, Deep learning for anomaly detection (2024), https://ff12.fastforwardlabs.com/
21. M.A.I. Mallick, R. Nath, Navigating the cyber security landscape: a comprehensive review of cyber-attacks, emerging trends, and recent developments. World Sci. News **190**(1), 1–69 (2024)
22. P. Pant, A. Kumar, L.K. Vashishtha, S. Dash, N.K. Ray, S.K. Sahu, A comparative study of deep learning techniques for network intrusion detection, in *2024 International Conference on Emerging Systems and Intelligent Computing (ESIC)* IEEE (2024), pp. 722–727
23. S.K. Sahu, S.K. Jena, A study of k-means and c-means clustering algorithms for intrusion detection product development. Int. J. Innov. Manag. Technol. **5**(3), 207–213 (2014)
24. S.K. Sahu, A. Katiyar, K.M. Kumari, G. Kumar, D.P. Mohapatra, An SVM-based ensemble approach for intrusion detection. Int. J. Inf. Technol. Web Eng. (IJITWE) **14**(1), 66–84 (2019)
25. S.K. Sahu, D.P. Mohapatra, A review on scalable learning approaches on intrusion detection dataset, in *Proceedings of ICRIC 2019: Recent Innovations in Computing* (2020), pp. 699–714
26. S.K. Sahu, D.P. Mohapatra, J.K. Rout, K.S. Sahoo, A.K. Luhach, An ensemble-based scalable approach for intrusion detection using big data framework. Big Data **9**(4), 303–321 (2021)
27. M. Saied, S. Guirguis, M. Madbouly, Review of artificial intelligence for enhancing intrusion detection in the internet of things. Eng. Appl. Artif. Intell. **127**, 107231 (2024)
28. I.H. Sarker, Machine learning: algorithms, real-world applications and research directions. SN Comput. Sci. **2**(3), 160 (2021)
29. I.H. Sarker, H. Janicke, A. Mohsin, A. Gill, L. Maglaras, Explainable AI for cybersecurity automation, intelligence and trustworthiness in digital twin: methods, taxonomy, challenges and prospects. ICT Express (2024)
30. N. Sessions, Bolstering cybersecurity: How large language models and generative AI are transforming digital security (2024), https://l1nq.com/xc5o0
31. M. Soori, B. Arezoo, R. Dastres, Artificial intelligence, machine learning and deep learning in advanced robotics, a review. Cogn. Robot. **3**, 54–70 (2023)
32. N. Tuptuk, S. Hailes, Security of smart manufacturing systems. J. Manuf. Syst. **47**, 93–106 (2018)
33. F. Ullah, M. Edwards, R. Ramdhany, R. Chitchyan, M.A. Babar, A. Rashid, Data exfiltration: a review of external attack vectors and countermeasures. J. Netw. Comput. Appl. **101**, 18–54 (2018)
34. L.K. Vashishtha, K. Chatterjee, S.K. Sahu, D.P. Mohapatra, A random forest-based ensemble technique for malware detection, in *International Conference on Information Systems and Management Science* (Springer, 2021), pp. 454–463

35. Y. Wang, M. Perry, D. Whitlock, J.W. Sutherland, Detecting anomalies in time series data from a manufacturing system using recurrent neural networks. J. Manuf. Syst. **62**, 823–834 (2022)
36. Z. Zhao, L. Alzubaidi, J. Zhang, Y. Duan, Y. Gu, A comparison review of transfer learning and self-supervised learning: definitions, applications, advantages and limitations. Expert Syst. Appl. 122807 (2023)
37. J. Zipfel, F. Verworner, M. Fischer, U. Wieland, M. Kraus, P. Zschech, Anomaly detection for industrial quality assurance: a comparative evaluation of unsupervised deep learning models. Comput. Ind. Eng. **177**, 109045 (2023)

Chapter 11
Sentiment Analysis on Movie Reviews Using the Convolutional LSTM (Co-LSTM) Model

Sireesha Moturi, S. N. Tirumala Rao, Srikanth Vemuru, M. Prasad, and M. Anusha

Abstract Inspection of people reviews published on social platforms is found to be essential for many businesses requisition. People's reviews published on social platform are growing in a rapid rate due to ubiquitous computing both interms of number and relevance, this leads the way to big data. A hybrid procedure has been proposed for sentiment analysis of reviews published on social platform. It consists of two deep learning techniques called as Convolutional Neural Network (CNN) and Long Short-Term Memory (LSTM). For local feature selection the increasingly effective technique is CNN, while for sequential inspection of lengthy text the recurrent neural networks LSTM frequently produce good results. The suggested Convolution LSTM techniques focally targeted on two goals in sentiment inspection. The first, one is that it is largely flexible in inspecting big social data, keeping scalability in mind, and secondly unlike traditional machine learning algorithms, it is not domain-specific. The outcomes of the experiment demonstrate that, in terms of accuracy and other metrics, the suggested ensemble model performs better than other Machine Learning techniques.

Keywords LSTM · Co-LSTM · Word embedding · RNN · CNN

S. Moturi (✉) · S. N. Tirumala Rao
Department of CSE, Narasaraopeta Engineering College, Narasaraopeta, Palnadu District, Andhra Pradesh, India
e-mail: sireeshamoturi@gmail.com

S. N. Tirumala Rao
e-mail: nagatirumalarao@gmail.com

S. Vemuru · M. Anusha
Department of CSE, KLEF, Vaddeswaram, Andhra Pradesh, India
e-mail: vsrikanth@kluniversity.in

M. Anusha
e-mail: anushaaa9@kluniversity.in

M. Prasad
Department of CSE, Shri Vishnu Engineering College for Women(A), Bhimavaram, Andhra Pradesh, India
e-mail: prasads.maddula@gmail.com

T. R. Benala et al. (eds.), *Boosting Software Development Using Machine Learning*, Artificial Intelligence-Enhanced Software and Systems Engineering 7,
https://doi.org/10.1007/978-3-031-88188-6_11

11.1 Introduction

Social networking sites have provided an excellent platform for big data analytics in several applications of real life [1]. Quite a huge amount of data is generated continuously when consumers upload their thoughts and share them with others on various social networking sites such as Facebook, Instagram, Twitter, and many others [2, 3]. Sentiment analysis plays a pivotal role in enabling software systems to understand and respond to user feedback, which directly influences product development and user satisfaction. Social data is one of those huge datasets generated in various social media platforms [4], comprising three big data characteristics: huge volume, velocity, and heterogeneity. Aside from those, it also possesses an incomparable quality called semantics, referring to the knowledge that it is created by hand, containing symbolic data with deeply rooted individual meaning [5]. Social media websites offer a fantastic platform for big data analysis in several real-world applications. For example, while consumers are posting their opinions or simply chatting with each other on multiple social media sites, such as Facebook, Instagram, Twitter, and many others, large amounts of data are constantly created. Social data generated from various social media platforms is one of the massive datasets containing huge volume, velocity, and heterogeneity characteristics. These characteristics are what make them the three big characteristics of data. Besides these, it has an unrivaled quality called semantics that pertains to knowledge; it is manually created and contains symbolic data with meaning deeply rooted in individuals. Undoubtedly, the complex feelings in social media platforms are a source for developing business processes to achieve organizational goals [4, 6, 7]. It has been used widely for monitoring the online influence of products or variations. However, because the amount of data in social network repositories is growing at an exponential pace, traditional methods for extracting sentiments from massive amounts of data fail. Referred to as intuitive computing, it's just one of the many growing experimental uses of sentiment analysis capable of automatically extracting user opinions from social media posts. It may also be considered a classification [8] goal, since sentiment analysis classifies the intent of a text into three classes: positive, negative, and neutral. In big data sentiment analysis of unstructured data, various generally suggested techniques can be identified, which can be separated into three categories: linguistic-based, lexicon-based, or machine learning based [9]. The subsequent segment of the paper is categorized as follows: In Segment 2, the inspiration for the amalgamation of LSTM and CNN has been considered [1, 6]. Segment 3 is about literature survey the approaches require for sentiment inspection [4, 10]. The strategy utilized for the study is introduced in Segment 4. Incremental approach of the suggested approach is debated in Segment 5. Appraisal specifications for the techniques were debated in Segment 6. The performance and outputs analysis have been debated in Segment 7. In Segment 8, the closure for the research paper is given.

11.2 Motivation

The motivation regarding the experimental work has been defined as follow:

In the current globe of computing, social platforms present in the internet is the large origin of user relationship and analysis. Sentiment inspection of such a large quantity of information helps in finding out and trail user opinion about brands, goods, or services. User commentary is crucially necessary in the process of conclusion creation [5, 11]. For instance, user opinion about a particular movie can help the new audience to get an opinion on the movie before buying the ticket to watch the movie. The similar technique is also helpful for product ratings and reviews since they assist consumers in determining the product for purchasing. One can analysis the sentiment of brand or goods in particular statistics placement for marketing [4, 12] to find out prospective users or business probable for fresh goods or favors in that place. Therefore, sentiment analysis assists to increase the marketing of an organization. Similarly, there were numerous utilizations of Sentiment Analysis, they are useful in our everyday tasks [2, 5, 9, 13].

The probable asset for sentiment analysis is social big data since it requires individual opinion on particular product or movie. For foretelling sentiment precisely, it utilizes most of dependencies and derision that requires to be utilized. Some data will also comprise small proverbs and texts in place where genuine opinion is ambitious to foretell. Numerous arithmetical training techniques that are previously present for sentiment inspection [14, 15]. Nevertheless, its accomplishment largely hangs on the status of attribute, taken from the opinion. It normally needs proficiency in attribute engineering, it is also very valuable in phrases of arithmetic space and time. The weight of actual attribute engineering can be decreased by neural network. Parallelism [9, 14] in derivation of localized correlations were utilized by convolution neural network and model from text athematic at time step didn't hang to the athematic at before time. We have endorsed convolution neural network for best attribute engineering for large social data [15] in this research paper. Nevertheless, for seizing the conditional information from present opinion since it didn't recollect previous condition so it could not be appropriate. For seizing the dependencies of terms inside opinion we acquired LSTM which is focally acceptable to seize secular conditional data, it is better acceptable.

11.3 Related Work

The related field in the study is sentiment analysis, which has recently gained momentum with the amount of data coming along through social media. A study proposed the Co-LSTM model that integrated convolutional layers with LSTM for the effective processing of huge volumes of social big data. With an accuracy that exploits both the temporal features of LSTM and the spatial features of CNNs for improving tasks related to sentiment classification, it proves to be a very apt tool for studying the dynamic nature of content on social media [1].

The paper [2] exhaustively reviews various techniques of sentiment analysis based on user reviews. The study emphasizes the fact that the domain of sentiment analysis rapidly evolves and encompasses the importance of hybrid models that make use of deep learning methods to deliver better performance in diverse scenarios such as customer evaluations and product reviews. As for sentiment analysis, a research study proposed a hybrid model based on BERT and Bi-LSTM, emphasizing Indonesian content from social media. Indeed, this stacking was able to determine correct sentiments and capture context efficiently for comments on the users themselves. This is an example of the interplay between transformer architecture and RNN in solving language problems [3].

Some reviews say that the applied a deep LSTM model in the analysis of sentiment to improve techniques for digital marketing. In their study, they processed huge and intricate marketing data using deep learning algorithms that allowed a better understanding of customer sentiment and improved decisions in the companies [4]. Knowledge graph embeddings used along with deep learning-based classification techniques have also been applied by studies in movie reviews. This resulted in improvements in text with large datasets using graph-based representation blended with deep learning [6]. A new hybrid word embedding and stacking ensemble method was proposed for sentiment analysis in intelligent customer service dialogues, thus providing a novel solution for dialogue-based sentiment analysis and highlighting the role of ensemble learning in increasing the robustness and accuracy of the model for real-time conversational data [8].

Based on the capability of hybrid and ensemble approaches to discover complex sentiment patterns, they have found a good place in recent research related to current sentiment analysis. In this paper, a hybrid of combining word embedding with stacking as an ensemble method for intelligent customer service conversation was proposed for sentiment analysis. This strategy is an enhancement of the accuracy of the classification of sentiment based on the amalgamation of the benefit of several approaches for the embedding of words along with ensemble methods, especially in real-time conversational data [10]. A new framework called FANS, which stands for Feature Selection in Sentiment Classification, was implemented; for its feature selection, an altered version of the Firefly algorithm was applied. This method tries to maximize the effectiveness of sentiment classification by selecting the most relevant features and enhancing the efficacy and precision of sentiment models [11].

Sentiment analysis of social media content was further pushed with the help of a model that categorized the sentiments of comments from Weibo using graph neural networks. This employs graph-based structures to efficiently catch the relationships between comments as they improve the contextual and relational elements of the discourse of social media for the model [12].

11.4 Background Details

11.4.1 Word Embedding Techniques

Word embeddings are essentially a word representation structure that bridges the gap between a machine's and human language recognition. They now understand how to characterise text in an n-dimensional space, where words with similar implications have almost identical characterisations. It denotes that two equal words are described by nearly identical vectors that are positioned almost exactly in a vector space. These are crucial for solving many Natural language processing issues [1, 3, 9]. Thus, when utilizing word embeddings, all separate terms are characterization as actual-evaluated vectors in a predefined vector volume. Each and every word is plotted to one vector and the vector values are grasped in a way that take after a neural network. Word2Vec is the most famous technique to grasp word embeddings by utilizing shallow neural network as we know the machine learning algorithms cannot process text so we need to figure out a way to transform these text data into numeric data. The techniques we used before like Count Vectorizer, Bag of Words and TF-IDF have been considered that can assist achieve use this objective.

Apart from this, we can use two more approach such as one-hot encoding, or we can use unique numbers to represent words in a vocabulary [8]. The latter technique is very much productive than one-hot encoding as rather than a sparse vector, we are having a dense one. Thus, this technique even works when our vocabulary is huge. Word2vec is an approach to productively create word embeddings by utilizing a two-layer neural network. Word2vec is not a single technique but a combination of two algorithms - Skip gram model and CBOW also called Continuous bag of words. GloVe also called as Global Vectors for Word Representation is another technique to create word embeddings [5, 6]. It is contingent on matrix factorization method on the word-context matrix. A huge matrix of co-occurrence data is built, and we count each "word", and how regularly we see this word in some "context" in a huge corpus.

11.4.2 Deep Learning Algorithms

The "deep" in deep learning directs to the number of hidden layers concerned in the model. Deep learning is a way of training artificial intelligence to identify particular data, such as speech or sentiment, and to make predictions based on preceding circumstance. Currently, there are six different types of neural networks. However, only two have gained a significant amount of popularity: convolution and recurrent [11, 14]. There are other techniques like LSTM, probabilistic neural network and feed forward etc.

11.4.3 Convolutional Neural Network (CNN)

It consists of three kinds of layers making up the CNN. These include the FC layer, also called the fully connected layer, the pooling layers, and the convolutional layers. When combined, these layers will provide a CNN architecture [8]. Two other important parameters apart from these three layers are the dropout layer and the activation function.

11.4.4 Recurrent Neural Network (RNN)

Recurrent Neural Network has its own memory which can remember all the previous characters. In the recurrent neural network, by the name, information goes in cycles, hence it is effective in capturing dependency with carry-out of sequential analysis. It takes two inputs at each step: information from the past states and information from the current state. Because of this, the neural network in question can master the tasks of character prediction and other similar, sequential data tasks like time series, speech, and audio. Figure 11.1 depicts the RNN's illustrated diagram.

11.4.5 Long Short-Term Memory (LSTM)

For sequential modeling focally on text data we use LSTM [1, 3] which is enlightened variety of RNN. It can be treated as a unique situation of RNN where only the necessary part of information is being send to the next layer rather than sending entire information. Vanishing gradient issues is one of the most important issue presents in RNN. To decrease the issues or errors the method called gradient descent

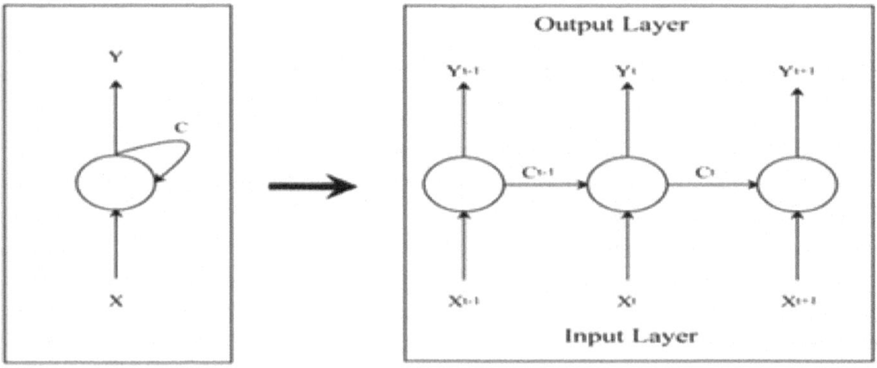

Fig. 11.1 Schematic flow diagram for recurrent neural network

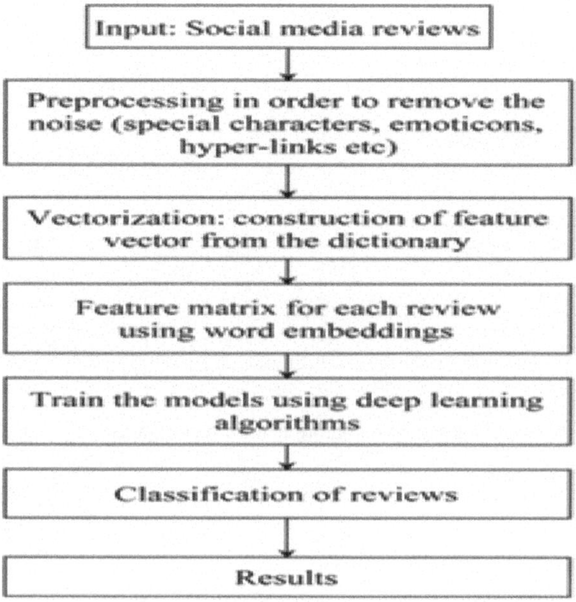

Fig. 11.2 Flow diagram of proposed approach

is frequently used in neural networks by retrofitting weight value of each and every neuron. Through back propagation the gradient loss diminishes at succeeding rate in RNN, It is also called as gradient vanishing issue. Consider the following sentence for example "I watch football, and I was best at goal keeping", the term 'goal keeping' rely on the term 'football', that is significantly behind the previous one in location. The gradient value vanishes remarkably and execution of RNN frequently diminishes with rise in separation between those dependent terms. LSTM [6] conquer this issue and execute better (Figs. 11.2, 11.3, 11.4, 11.5, 11.6 and 11.7).

11.5 Proposed Technique for Sentiment Analysis

A furnished technique travel across the following layers, those are LSTM, Convolution, and Word embedding layer. The illustrative picture of the suggested technique for sentiment inspection is furnished in Fig. 11.2. Word embedding is solicited in the first layer to implant the terms in opinion that can remove domain dependencies of opinion attributes [6, 11]. The next layer aspect utilizes the pooling layer and the convolution layer for the purpose of figure out prime neighbor and deep attributes present in the sentence. LSTM network is the third layer that solicit on result acquired from this layer to seize dependency from right to left in sequence. The amalgamation of all the three layers assists and perceive the sentiment of the sentence. LSTM will

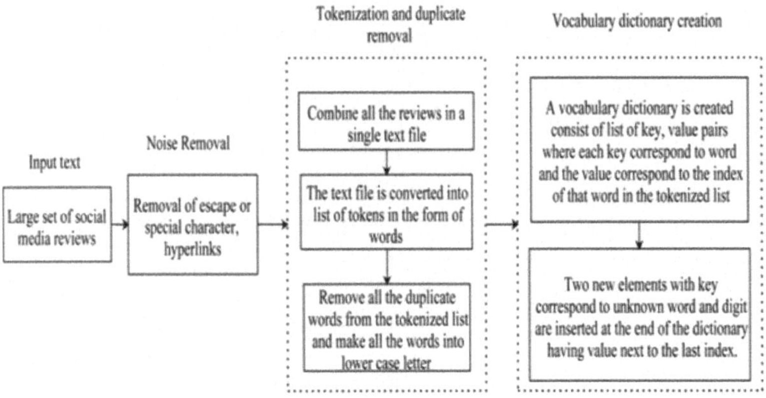

Fig. 11.3 Vocabulary dictionary creation before feature vectorization

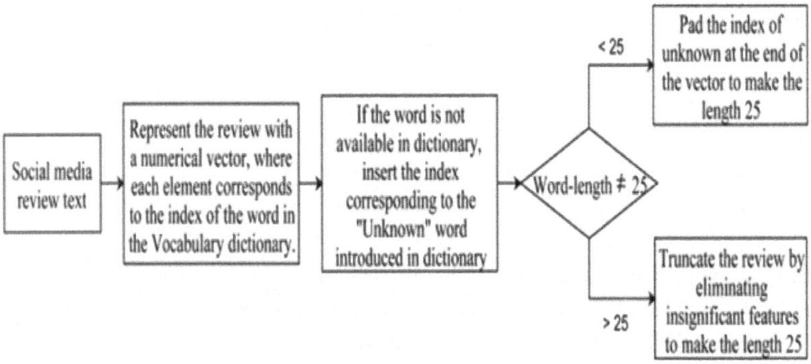

Fig. 11.4 Process of feature vectorization

produce result that are given to sigmoid layer which is fully connected to compute output using contemplation loss function which is called binary cross entropy [3, 14]. Classifiers final architectonic is depicted in Fig. 11.7. Proposed technique steps are illustrated as shown below.

11.5.1 Preprocessing Reviews

preprocessing reviews and opinions present in social platforms [2] are frequently in the structure of text that consist meaningless information such as hyperlinks, symbols, and special characters etc. Those meaningless data are removed by using regular expression. All opinions are split into tokens that are in the form of terms in preprocessing stage. To build a special characterization of each and all word the similar matching terms were removed. With rare terms as keys and term index

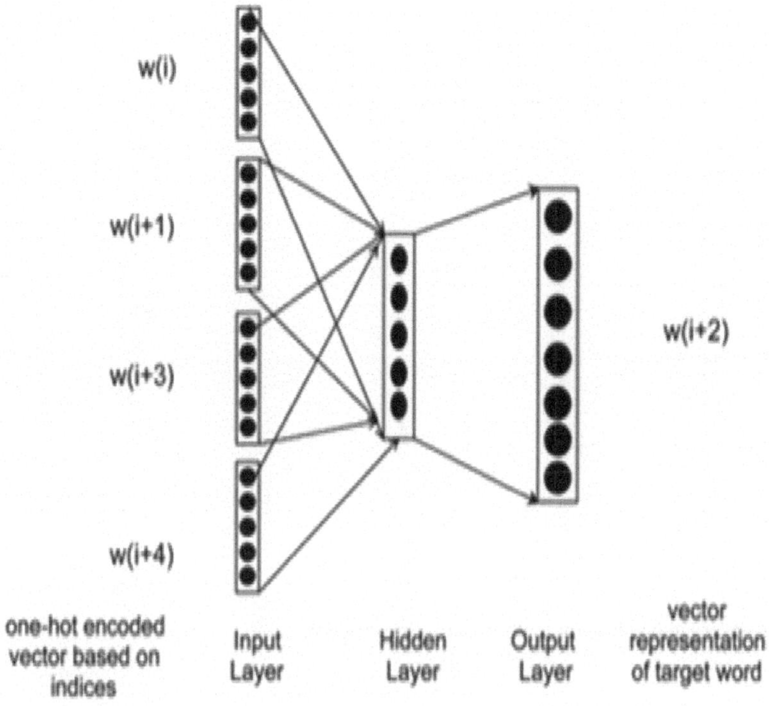

Fig. 11.5 Semantic diagram of word embedding model

as values a vocabulary dictionary is built. To portray terms that are not present in dictionary, two new terms "unknown" and "digits" were instigated, respectively. Procedure for vocabulary dictionary establishment is depicted in Fig. 11.3.

Text vectorization [4] procedure will be implemented following the preprocessing stage. Every component of the vector portrayal of opinion communicates to the indices of the terms in the vocabulary dictionary. Vector range has been rooted to 20. Since many of the opinions are having word-range less than 20, the indices of the freshly presented term unknown was fill out at the last to construct range 25.

If thc term-range of any opinion outplays 20, the fewer notable attributes are eliminated, i.e., the term-range of the opinion is abbreviated to 20. The unimportant terms are figured out with the technique of stop word elimination and lemmatization using the Natural language Toolkit package present in python. Many of the terms in English language have different substitute terms with equal meaning. The technique for changing substitute form to base form is called Lemmatization from which essentially decreased the frequency of terms [3]. The attribute vectorization technique is presented in Fig. 11.4.

To remove features and decrease estimation complexity dimensional reduction is needed in some cases. Instinctual instance could be "magnificently astonishing" [1] could also be mapped as only "astonishing" since it decreases the proportions

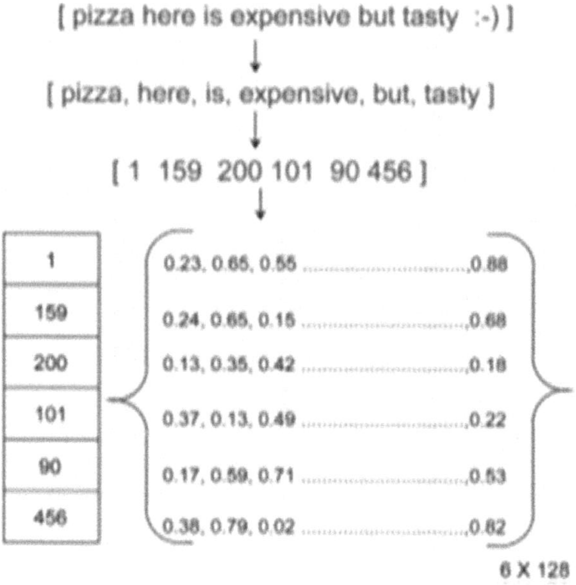

Fig. 11.6 Wordembedding for feature matrix construction

of input instead of dropping semantic data. For dimensional reduction of attribute matric we have taken principal component analysis, then it is traversed over to LSTM and CNN models as input. An ideal framework of pooling and convolution layers were also been considered for the purpose of feature removal.

11.5.2 Word-Embedding Approach

Using backpropagation technique every term in the record of text is implanted into a vector of dimension 130 where trained. For vector representation Word2Vec procedure have been utilized for training term implanting since it is easy and more coherent [11]. To portray the opinions in textual structure into digital vector array the word embedding technique is used which can later procedure through neural networks. The vocabulary dictionary is built preceding to portrayal, for considered datasets. Every word in vocabulary dictionary is correlated using index that portrays location of terms in dictionary. Since the location of each term is rare in nature, they have grasped that for vector portrayal of each opinion present in the dataset [6]. To portray the words in vocabulary dictionary the index is utilized.

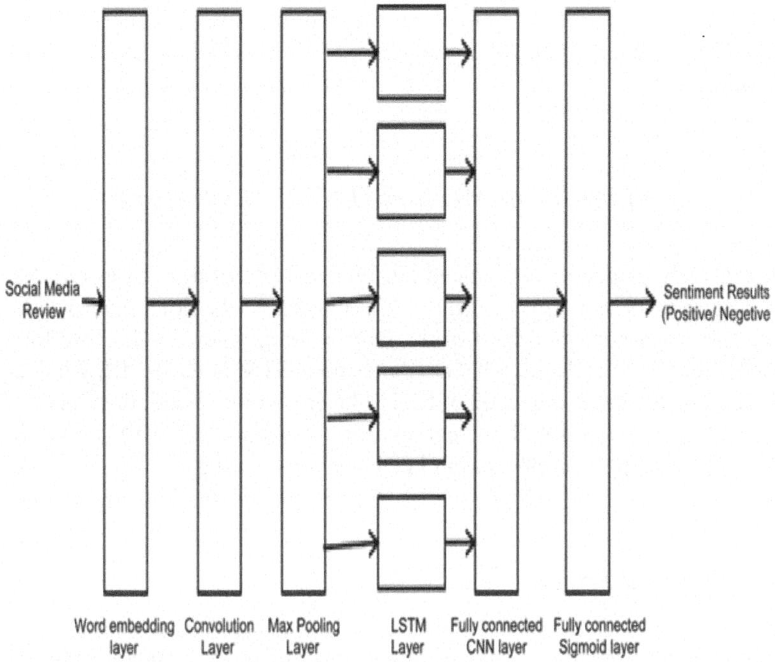

Fig. 11.7 Default caption

11.5.3 Convolutional Neural Layer

The convolution layer is the first layer that extracts different information from the input text. This layer carries out a convolutional arithmetic operation between the filter of size MxM and the input text. While the filter slides over the loaded text, a dot product between the filter and portions of the input text that match the magnitude of the filter MxM is computed.

The resultant of this process is termed a feature matrix, which provides information regarding sentiment and dependencies of text. Further, this feature matrix acts as an input for subsequent layers in order to find more features of the input text. We use the sigmoid function in order to get class label. It captures the natural hidden features as shown in Fig. 11.5.

11.5.4 Max-Pooling Layer

In the Max Pooling layer, at each position of a filter, the feature matrix will extract the biggest term. Average pooling determines the average of the phrases inside a pre-defined section of the text. Sum Pooling calculates the total sum of the terms within the pre-defined section. Sometimes the Pooling Layer is in between the Fully Connected Layer and the Convolutional Layer [14]. This is done separately for each convolution filter.

11.5.5 Long Short-Term Memory (LSTM) Technique

To consecutively inspect the produced attribute vectors from right to left the result of max polling layer was traversed into LSTM layer. As the crucial native attributes has been taken out near the result of the max-pooling layer, To disclose long-term dependencies and to disclose global attributes the LSTM layer [9, 10] will examine. To decrease the attributes the result of LSTM layer is compressed, then it is traversed through the FC layer to foretell the real opinion. Thousands of LSTM networks were appealed with nine percent dropout to keep away overfitting.

11.5.6 Sigmoid Layer

Thus, the attribute vector received from the result of the previous LSTM [3] layer feeds the FC sigmoid layer to estimate the probability distribution for every cluster. The Sigmoid activation function normalizes the classifier confidence score between one and zero. When the output is known, a binary cross entropy is used as a loss function to compute the coherence between the predicted and actual opinions.

11.6 Implementation

11.6.1 Dataset Used for Research

Reviews of Movies: The Big Movie Review [1] Dataset (frequently also called as IMDB dataset) consist of 50,000 largely biased movie reviews (bad or good) for testing and the equal quantity again for the training. Here, the question is whether the movie review obtained is sentimentally favorable or not. The data was collected by a scientist at Stanford, and used in a publication where a mix of 75:25 of the data was utilized for training versus testing.

Table 11.1 Comparing accuracies of different models

Models	Precision	Recall	F-measure	Accuracy
[1] CNN	0.8000	0.8294	0.8144	0.8200
[1] RNN	0.7494	0.7810	0.7649	0.7725
[1] CO-LSTM	0.8354	0.8350	0.8302	0.8313
*CNN	0.8200	0.8492	0.8231	0.8349
*RNN	0.7498	0.7810	0.7848	0.7745
*CO-LSTM	**0.8794**	**0.8740**	**0.8710**	**0.8759**

Bold refers to the proposed work

11.6.2 Result Analysis and Discussion

Machine Learning techniques are less productive when compared to the Deep learning [8, 14] techniques. LSTM and CNN neural network algorithms. Even though substantial study have been accomplished utilizing convolutional techniques. Huge amount of research have been accomplished using deep learning techniques too in current years. In order to expand validation the model was applied on additional datasets like sentiment analysis in healthcare and social media reviews. Domain-specific evaluation highlighted that the model generalizes well with minimal fine-tuning. Based on our findings, resource consumption of Co-LSTM increases exponentially after a certain threshold of the scale of the dataset and consequently, it becomes difficult to work with a large dataset. This paves the way for further investigation of models that can run in parallel and are architected to be highly efficient in memory usage (Table 11.1).

Correlative nspection of the classification techniques based on recall, f-measure, precision, and accuracy for the reviews of movie dataset is shown in Table. From proposed Co-LSTM technique we can note that f-measure and accuracy submit best outputs when compared to remaining techniques. Better precision and recall values were submitted by the CNN and RNN [1] techniques for reviews of movie dataset because they are influenced more towards the negative opinion. In terms of accuracy the top two techniques for reviews of movie dataset are identified to be Co-LSTM with 87.59% and CNN with 83.49% respectively. Compared to other techniques we got poor results for all the metrics in RNN due to gradient descendent problem (Fig. 11.8).

11.7 Conclusion

We have proposed a hybrid technique which is a neural network that incorporates both LSTM and CNN to foretell the opinion of user review of any product or movie irrespective of domain dependency, This is the huge advantage that it is not restricted

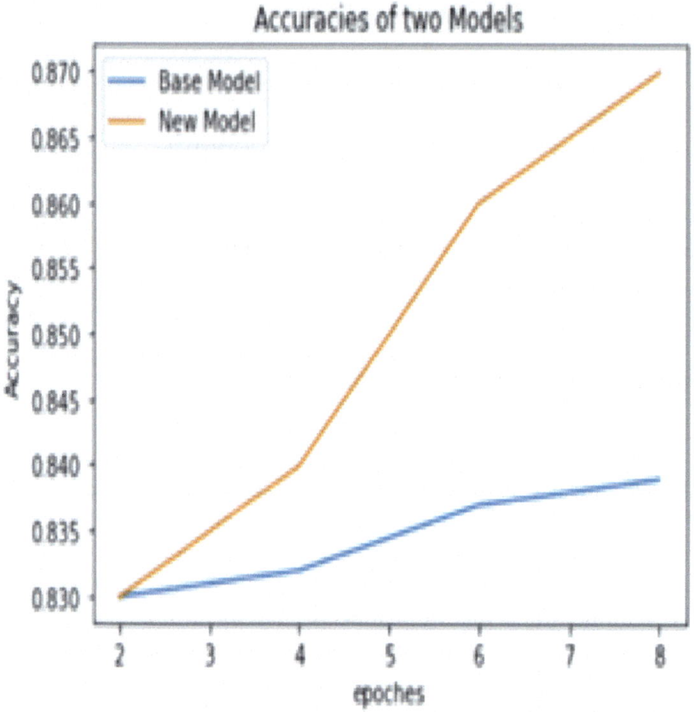

Fig. 11.8 Comparing accuracies

to specific domain. This technique can be trained for goods opinions as well as service opinions but not decreasing the efficiency. We don't need any complex manual attribute engineering so we can be steer clear of expertise of particular domain. This is possible because we have used a pre-trained word embedding technique for embedding attribute vector. By using Convolution layer we have find out the necessary attributes only from the vectors before LSTM layer, so it has hugely decreased training time and make it feasible. By studying sequential patterns in the opinions instead of just considering terms or expressions we used LSTM layer at the last stage.

11.8 Limitations and Future Work

The resource consumption of Co-LSTM increases exponentially after a certain threshold of the scale of the dataset and consequently, it becomes difficult to work with a large dataset. To overcome this challenge exploring frameworks like edge computing for resource constrained environments could further enhance its deployment readiness and Future work will explore lightweight model architectures and hybrid approaches, such as combining Co-LSTM with attention mechanisms.

References

1. R.K. Behera, M. Jena, S.K. Rath, S. Misra, Co-LSTM: convolutional LSTM model for sentiment analysis in social big data. Inf. Process. Manag. **58**(1) (2021). art. no. 102435. https://doi.org/10.1016/j.ipm.2020.102435
2. S. Naseem, T. Mahmood, M. Asif, J. Rashid, M. Umair, M. Shah, Survey on sentiment analysis of user reviews, in *Proceedings of the 2021 International Conference on Innovative Computing (ICIC)*, Lahore, Pakistan (2021), pp. 1–6. https://doi.org/10.1109/ICIC53490.2021.9693029
3. A. Chowanda, Y. Muliono, Indonesian sentiment analysis model from social media by stacking BERT and BI-LSTM, in *Proceedings of the 3rd International Conference on Artificial Intelligence and Data Sciences (AiDAS)*, Ipoh, Malaysia (2022), pp. 278–282. https://doi.org/10.1109/AiDAS56890.2022.9918717
4. M.B.A. Lasi, A.B.A. Hamid, A.H. Jantan, S.B. Goyal, N.N. Tarmidzi, Improving digital marketing using sentiment analysis with deep LSTM, in ed. by A. Swaroop, Z. Polkowski, S.D. Correia, B. Virdee. *Proceedings of Data Analytics and Management. ICDAM 2023. Lecture Notes in Networks and Systems*, vol. 785 (Springer, Singapore, 2024), pp. 191–202. https://doi.org/10.1007/978-981-99-6544-1_17
5. G.R. Trivedi, J.V. Bolla, M. Sireesha, A Bitcoin transaction network using cache-based pattern matching rules, in *Proceedings of the 5th International Conference on Smart Systems and Inventive Technology (ICSSIT)*, Tirunelveli, India (2023), pp. 676–680. https://doi.org/10.1109/ICSSIT55814.2023.10061064
6. Y. Mao, Y. Zhang, L. Jiao, H. Zhang, Document-level sentiment analysis using attention-based bi-directional long short-term memory network and two-dimensional convolutional neural network. Electronics **11**(12) (2022). art. no. 1906. https://doi.org/10.3390/electronics11121906
7. M. Sireesha, S. Vemuru, S.N.T. Rao, Classification model for prediction of heart disease using correlation coefficient technique. Int. J. Adv. Trends Comput. Sci. Eng. **9**(2), 2116–2123 (2020)
8. K. Hussain, I. Ihsan, Sentiment analysis in movie reviews using knowledge graph embeddings and deep learning classification. J. Artif. Intell. Appl. **6**, 21–29 2023. ISSN: 2710-1606. https://doi.org/10.56979/601/2023
9. S. Moturi, S. Vemuru, S.N.T. Rao, Two-phase parallel framework for weighted coalesce rule mining: a fast heart disease and breast cancer prediction paradigm. Biomed. Eng.: Appl., Basis, Commun. **34**(4) (2024). https://doi.org/10.4015/S1016237222500107
10. D. Chen, H. Zhengwei, M. Jintao et al., Sentiment analysis for intelligent customer service dialogue using hybrid word embedding and stacking ensemble. Soft Comput. (2024). https://doi.org/10.1007/s00500-024-09899-2
11. R. Asgarnezhad, A. Monajemi, FANS: a framework for feature selection in sentiment classification using a modified firefly algorithm. Evol. Intell. **17**, 2279–2291 (2024). https://doi.org/10.1007/s12065-023-00887-3
12. Y. Li, N. Li, Sentiment analysis of Weibo comments based on graph neural network. IEEE Access **10**, 23497–23510 (2022). https://doi.org/10.1109/ACCESS.2022.3154107
13. A. Seva, S.N. Tirumala Rao, M. Sireesha, Prediction of liver disease with random forest classifier through SMOTE-ENN balancing, in *Proceedings of the IEEE 13th International Conference on Communication Systems and Network Technologies (CSNT)*, Jabalpur, India (2024), pp. 928–933. https://doi.org/10.1109/CSNT60213.2024.10546170
14. B. Greeshma, M. Sireesha, S.N. Tirumala Rao, Detection of arrhythmia using convolutional neural networks, in ed. by S. Shakya, K.L. Du, W. Haoxiang. *Proceedings of the Second International Conference on Sustainable Expert Systems. Lecture Notes in Networks and Systems*,

vol. 351 (Springer, Singapore, 2022), pp. 1–10. https://doi.org/10.1007/978-981-16-g7657-4_3
15. S. Moturi, S.N. Tirumala Rao, S. Vemuru, Grey wolf assisted dragonfly-based weighted rule generation for predicting heart disease and breast cancer. Comput. Med. Imaging Graph. **91** (2021). art. no. 101936. https://doi.org/10.1016/j.compmedimag.2021.101936

Chapter 12
An Overview of AI Workload Optimization Techniques

Ravi Panchumarthy and Tirimula Rao Benala

Abstract Artificial intelligence (AI) workloads such as machine learning (ML) model training and inference have unique computational requirements that can benefit greatly from optimization. Owing to the recent surge in the adoption of AI across industries, new challenges related to the management of large-scale ML workloads have emerged. This chapter provides a high-level survey of techniques for optimizing AI applications for productivity, cost efficiency, and performance, without delving into technical intricacies. Commonly used optimization approaches are categorized into five broad dimensions: hardware, software, data, model, and hybrid optimization. Concepts such as hardware acceleration using specialized chips, software frameworks, and libraries for AI, data management techniques, model simplification methods, and synergistic approaches combining multiple strategies are briefly discussed. This overview is designed to offer technology decision-makers a foundation for understanding the contemporary landscape of AI optimization; it can enable them to grasp potential benefits and trade-offs of various techniques for specific use cases without requiring in-depth technical knowledge, ultimately leading to improved efficiency and performance in AI operations.

Keywords Artificial intelligence · Workload optimization · Optimization overview

R. Panchumarthy
Intel Corporation, Hillsboro, OR, USA
e-mail: ravi.panchumarthy@intel.com

T. R. Benala (✉)
Department of Information Technology, JNTU-GV College of Engineering Vizianagaram, Jawaharlal Nehru Technological University Gurajada Vizianagaram, Dwarapudi, Vizianagaram, Andhra Pradesh, India
e-mail: btirimula.it@jntugvcev.edu.in

T. R. Benala et al. (eds.), *Boosting Software Development Using Machine Learning*, Artificial Intelligence-Enhanced Software and Systems Engineering 7,
https://doi.org/10.1007/978-3-031-88188-6_12

12.1 Introduction

Artificial Intelligence (AI) is reshaping industries such as healthcare, finance, manufacturing, and retail, unlocking new opportunities and driving innovation. Applications in computer vision, natural language processing, and decision-making illustrate the transformative power of AI across diverse domains [1–5]. However, the rapid growth of AI adoption has introduced computationally intensive workloads, especially in training and deploying large-scale machine learning (ML) models [6, 7].

These workloads often involve massive datasets, complex algorithms, and repetitive training cycles, requiring substantial computational resources, memory bandwidth, and energy [8, 9]. The associated challenges are amplified by the increasing demand for real-time or near-real-time inference, which necessitates efficient infrastructure, resource allocation, and optimization techniques. Without proper optimization, AI workloads are likely to become inefficient, slow, and prohibitively expensive to manage [10, 11].

Optimizing AI workloads is a multidimensional challenge, necessitating a holistic approach that requires optimization of hardware, software, data, and models [12–14]. This chapter presents a survey of techniques used for optimization of AI workload; these are categorized into five broad dimensions: hardware, software, data, model, and hybrid optimization. Hardware optimization utilizes specialized chips such as graphics processing units (GPUs), field-programmable gate arrays (FPGAs), and dedicated application-specific integrated circuits (ASICs) for accelerating neural network processing. Software optimization focuses on frameworks, compilers, and libraries tailored for AI computations. Data optimization reduces overheads via augmentation, compression, and pruning, improving model quality. Model optimization employs methods such as quantization, pruning, and knowledge distillation to minimize complexity. Hybrid optimization synergistically combines hardware, software, data, and model techniques to maximize end-to-end gains. By exploring these categories, this chapter serves as a guide for technology decision-makers to grasp the benefits and trade-offs of different optimization techniques and thus make informed decisions for achieving improved productivity, cost efficiency, and performance in their AI workloads, without the need for technical expertise.

12.2 Hardware Optimizations

Rapid advancements of AI and ML are closely tied to the developments in hardware technology. As AI workloads have become increasingly complex and computationally intensive, traditional general-purpose processors have struggled to keep pace with the demands of training and deploying large-scale models. This section explores the various hardware optimization techniques used for addressing these challenges, focusing on specialized chips designed for AI processing. It also compares different acceleration techniques and emerging technologies that can reshape the landscape of AI hardware.

12.2.1 Specialized Chips for AI Processing

Owing to the need for efficient processing of AI workloads, specialized hardware accelerators have been developed [15]. These devices are designed to optimize the specific computational patterns found in AI algorithms, particularly the matrix multiplications and convolutions that form the backbone of many neural network architectures.

GPUs [16–20]: Originally designed for rendering complex 3D graphics, GPUs have become the workhorse of modern AI computations. Their highly parallel architecture, comprising thousands of small, efficient cores, is well-suited to the parallel nature of many AI algorithms. The introduction of NVIDIA's compute unified device architecture (CUDA) in 2006 marked a turning point, enabling developers to harness the power of GPUs for general-purpose computing, including AI workloads.

GPUs excel in both training and inference tasks, offering significantly faster speeds than CPUs for deep learning workloads. The latest generations of NVIDIA GPUs incorporate specialized tensor cores that further accelerate matrix multiplication operations, which are common in AI workloads. These advancements have enabled the training of increasingly large and complex models, such as GPT-3 and its successors. However, GPUs also face challenges, particularly in terms of power consumption and memory bandwidth. Moreover, multi-GPU systems are often employed to scale up performance, but this introduces complexities in data parallelism and model parallelism strategies.

FPGAs [21–24]: FPGAs offer a unique blend of flexibility and efficiency for AI acceleration. These devices comprise an array of programmable logic blocks that can be reconfigured for implementing custom digital circuits. Owing to this reconfigurability, FPGAs can be optimized for specific AI models or tasks, offering better performance and higher energy efficiency compared with fixed-architecture devices. In particular, FPGAs have shown promise in inference tasks, where the model architecture is fixed and can be efficiently mapped to the FPGA's resources. For example, Microsoft, after deploying FPGAs in its data centers to accelerate Bing search rankings and other AI workloads, has reported significant improvements in performance and energy efficiency.

However, FPGAs also face some challenges, such as widespread adoption for AI workloads. Traditional programming of FPGAs requires expertise in hardware description languages; recent advancements of high-level synthesis tools, however, are making them more accessible to software developers.

ASICs [23–27]: ASICs represent the pinnacle of hardware specialization for AI workloads. These custom-designed chips are tailored for specific AI tasks, offering the best possible performance and energy efficiency. The most prominent example of AI-specific ASICs is Google's Tensor Processing Unit (TPU) [28], first announced in 2016 and currently in its fourth generation.

TPUs are designed to accelerate both the training and inference of neural networks, with focus on Google's TensorFlow framework. They feature a systolic array architecture that is highly efficient for matrix multiplications and convolutions. In benchmark tests, TPUs have demonstrated significant performance and energy efficiency gains over GPUs for certain workloads, especially in large-scale ML tasks.

Other technology giants have followed suit in developing their own AI ASICs. Intel has introduced the Gaudi AI training processor [29], which is designed to accelerate the training of large-scale deep learning models. AWS has developed its own AI-focused chips [30]—Trainium for training and Inferentia for inference—to power its cloud-based AI services. These specialized chips offer compelling performance and efficiency benefits, but they also introduce challenges in terms of programming complexity and vendor lock-in. Apple's Neural Engine [31], when integrated into mobile and desktop processors, accelerates ML tasks on consumer devices. Tesla [32–34] has developed custom AI chips for autonomous driving, emphasizing the importance of specialized hardware in edge AI applications.

Several other AI-focused ASICs have emerged in recent years [17]. Cerebras Systems [35] has developed the wafer-scale engine, a massive ASIC that integrates over 400,000 AI-optimized cores on a single wafer, enabling unprecedented performance and scalability for training large-scale neural networks. Graphcore [36] has introduced its intelligence processing unit, a unique ASIC architecture designed specifically for highly parallel ML workloads. Furthermore, startups such as Groq and SambaNova have unveiled innovative ASIC designs [37] that promise to push the boundaries of AI acceleration; these focus on delivering high performance and energy efficiency.

Neural processing units (NPUs) are highly specialized ASICs designed to accelerate neural network operations and AI tasks. They are optimized for data-driven parallel computing, particularly for inference tasks in ML and AI. They are tailored to perform huge amounts of matrix and vector computations typical in deep learning models. NPUs are often integrated into the system-on-chip (SoC) alongside CPUs and GPUs. A few examples of NPUs are Intel's Core Ultra processors, AMD's Ryzen AI series, Apple's Processors, and Qualcomm's Snapdragon X Elite and Plus processors. NPUs are critical components in modern computing, designed to accelerate AI and ML tasks efficiently, reduce power consumption, and enhance the overall performance of devices by offloading AI workloads from CPUs and GPUs [38, 39].

12.2.2 Comparison of Hardware Acceleration Techniques

The choice between different hardware acceleration techniques involves complex trade-offs between performance, energy efficiency, flexibility, and cost. GPUs offer a good balance between performance and programmability, making them the most widely used accelerators for AI workloads. FPGAs provide greater energy efficiency and the ability to tailor the hardware to specific models, but this comes at the cost of increased development complexity. ASICs offer the highest performance and efficiency for specific workloads, but they lack flexibility and involve high upfront costs. Benchmark studies have shown that the relative performance of these different accelerators can vary significantly depending on the specific AI task, model architecture, and scale of deployment. For example, GPUs often excel in training large models, whereas ASICs and FPGAs can offer superior performance and efficiency for inference tasks with well-defined models [40]. Energy efficiency is an increasingly important consideration, particularly for large-scale deployments in data centers. The total cost of ownership (TCO) for AI hardware must account not only for the initial hardware costs but also for ongoing energy and cooling expenses. In this regard, specialized ASICs and well-optimized FPGA implementations can offer significant advantages over general-purpose GPUs for certain workloads [41].

12.2.3 Emerging Hardware Technologies for AI

With the increasing demand for AI computations, researchers and companies are exploring novel hardware architectures that could offer order-of-magnitude improvements in performance and efficiency.

Neuromorphic computing [42] is one such approach; it attempts to mimic the structure and function of biological neural networks in hardware. Projects such as Intel's Loihi chip [43, 44] aim to create highly efficient, event-driven processing systems that could excel at certain types of AI workloads, particularly those involving temporal data or sparse computations. IBM True North [27] and SpiNNaker [25] are other examples of neuromorphic computing platforms that have demonstrated promising results in areas such as low-power edge inference and large-scale neural network simulations.

Photonic computing [45] leverages light for information processing, promising ultra-low latency, and high energy efficiency. While still in early stages, photonic neural networks have shown potential for accelerating certain AI operations at the speed of light.

Quantum computing [31], though still in nascent stages, has the potential to revolutionize certain AI algorithms by leveraging quantum phenomena to perform computations that are intractable on classical computers.

In-memory computing and processing-in-memory architectures [32, 33] aim to overcome the von Neumann bottleneck by performing computations directly in memory. This approach can significantly reduce data movement, which is a major source of energy consumption in contemporary AI hardware. Recent advancements of non-volatile memory technologies, such as resistive RAM (ReRAM) and phase-change memory (PCM), are making in-memory computing increasingly viable for AI workloads.

These emerging technologies, while still in various stages of development, hold the promise of transforming the landscape of AI hardware in the coming years, offering new opportunities for improved performance, energy efficiency, and scalability.

12.3 Software Optimizations

Though hardware advancements have been instrumental in accelerating AI workloads, software optimizations are equally crucial for maximizing the efficiency and performance of AI systems. This section examines various software optimization techniques, encompassing AI-specific frameworks and libraries, compiler optimizations, runtime optimizations, and strategies for distributed computing and parallelization.

12.3.1 AI-Specific Frameworks and Libraries

The creation of specialized software frameworks and libraries has been crucial for enhancing the accessibility and efficiency of AI systems. These tools abstract the complexities associated with implementing and optimizing AI algorithms, enabling researchers and developers to focus on model design and application-specific challenges.

Deep learning frameworks [17, 34, 35]: Prominent deep learning frameworks, including TensorFlow, PyTorch, and MXNet, have emerged as the foundation for contemporary AI development. These tools offer high-level application programming interfaces that facilitate model construction, automatic differentiation for gradient calculation, and optimized implementations of ubiquitous neural network operations. By providing building blocks for constructing complex neural networks, these frameworks have significantly reduced the barriers to entry for AI development and enabled rapid prototyping and experimentation.

Ongoing efforts to optimize these frameworks for performance and energy efficiency have further enhanced their suitability for deployment on resource-constrained edge devices [36, 37]. For example, TensorFlow Lite and PyTorch Lite provide lightweight versions of the frameworks optimized for mobile and embedded devices, enabling efficient on-device inference. Similarly, Intel OpenVINO and Microsoft ONNX Runtime offer optimizations and runtime support to deploy deep learning models on a variety of hardware platforms, including edge devices. These are cross-platform inference engines that can efficiently execute models exported from different deep learning frameworks, making it easier to deploy AI applications on a wide range of hardware [38, 39].

Optimized linear algebra libraries [40, 41]: At a lower level, highly optimized linear algebra libraries form the backbone of efficient AI computations. Libraries such as NVIDIA's cuBLAS, Intel's oneDNN, and OpenBLAS provide hardware-specific implementations of basic linear algebra subprograms (BLAS), enhancing performance via techniques such as cache-aware algorithms, vectorization, and threading.

NVIDIA's cuDNN library provides highly optimized implementations of common deep learning operations for NVIDIA GPUs, significantly accelerating training and inference tasks [42, 43].

Intel's oneAPI Deep Neural Network Library (oneDNN) [44] is a comprehensive collection of highly optimized computational primitives that can be deployed across a diverse range of hardware platforms, including CPUs, GPUs, and FPGAs. By furnishing these optimized primitives, oneDNN empowers deep learning developers to achieve efficient and high-performance execution of their models on various hardware architectures.

12.3.2 Compiler Optimizations

Compiler technologies are crucial for translating high-level AI models into efficient executable code. AI-Specific compiler optimizations aim to leverage the distinctive attributes of AI workloads for generating more efficient executable representations.

Many AI frameworks represent computations as dataflow graphs. Compiler optimizations [45] at the graph level can significantly improve performance by eliminating redundant computations; fusing adjacent operations into more efficient kernels [46]; partitioning the graph for heterogeneous execution on different hardware accelerators [44]; and generating hardware-specific code for performance-critical kernels.

TensorFlow's XLA compiler [47, 48] can perform graph-level optimizations, such as layout transformations, operator fusion, and constant folding, to generate highly efficient codes for both training and inference tasks.

At a more granular level, compilers can generate highly efficient computational kernels customized to specific hardware platforms. Techniques such as loop restructuring, data vectorization, and memory access optimizations can significantly enhance the performance of individual AI operations. Software projects such as TVM [49] and Intel OpenVINO [50, 51] offer a unified optimization framework that can target a diverse range of hardware backends, spanning from mobile CPUs to specialized AI accelerators, thereby further advancing the capabilities of these compiler-level optimizations.

12.3.3 Runtime Optimizations

The efficient execution of AI workloads is dependent on model or data optimizations as well as on runtime optimizations and intelligent scheduling of computations to maximize hardware efficiency.

Dynamic tensor rematerialization [52]: With an increase in the size of AI models, they often exceed the available memory, especially on resource-constrained devices such as edge hardware. Dynamic tensor rematerialization can address this challenge by selectively recomputing certain tensors, instead of storing them in memory. This technique effectively trades computation for memory, enabling the training and inference of larger models on hardware with limited memory capacity, which is crucial under resource-constrained environments.

Kernel fusion and operator scheduling [53, 54]: Another key runtime optimization involves kernel fusion, where multiple operators are dynamically combined into a single kernel. This reduces the memory bandwidth requirements and enhances cache utilization, enabling faster execution. Furthermore, intelligent operator scheduling can optimize the order in which operations are executed, ensuring better hardware utilization and minimizing data movement, thus reducing processing delays.

Memory management optimizations [55–57]: Efficient memory management is critical for the smooth execution of AI workloads. Techniques such as memory pooling, pre-allocation of resources, and intelligent data prefetching help reduce memory overheads. By minimizing the need for frequent memory allocations and ensuring that data are readily available, these strategies can significantly improve system performance and reduce latency, particularly in large-scale AI applications.

12.3.4 Distributed Computing Strategies

With the increasing size and complexity of AI models, distributed computing has become essential for efficient training and deployment at scale. By leveraging the power of multiple devices and nodes, distributed strategies significantly accelerate processing, enabling the training of models that would otherwise be impossible on a single machine.

Data Parallelism [58–60]: Data parallelism divides large datasets across multiple compute nodes, with each node processing a fraction of data independently. After processing, the results from all nodes are aggregated, and model parameters across the network are synchronized. This method can significantly boost training for large datasets, but it requires efficient synchronization techniques to ensure that the model updates are correctly coordinated, preventing bottlenecks or inconsistencies in the model.

Model parallelism [56, 61, 62]: When models are excessively large to fit into the memory of a single device, model parallelism partitions the model itself across multiple devices. This strategy is particularly important for training expansive models, such as GPT-3 or other large language models. Techniques such as pipeline parallelism and tensor model parallelism can ensure that different parts of the model are trained in parallel; this can allow the model to scale efficiently across hundreds or even thousands of GPUs, while managing inter-device communication to minimize latency.

Distributed optimization algorithms [63–65]: Distributed optimization algorithms, such as parameter servers or ring all-reduce, enable efficient synchronization of model updates across multiple compute nodes. These algorithms are crucial for balancing the trade-off between communication overhead and convergence speed. By optimizing how updates are communicated and applied, distributed optimization can ensure that large-scale training can proceed efficiently without significant performance degradation from synchronization delays.

Federated learning [66–68]: Federated learning is an emerging paradigm that facilitates model training on distributed, decentralized data. This approach is beneficial in scenarios where data privacy is a critical concern because it enables model training without the need to centralize and aggregate sensitive data from multiple sources. Because federated learning allows the model to be trained on decentralized data, it preserves the privacy of individual user data. This approach is especially valuable in applications where regulatory requirements or user preferences require sensitive data to be protected from central servers or cloud services.

To optimize the performance for AI workloads, a close collaboration between software and hardware optimizations is a prerequisite. Software optimizations are instrumental in maximizing the efficiency and performance of AI workloads. Diverse techniques, ranging from high-level frameworks to low-level compiler optimizations and distributed computing strategies, are employed to expand the frontiers of AI.

12.4 Data Optimizations

Hardware and software optimization is crucial for improving the efficiency of AI workloads. Data optimization techniques also play a vital role in enhancing model performance and reducing computational requirements. This section examines various data optimization approaches, encompassing data augmentation, compression, and pruning, along with their impact on model quality and training efficiency.

12.4.1 Data Augmentation Techniques

Data augmentation is employed to enhance the diversity of training data without the need to acquire new datasets. This approach is beneficial in scenarios where procuring additional labeled data is costly or burdensome [69].

For computer vision tasks, common image data augmentation techniques include geometric transformations (rotation, flipping, scaling, cropping), color space transformations (brightness, contrast, and saturation), noise injection methods (Gaussian, salt-and-pepper noise), mixing augmentations (CutMix [70], MixUp [71]), and so on. These techniques can significantly enhance model generalization and robustness to variations in input data [72].

In the context of natural language processing tasks, text data augmentation techniques have demonstrated improved performance of NLP models on a wide range of tasks, from text classification to machine translation [73, 75], including the following:

- Synonym substitution, where words are replaced with semantically similar alternatives
- Random operations such as inserting, deleting, or rearranging words
- Back translation, where text is translated to another language and then back to the original
- Contextual augmentation [74], which leverages language models to generate synthetic yet plausible text.

More advanced data augmentation methods have emerged in recent years. AutoAugment employs reinforcement learning to automatically identify optimal augmentation strategies. RandAugment simplifies the search space of AutoAugment, enabling more efficient data augmentation. Generative adversarial networks can be leveraged to generate synthetic training data, which is particularly beneficial in scenarios with limited available data [76].

12.4.2 Data Compression Methods

With increasing size and complexity of AI models and datasets, the need for efficient data storage and transmission is becoming increasingly vital. Data compression techniques can substantially alleviate storage requirements and accelerate the data loading process during model training.

Lossless data compression methods preserve the entirety of original data and are valuable for applications where data integrity is paramount. These techniques include general-purpose algorithms such as GZIP, LZMA, and Brotli, as well as specialized formats such as Parquet and ORC for tabular data, and domain-specific approaches such as FLIF for images and FLAC for audio [3].

In the context of AI workloads, lossy compression techniques can be employed when a certain degree of data loss is acceptable, provided that it significantly reduces the data size without substantially compromising model performance. For instance, image data can be compressed using formats such as JPEG and WebP with carefully tuned quality settings, while audio data can be compressed using codecs such as MP3 and Opus with appropriate bitrate configurations [77, 78]. Furthermore, vector quantization has proven particularly useful for compressing embeddings in natural language processing tasks [79], where a trade-off between data size and model performance can be leveraged.

Advancements in AI have enabled the development of learned compression techniques [80] that can surpass the performance of traditional compression methods. These approaches leverage neural networks to learn optimal compression strategies customized to specific data domains. For instance, Ballé et al. [81] demonstrated the efficacy of learned image compression, while DeepZip was applied for compressing genomic data [82].

12.4.3 Data Pruning and Filtering Approaches

Data pruning and filtering approaches endeavor to identify and remove superfluous or detrimental data points, thereby reducing dataset size and enhancing model performance [83].

Redundancy reduction techniques identify and discard duplicate or highly similar data instances, effectively reducing dataset size without causing significant information loss. Data denoising methods aim to identify and remove corrupted or mislabeled samples that can negatively impact model training and generalization [84].

Recent advances in meta-learning have enabled the development of more sophisticated data pruning techniques. Meta-Dropout proposes a meta-learning approach to selectively retain important training examples, while GLISTER leverages gradient-based influence scores to identify the most informative data points [61, 85, 86].

Noise reduction techniques, such as those developed by Karimi et al. and Lee et al. [87], can enhance model performance by filtering low-quality or uninformative training samples. Outlier detection and confidence-based filtering are other promising approaches in this domain, and there is potential to improve model quality and efficiency [88, 89].

Curriculum learning involves presenting training data in a structured order, often progressing from simpler to more complex examples. This approach can accelerate the learning process and enhance model generalization. Self-paced learning automatically selects the order of training examples based on model performance. Difficulty-based ordering, whether manually or automatically assessed, structures the training curriculum by organizing samples according to the level of complexity [90–92].

Core-set selection methodologies seek to identify a compact subset of the data that can effectively encapsulate the essence of the entire dataset. These methods encompass a range of techniques, such as geometric techniques that select data points intended to span the underlying data manifold; diversity-based methods aiming to ensure a representative distribution across the data space; and influence-based pruning strategies that identify and retain the most impactful training samples. These techniques can significantly reduce dataset size while preserving model performance, which is beneficial for memory-constrained environments or scenarios with limited computational resources [93, 94].

12.4.4 Challenges and Considerations

Data optimization is a key factor for enhancing the efficiency and scalability of AI models. By addressing challenges such as overfitting, data imbalance, and computational bottlenecks, these techniques can significantly improve model generalization, reduce training time, and optimize resource utilization. However, despite the clear benefits, data optimization presents several challenges that require careful consideration [95, 96].

One of the primary challenges is balancing data compression with information loss. Reducing the size of datasets can improve efficiency, but excessive compression may cause the risk of information loss, potentially degrading model performance. Striking the right balance between minimizing data size and maintaining the integrity of crucial features is critical. While advanced techniques like data augmentation can improve model robustness, they often come with increased computational overheads. These methods may lengthen processing times, necessitating efficient implementations to justify the trade-offs.

Another significant concern is the potential introduction of biases during data optimization. Techniques such as pruning or filtering must be applied carefully to avoid distortion of the underlying data distribution. Introducing or amplifying biases can negatively impact model fairness and accuracy, making it essential to rigorously evaluate the effects of any data-related optimizations. Furthermore, the effectiveness of optimization strategies can vary across different tasks, and a strategy suitable for one application may not fit into the other one. Thus, tailored approaches may be required to fine-tune the optimization methods based on the specific requirements of each task.

12.5 Model Optimizations

With increasing size and complexity of AI models, optimizing their structure and parameters becomes crucial for efficient deployment and execution. This section explores various model optimization techniques, including quantization, pruning, knowledge distillation, and neural architecture search. These methods aim to reduce model size, improve inference speed, and maintain or even enhance model performance [96, 97].

12.5.1 Quantization Techniques

Quantization [98, 99], a technique that reduces the precision of model weights and activations, is a key strategy for optimizing AI models. By converting high-precision floating-point values to lower bit-width representations, quantization can significantly reduce model size and inference latency.

Post-training quantization [100] is applied after a model is trained. It can be dynamic range quantization, which determines quantization parameters on-the-fly during inference, or static quantization, which uses a representative dataset to pre-determine optimal quantization parameters. Weight-only quantization quantizes only the weights, leaving activations in floating-point. While these methods can significantly reduce model size and inference latency, they may result in the loss of accuracy to some extent.

Quantization-aware training [101] incorporates quantization effects into the training process. It simulates low-precision operations during training, allowing the model to learn optimal quantization parameters. This approach often yields better results than post-training quantization, especially for lower bit-widths. Differentiable quantization enables end-to-end training of quantized models.

Mixed-precision quantization [102] employs different levels of quantization for different parts of the model. It can be layer-wise, mixed precision, assigning different bit-widths to different layers based on their sensitivity, or channel-wise quantization, which applies different quantization parameters to different channels within a layer. Hybrid float/fixed-point models use a mix of floating-point and fixed-point representations. These methods offer a flexible approach for balancing model size, speed, and accuracy.

12.5.2 Model Pruning and Sparsification

Model pruning [103–106] is a powerful technique for optimizing AI models by eliminating unnecessary weights or neurons, effectively reducing the size and computational complexity of the model without compromising performance. By removing redundant parameters, pruning not only accelerates inference but also lowers memory requirements, making it useful for deploying models on resource-constrained devices.

Weight pruning techniques: Weight pruning focuses on removing individual weights within the model based on their contribution to the overall performance. The most common weight pruning techniques are as follows:

- **Magnitude-based pruning**: This method removes weights that fall below a predefined threshold, assuming that smaller weights have a lesser impact on the output of the model.

- **Gradient-based pruning**: Weights are pruned based on their influence on the loss function of the model, focusing on removing those that contribute the least to minimizing the loss.
- **Regularization-based pruning**: This technique integrates sparsity-inducing regularizers, such as L1 or L2 regularization, during the training process to encourage the model to produce sparse representations naturally.

Structured pruning: Unlike unstructured pruning, which targets individual weights, structured pruning eliminates entire structures, such as filters, neurons, or blocks, resulting in more hardware-efficient models. This approach is particularly beneficial for AI accelerators and hardware systems optimized for sparse computation. These techniques are as follows:

- **Filter pruning in CNNs**: Removes entire convolutional filters, reducing the number of feature maps and computations required in convolutional neural networks.
- **Neuron pruning in fully connected layers**: Eliminates entire neurons along with their connections, significantly reducing model size in dense layers.
- **Block-based pruning**: Prunes groups of weights together in a structured fashion, leading to more efficient execution on parallel processing hardware.

Dynamic pruning techniques: Dynamic pruning adapts the model's structure during inference, tailoring computations to the specific input for enhancing the efficiency. While dynamic pruning offers significant efficiency gains through dynamic adjusting, it often requires more complex implementation and fine-tuning to ensure consistent performance across varying inputs. These techniques include:

- **Runtime neural pruning**: Dynamically prunes parts of the network during inference, selecting which portions of the model to activate based on the input's requirements.
- **Conditional computation**: Activates only the necessary parts of the model for each input, reducing computation by processing only the most relevant network components.

12.5.3 Knowledge Distillation

Knowledge distillation [107–109] transfers knowledge from a large, complex model (the teacher) to a smaller, simpler model (the student). This process can significantly reduce model size and inference latency while preserving much of the original model's performance.

Traditional knowledge distillation techniques aim to transfer knowledge from a larger, more complex model (the teacher) to a smaller, simpler model (the student). This is achieved through the following:

- **Soft targets**: Use the teacher's output probabilities, rather than just the final classification labels, to guide the student's training.

- **Temperature scaling**: Adjusts the "softness" of the teacher's probability distribution to control the degree of knowledge transfer.
- **Attention transfer**: Transfers the teacher's attention maps or feature importance to that of the student model.

More advanced distillation techniques include:

- **Multi-teacher distillation**: Combines knowledge from multiple teacher models to create a more comprehensive and robust student model.
- **Online distillation**: Performs distillation during the training process of both teacher and student models, allowing for continuous knowledge transfer and refinement.
- **Self-distillation**: Uses earlier checkpoints or versions of the same model for distillation, enabling the student model to learn from its own progression and evolution, resulting in better generalization and performance.

For specific tasks, knowledge distillation can be adjusted to preserve key aspects of the original model's performance:

- **Feature-based distillation**: Transfers the teacher model's intermediate feature representations to the student.
- **Relation-based distillation**: Preserves the relational information between samples in the student model.
- **Task-adaptive distillation**: Customizes the distillation process to better suit the requirements of specific downstream tasks.

12.5.4 Neural Architecture Search and AutoML

Neural architecture search (NAS) [110–112] automates the process of finding optimal neural network architectures, freeing researchers and engineers from the time-consuming task of manual design. NAS employs various search algorithms, including reinforcement learning-based NAS, evolutionary algorithms, and gradient-based approaches.

To address real-world constraints, NAS has evolved to incorporate efficiency considerations. Hardware-aware NAS considers hardware limitations during the search process, while multi-objective NAS optimizes for multiple objectives such as accuracy, latency, and model size. Once-for-all networks, a recent innovation, train a single large network that can be adapted to different hardware platforms.

AutoML [113, 114], a broader field encompassing NAS, automates various aspects of model optimization. It includes automated hyperparameter optimization, a combination of architecture and hyperparameter search, and meta-learning for architecture search. By leveraging NAS and AutoML techniques, researchers and engineers can accelerate the development of high-performing AI models tailored to specific requirements.

12.5.5 Emerging Techniques and Challenges in Model Optimization

With the constant evolution of AI technologies, novel model optimization techniques are emerging. Neural ordinary differential equations, an alternative to traditional layer-based models [111, 115], offer potential benefits in terms of parameter and memory efficiency. Federated learning [116, 117], a privacy-preserving approach to distributed model training, necessitates specialized optimization methods to address challenges such as communication efficiency and data heterogeneity. Neuro-symbolic AI [118], which combines neural networks with symbolic AI, aims to create more efficient and interpretable models.

While model optimization presents numerous advantages, it also introduces challenges. Striking a balance between model performance, size, and speed requirements is a constant trade-off. Ensuring that optimized models maintain satisfactory performance on unseen data is another critical concern. In addition, adapting optimization techniques to diverse hardware platforms can be complex, and managing the increased complexity of optimized model pipelines poses a significant challenge.

12.5.6 Impact on AI Workloads

The application of model optimization techniques [35, 114, 119, 120] can significantly enhance the performance and efficiency of AI workloads, making them more practical and accessible across diverse platforms and environments.

One major benefit is the reduction in computational requirements. By streamlining models, optimization enables the deployment of AI on resource-constrained devices such as smartphones, IoT devices, and embedded systems. This broadens the range of applications where AI can be effectively utilized, making it more accessible to industries with limited hardware resources. These techniques contribute to improved inference speed, a critical factor for real-time applications such as autonomous driving, robotics, and natural language processing, where quick and accurate decision-making is essential. Lastly, energy efficiency is a key outcome of model optimizations.

These optimizations are crucial in making AI models more efficient and scalable across various hardware platforms, from powerful servers to portable devices. As AI continues to integrate into diverse industries, the importance of model optimization will continue to increase. Future research is likely to focus on automated, adaptive optimization techniques that can dynamically adjust models to meet specific deployment needs and constraints, ensuring optimal performance across different use cases.

12.6 Hybrid Optimization Approaches

As AI systems become increasingly complex, we understand that there is no single optimization technique that can address all the challenges in improving AI workload efficiency. Hybrid optimization approaches, which combine multiple optimization strategies across hardware, software, data, and model levels, have emerged as a powerful means to achieve synergistic performance improvements. This section explores the integration of various optimization techniques, their synergies, and associated challenges and opportunities.

12.6.1 Synergies Between Hardware, Software, Data, and Model Optimizations

The integration of multiple optimization techniques, commonly referred to as hybrid optimization approaches, can significantly enhance system performance. Specifically, the system performance improves substantially with the combination of hardware and software optimizations, such as ASIC-aware NAS, software-defined hardware, and compiler-hardware co-optimization [121, 122].

Integrating data optimization techniques with model optimization methods can lead to improvements in both training efficiency and model performance. Examples of such synergies include data-efficient NAS, joint optimization of data augmentation and model pruning, and adaptive quantization approaches that leverage data characteristics [123].

Integrating software and model optimization techniques can also enable the development of more efficient AI systems. Examples of such integrations include model-aware operator fusion, which optimizes computational graphs; dynamic model adaptation, which adjusts model complexity at runtime; and compiler-driven model transformations, which apply model-specific optimizations during the compilation process. Collectively, these approaches can enhance the overall efficiency of AI systems [124, 125].

12.6.2 End-to-End Optimization Strategies

Holistic end-to-end optimization strategies seek to comprehensively optimize the entire AI workflow, from data preprocessing to model deployment. AutoML can be expanded beyond model architecture to encompass the entire AI ecosystem, enabling the joint optimization of hardware, software, and model components, as well as the application of transfer learning techniques and multi-objective optimization approaches [113, 114, 126].

Differentiable approaches to NAS can be expanded to incorporate hardware and software factors, directly integrating hardware constraints into the architecture search process and optimizing software frameworks and hardware designs simultaneously [127, 128].

Federated learning, a privacy-preserving approach to distributed model training, can benefit from the integration of various optimization techniques, exemplified by strategies such as communication-efficient federated learning and privacy-preserving hybrid optimization [116].

12.6.3 Case Studies of Successful Hybrid Optimization Implementations

Several real-world case studies have demonstrated the effectiveness of hybrid optimization approaches in improving the performance and efficiency of AI workloads.

Google's TPU and TensorFlow Ecosystem [28, 129, 130]: Google's TPUs represent the pinnacle of hardware–software co-design, specifically tuned for high-performance ML workloads. Its integration with TensorFlow, Google's open-source AI platform, highlights the efficiency and scalability of this hybrid approach:

- **TPU-Specific optimizations in TensorFlow**: TensorFlow includes custom operators and graph-level optimizations designed for TPUs. These enhancements streamline data flow and execution, allowing models to fully utilize the specialized architecture of TPUs.
- **Quantization-aware training**: Google pioneered the integration of quantization during the training process, ensuring that models can handle lower-precision data (e.g., INT8) without compromising accuracy. This allows faster and more memory-efficient inference on TPUs.
- **Accelerated linear algebra (XLA) compiler**: XLA further optimizes TensorFlow's computations by transforming high-level operations into highly efficient machine codes, leveraging TPU architecture and enhancing performance across various hardware platforms. This compilation approach helps reduce execution time and memory overhead for both training and inference workloads [17].

NVIDIA's CUDA and cuDNN Libraries [17, 122, 131]: NVIDIA's strategy is built for maximizing GPU performance for deep learning by offering a tightly integrated software ecosystem that complements its hardware accelerators. The combination of CUDA, cuDNN, and complementary tools such as TensorRT illustrates a holistic approach to hybrid optimization:

- **GPU-Accelerated deep learning primitives**: Using cuDNN, NVIDIA provides highly optimized implementations of core deep learning operations (e.g., convolutions and pooling) tailored for its GPUs. These libraries reduce the overhead of manual optimization and ensure that models take full advantage of GPU parallelism.

- **TensorRT Inference Optimization**: TensorRT is NVIDIA's inference SDK that applies a hybrid approach to inference, utilizing a combination of layer fusion, kernel auto-tuning, and quantization to optimize deep learning models for deployment. This creates efficiency gains across various inference environments, from edge devices to data centers.
- **Multi-instance GPU (MIG) in A100 GPUs**: MIG allows NVIDIA A100 GPUs to be partitioned into multiple independent instances, each of which can handle separate tasks concurrently. This optimizes resource utilization by allowing multiple models to be deployed on the same hardware, enabling scalability in multi-tenant environments.

Intel's OpenVINO and OneDNN Libraries [31, 50, 61, 132]: Intel's approach to hybrid optimization is exemplified by the OpenVINO Toolkit, which leverages both hardware-specific optimizations and software-level enhancements to maximize performance across Intel's diverse range of processors and accelerators:

- **Model optimizer**: The model optimizer pre-processes trained models to enable them to run more efficiently at deployment. Neural network compression framework allows various compression techniques such as pruning, quantization, and graph simplifications, which aid in reducing memory usage and computational complexity without compromising model accuracy.
- **OpenVINO Runtime**: OpenVINO Runtime provides a flexible and efficient framework for executing AI models, selecting the most suitable hardware target (CPU, GPU, NPU, or FPGA) and applying hardware-specific optimizations in real time. This flexibility allows models to be executed seamlessly across different × 86 and ARM architectures while maximizing performance. The engine dynamically selects the most efficient hardware for each task, allowing models to be deployed across edge and cloud environments with consistent performance gains.
- **Deep neural network library (OneDNN)**: OneDNN provides highly efficient implementations of key deep learning primitives, such as convolution and matrix multiplication, optimized for Intel's hardware. By focusing on low-level operations, OneDNN enhances computational efficiency and scalability, particularly in AI workloads.

These case studies highlight the importance of integrating hardware, software, and model optimizations to achieve significant performance improvements in AI workloads. By leveraging specialized hardware, optimized software libraries, and advanced techniques such as quantization and mixed-precision, these companies have successfully addressed the computational demands of modern AI applications.

12.6.4 Challenges in Implementing Hybrid Optimization Approaches

Hybrid optimization approaches present significant potential for enhancing the efficiency of AI workloads; however, they also introduce complex challenges that require careful management [3]. The primary challenge is the increased system complexity arising from the coordination of multiple optimization techniques across hardware, software, data, and model levels. This inherent coordination can lead to intricate interdependencies, where optimization strategies may be in mutual conflict. For instance, hardware-specific enhancements could clash with algorithmic changes, resulting in reduced performance or unexpected behavior. Furthermore, debugging and troubleshooting these highly optimized systems becomes more complex, as identifying the root cause of issues often requires extensive cross-domain expertise.

Another challenge is ensuring that hybrid optimizations generalize well to a wide range of tasks and hardware environments [133]. Many optimizations are tailored to specific hardware configurations, models, or datasets, which can lead to overfitting to certain scenarios. This presents a risk when deploying AI models in diverse or unfamiliar environments. Achieving a balance between maximizing performance on a particular hardware platform and maintaining portability across different systems is a persistent challenge. Additionally, optimized models must remain compatible with various software frameworks and tools, which further complicates the task of ensuring generalization and broad applicability.

The development and maintenance overhead associated with hybrid optimization is another significant hurdle [12, 14, 35]. Implementing and testing these multi-faceted strategies requires a long development time, particularly as optimization techniques evolve to keep pace with advances in hardware and AI research. Furthermore, the skills and expertise required to develop and maintain these systems are often interdisciplinary, encompassing deep knowledge of both hardware and software, along with a robust understanding of AI models and techniques. This need for diverse expertise adds to the complexity and cost of managing hybrid optimization approaches in real-world applications.

12.6.5 Future Trends

With the progressive advancement of AI, hybrid optimization approaches continue to evolve to meet new challenges, with several key trends shaping the future. One significant development is the rise of automated hybrid optimization, where AI systems are increasingly using AI techniques to optimize themselves. This recursive loop of AI-driven optimization could lead to self-adapting systems capable of dynamically adjusting their strategies based on changing conditions and requirements. Such systems can streamline the optimization process, reducing the need for manual interventions and allowing AI to fine-tune its performance autonomously across multiple levels of hardware and software.

Another exciting trend lies in the integration of quantum and classical computing within hybrid systems. Quantum-accelerated ML [134–136] is emerging as a way to leverage the unique capabilities of quantum computers to speed up specific components of classical AI workloads. At the same time, quantum-inspired algorithms are influencing classical optimization methods, offering new avenues for improving performance even on traditional hardware. These quantum–classical hybrids could revolutionize how AI workloads are optimized, particularly for complex tasks where quantum computing's advantages for solving certain types of problems could complement classical AI's strengths.

Neuromorphic computing [45, 137–139], which mimics the architecture of the human brain, is gaining traction as a potential pathway for developing highly energy-efficient AI systems. By incorporating brain-inspired optimization techniques into hybrid approaches, researchers are exploring ways to develop AI systems that can perform sophisticated tasks while consuming minimal power. This is relevant in edge computing scenarios where power efficiency is critical. The fusion of neuromorphic principles with traditional hardware and software optimizations could result in AI systems that not only excel in performance but also minimize their environmental footprint.

Hybrid optimization represents the forefront of AI workload enhancements [17, 140, 141], enabling the creation of systems that outperform contemporary singular approaches. By intelligently combining optimization strategies across hardware, software, data, and model levels, researchers and engineers can develop AI systems that are more efficient and scalable. Future research might focus on automating and enhancing these hybrid techniques, driving the development of systems that can continuously adapt and optimize themselves across all layers of the AI stack—from hardware to high-level algorithms.

12.7 Practical Considerations for Decision-Makers

As AI transforms industries, AI engineers and decision-makers face the challenge of efficiently implementing and managing AI workloads. This section outlines practical considerations for selecting optimization techniques, understanding trade-offs, and adhering to best practices [12].

Selecting the right optimization techniques for specific use cases is critical to maximize AI workload efficiency. Key factors to consider include workload characteristics such as model complexity, data volume, and the nature of the task—whether inference or training. Larger models may benefit from pruning or quantization, while high-volume datasets might require data-level optimizations. Hardware capabilities and budget constraints are also pivotal in determining the feasibility of certain optimizations, particularly when specialized hardware like GPUs or FPGAs are involved.

Performance requirements such as latency sensitivity or throughput demands also guide the choice of techniques, ensuring that real-time applications maintain low inference times, and batch processing benefits from parallelization. Finally, scalability is essential; optimizations should account for future growth, allowing flexibility for updates as models and data evolve, while also considering team expertise and the availability of support resources [120, 125, 142, 143].

Understanding trade-offs is crucial when implementing optimization strategies [41, 57, 144]. One common challenge is balancing accuracy with efficiency. Although techniques such as quantization or pruning can reduce model size and improve speed, they may also reduce accuracy. Another trade-off is development time versus runtime performance; custom ASIC development offers exceptional performance, but it requires a significant upfront investment; pre-optimized libraries are quicker to implement but offer less fine-grained control. Flexibility versus specialization is another consideration, where specialized optimizations may boost performance for specific workloads, but they limit adaptability for future changes. Similarly, high-performance hardware accelerators can provide significant boosts at a higher cost; software-based optimizations might be more affordable, but they offer lower performance gains. Complexity versus maintainability must also be weighed; advanced techniques might provide better results, but they increase maintenance burdens, whereas simpler optimizations are easier to maintain but may not deliver peak performance.

To ensure successful AI workload optimizations [3, 12, 13, 145], certain best practices should be followed. First, a baseline should be established by benchmarking the current workload performance, which can help accurately quantify the impact of optimizations. Optimizations should be prioritized by focusing on the most significant bottlenecks, using profiling tools to identify performance hotspots. Incremental implementation is the key, applying optimizations in stages and quantifying the effects of each change to simplify troubleshooting and fine-tune performance. Thorough validation is also essential; optimized models should be tested across various scenarios to ensure they maintain accuracy and reliability. Monitoring optimized workloads in production is equally important, allowing for iterative adjustments as workload characteristics evolve. Extensive documentation of applied techniques and their impacts aids in future troubleshooting and knowledge transfer.

Furthermore, it is vital to consider the entire AI pipeline, rather than focusing only on individual model optimizations. Optimizing data pre- and post-processing steps can significantly enhance overall performance. Moreover, staying informed about the latest AI optimization techniques and hardware advancements will help ensure the use of latest strategies in projects. Collaborating across disciplines, between data scientists, software engineers, and hardware specialists, can lead to more comprehensive optimization efforts. Finally, scalability should always be considered, ensuring that chosen techniques will scale efficiently despite increasing data volumes and model complexities.

12.8 Conclusion

As we conclude this comprehensive exploration of AI workload optimization, it is important to reflect on the key insights gained and recognize the pivotal role that optimization plays in the rapidly evolving landscape of AI. This chapter summarized various dimensions of AI workload optimization, each offering distinct advantages and challenges.

We examined **hardware optimizations**, where specialized chips such as GPUs, FPGAs, and ASICs accelerate AI computations, providing substantial performance gains. However, these solutions often require substantial investment and expertise to ensure effective implementation. **Software optimizations** highlight the role of AI-specific frameworks, compiler techniques, and efficient runtime strategies, offering flexibility and often integrating with existing hardware. In terms of **data optimization**, techniques such as data augmentation, compression, and pruning were discussed, elaborating on their critical role in managing the increasing volumes of data while preserving model quality and improving training efficiency.

Model optimizations such as quantization, pruning, and knowledge distillation demonstrated how model size and inference speed can be improved with minimal impact on accuracy. Finally, **hybrid approaches** revealed the power of combining these techniques, illustrating how an integrated strategy can yield outcomes superior to those of isolated optimizations.

The key takeaway is that there is no one-size-fits-all solution in AI workload optimization. The most effective approach involves a meticulous combination of techniques tailored to the specific requirements and constraints for each use case.

As AI continues to permeate industries and societal applications, ensuring efficient scaling of AI operations becomes essential. Optimization plays a vital role in this scaling process in several ways:

- **Cost efficiency**: By reducing computational requirements and improving resource utilization, optimizations allow organizations to manage the escalating costs associated with large-scale AI deployments.
- **Energy efficiency**: Optimized workloads consume less power, promoting sustainable AI practices and reducing the environmental impact of AI operations.
- **Improved accessibility**: Optimization techniques that minimize model size and computational demands make AI more accessible across a range of devices and applications, including edge and mobile computing.
- **Enhanced performance**: Optimized systems process more data, ensure faster decision-making, and handle increasingly complex tasks, unlocking new possibilities for AI applications.
- **Scalability**: Well-executed optimizations enable organizations to scale AI operations effectively, managing increasing data volumes and sophisticated models without proportionally increasing resource demands.

AI workload optimization is a key enabler of the future of AI. By continuously pushing the boundaries of efficiency and performance, we can unlock AI's full potential to solve complex problems, drive innovation, and improve lives worldwide. The journey of optimization is ongoing, and it is up to researchers, engineers, and decision-makers to drive this field forward, ensuring AI scales to meet the challenges and opportunities of tomorrow.

References

1. Y. Weng, J. Wu, T. Kelly, W. Johnson, Comprehensive overview of artificial intelligence applications in modern industries (2024). arXiv preprint arXiv:2409.13059
2. N. Maslej, L. Fattorini, R. Perrault, V. Parli, A. Reuel, E. Brynjolfsson, J. Etchemendy, K. Ligett, T. Lyons, J. Manyika, J.C. Niebles, Y. Shoham, R. Wald, J. Clark, Artificial Intelligence Index Report 2024 (2024)
3. B. Walsh, *Productionizing AI: How to Deliver AI B2B Solutions with Cloud and Python* (Apress, 2023)
4. P. Cohan, *Brain Rush* (Apress, 2024)
5. M.H. Calp, Evaluation of multidisciplinary effects of artificial intelligence with optimization perspective (2019). arXiv preprint arXiv:1902.01362
6. N.C. Thompson, K. Greenewald, K. Lee, G.F. Manso, The computational limits of deep learning **10** (2020). arXiv preprint arXiv:2007.05558
7. C.J. Wu, R. Raghavendra, U. Gupta, B. Acun, N. Ardalani, K. Maeng, G. Chang, F. Aga, J. Huang, C. Bai, K. Hazelwood, Sustainable AI: environmental implications, challenges and opportunities. Proc. Mach. Learn. Syst. **4**, 795–813 (2022)
8. Y. Wang, Q. Wang, S. Shi, X. He, Z. Tang, K. Zhao, X. Chu, Benchmarking the performance and energy efficiency of AI accelerators for AI training, in *2020 20th IEEE/ACM International Symposium on Cluster, Cloud and Internet Computing (CCGRID).* (IEEE, 2020), pp. 744–751
9. M. Naumov, J. Kim, D. Mudigere, S. Sridharan, X. Wang, W. Zhao, S. Yilmaz, C. Kim, H. Yuen, M. Ozdal, K. Nair, M. Smelyanskiy, Deep learning training in facebook data centers: design of scale-up and scale-out systems (2020). arXiv preprint arXiv:2003.09518
10. D. Richins, D. Doshi, M. Blackmore, A.T. Nair, N. Pathapati, A. Patel, B. Daguman, D. Dobrijalowski, R. Illikkal, K. Long, D. Zimmerman, V.J. Reddi, Ai tax: the hidden cost of AI data center applications. ACM Trans. Comput. Syst. (TOCS) **37**(1–4), 1–32 (2021)
11. R. Caspart, S. Ziegler, A. Weyrauch, H. Obermaier, S. Raffeiner, L.P. Schuhmacher, J. Scholtyssek, D. Trofimova, M. Nolden, I. Reinartz, C. Debus, Precise energy consumption measurements of heterogeneous artificial intelligence workloads, in *International Conference on High Performance Computing* (Springer International Publishing, Cham, 2022), pp. 108–121
12. J. Bosch, H.H. Olsson, I. Crnkovic, Engineering AI systems: a research agenda, in *Artificial Intelligence Paradigms for Smart Cyber-Physical Systems* (2021), pp. 1–19
13. H. Carvalho, P. Zaykov, A. Ukaye, Leveraging the HW/SW Optimizations and Ecosystems that Drive the AI Revolution (2022). arXiv preprint arXiv:2208.02808
14. C.J. Wu, B. Acun, R. Raghavendra, K. Hazelwood, *Beyond Efficiency: Scaling AI Sustainably* (IEEE Micro, 2024)
15. P. Gupta, N.K. Sehgal, J.M. Acken, Hardware based AI and ML, in *Introduction to Machine Learning with Security: Theory and Practice Using Python in the Cloud.* (Springer International Publishing, Cham, 2024), pp.247–270
16. S. Che, M. Boyer, J. Meng, D. Tarjan, J.W. Sheaffer, K. Skadron, A performance study of general-purpose applications on graphics processors using CUDA. J. Parallel Distrib. Comput. **68**(10), 1370–1380 (2008)

17. A. Reuther, P. Michaleas, M. Jones, V. Gadepally, S. Samsi, J. Kepner, AI and ML accelerator survey and trends, in *2022 IEEE High Performance Extreme Computing Conference (HPEC)*. (IEEE, 2022), pp. 1–10
18. J.D. Owens, M. Houston, D. Luebke, S. Green, J.E. Stone, J.C. Phillips, GPU computing. Proc. IEEE **96**(5), 879–899 (2008)
19. J.D. Owens, D. Luebke, N. Govindaraju, M. Harris, J. Krüger, A.E. Lefohn, T.J. Purcell, A survey of general-purpose computation on graphics hardware, in *Computer Graphics Forum*, vol. 26, no. 1 (Blackwell Publishing Ltd., Oxford, UK, 2007), pp. 80–113
20. S. Pal, E. Ebrahimi, A. Zulfiqar, Y. Fu, V. Zhang, S. Migacz, D. Nellans, P. Gupta, Optimizing multi-GPU parallelization strategies for deep learning training. IEEE Micro **39**(5), 91–101 (2019)
21. A. Putnam, A.M. Caulfield, E.S. Chung, D. Chiou, K. Constantinides, J. Demme, H. Esmaeilzadeh, J. Fowers, G.P. Gopal, J. Gray, D. Burger, A reconfigurable fabric for accelerating large-scale datacenter services. ACM SIGARCH Comput. Arch. News **42**(3), 13–24 (2014)
22. C. Bobda, J.M. Mbongue, P. Chow, M. Ewais, N. Tarafdar, J.C. Vega, R. Tessier, The future of FPGA acceleration in datacenters and the cloud. ACM Trans. Reconfigurable Technol. Syst. (TRETS) **15**(3), 1–42 (2022)
23. A. Putnam, A.M. Caulfield, E.S. Chung, D. Chiou, K. Constantinides, J. Demme, K. Eguro, D. Koch, S. Handagala, M. Leeser, D. Burger, A reconfigurable fabric for accelerating large-scale datacenter services. ACM SIGARCH Comput. Arch. News **42**(3), 13–24 (2014)
24. A. Boutros, E. Nurvitadhi, R. Ma, S. Gribok, Z. Zhao, J.C. Hoe, V. Betz, M. Langhammer, Beyond peak performance: comparing the real performance of AI-optimized FPGAs and GPUs, in *2020 International Conference on Field-Programmable Technology (ICFPT)* (IEEE, 2020), pp. 10–19
25. A. HajiRassouliha, A.J. Taberner, M.P. Nash, P.M. Nielsen, Suitability of recent hardware accelerators (DSPs, FPGAs, and GPUs) for computer vision and image processing algorithms. Signal Process.: Image Commun. **68**, 101–119 (2018)
26. J. Dean, 1.1 the deep learning revolution and its implications for computer architecture and chip design, in *2020 IEEE International Solid-State Circuits Conference-(ISSCC)* (IEEE, 2020), pp. 8–14
27. Y. Chen, Y. Xie, L. Song, F. Chen, T. Tang, A survey of accelerator architectures for deep neural networks. Engineering **6**(3), 264–274 (2020)
28. N. Jouppi, C. Young, N. Patil, D. Patterson, Motivation for and evaluation of the first tensor processing unit. IEEE Micro **38**(3), 10–19 (2018)
29. E. Medina, E. Dagan, Habana labs purpose-built AI inference and training processor architectures: scaling AI training systems using standard Ethernet with Gaudi processor. IEEE Micro **40**(2), 17–24 (2020)
30. T. Xue, X. Li, R. Smirnov, T. Azim, A. Sadrieh, B. Pahlavan, NinjaLLM: fast, scalable and cost-effective RAG using Amazon SageMaker and AWS Trainium and Inferentia2 (2024). arXiv preprint arXiv:2407.12057
31. Q. Liang, P. Shenoy, D. Irwin, *AI on the Edge: Rethinking AI-based IoT Applications Using Specialized Edge Architectures* (2020). arXiv e-prints, arXiv-2003
32. E. Talpes, D.D. Sarma, D. Williams, S. Arora, T. Kunjan, B. Floering, A. Jalote, C. Hsiong, C. Poorna, V. Samant, P. Banon, The microarchitecture of dojo, tesla's exa-scale computer. IEEE Micro **43**(3), 31–39 (2023)
33. An, A. I. (2020). AI Chips: What They Are and Why They Matter.
34. A. Reuther, P. Michaleas, M. Jones, V. Gadepally, S. Samsi, J. Kepner, AI accelerator survey and trends, in *2021 IEEE High Performance Extreme Computing Conference (HPEC)* (IEEE, 2021), pp. 1–9
35. S. Lie, Cerebras architecture deep dive: first look inside the hardware/software co-design for deep learning. IEEE Micro **43**(3), 18–30 (2023)
36. Z. Jia, B. Tillman, M. Maggioni, D.P. Scarpazza, Dissecting the graphcore ipu architecture via microbenchmarking (2019). arXiv preprint arXiv:1912.03413

37. M. Emani, Z. Xie, S. Raskar, V. Sastry, W. Arnold, B. Wilson, R. Thakur, V. Vishwanath, Z. Liu, M.E. Papka, M. Boyd, A comprehensive evaluation of novel AI accelerators for deep learning workloads, in *2022 IEEE/ACM International Workshop on Performance Modeling, Benchmarking And Simulation of High Performance Computer Systems (PMBS)* (IEEE, 2022), pp. 13–25
38. S. Wang, A. Pathania, T. Mitra, Neural network inference on mobile SoCs. IEEE Design & Test **37**(5), 50–57 (2020)
39. H. Peng, C. Ding, T. Geng, S. Choudhury, K. Barker, A. Li, Evaluating emerging ai/ml accelerators: Ipu, rdu, and nvidia/amd gpus, in *Companion of the 15th ACM/SPEC International Conference on Performance Engineering* (2024), pp. 14–20
40. A. Reuther, P. Michaleas, M. Jones, V. Gadepally, S. Samsi, J. Kepner, Survey and benchmarking of machine learning accelerators, in *2019 IEEE High Performance Extreme Computing Conference (HPEC)* (IEEE, 2019), pp. 1–9
41. S. Shankar, A. Reuther, Trends in energy estimates for computing in ai/machine learning accelerators, supercomputers, and compute-intensive applications, in *2022 IEEE High Performance Extreme Computing Conference (HPEC)* (IEEE, 2022), pp. 1–8
42. C.D. Schuman, S.R. Kulkarni, M. Parsa, J.P. Mitchell, P. Date, B. Kay, Opportunities for neuromorphic computing algorithms and applications. Nat. Comput. Sci. **2**(1), 10–19 (2022)
43. M. Davies, Lessons from Loihi: progress in neuromorphic computing, in *2021 Symposium on VLSI Circuits* (IEEE, 2021), pp. 1–2
44. Y.S. seYang, Y. Kim, Recent trend of neuromorphic computing hardware: Intel's neuromorphic system perspective, in *2020 International SoC Design Conference (ISOCC)* (IEEE, 2020), pp. 218–219
45. B.J. Shastri, A.N. Tait, T. Ferreira de Lima, W.H. Pernice, H. Bhaskaran, C.D. Wright, P.R. Prucnal, Photonics for artificial intelligence and neuromorphic computing. Nat. Photonics **15**(2), 102–114 (2021)
46. Y. Kwak, W.J. Yun, S. Jung, & J. Kim, Quantum neural networks: concepts, applications, and challenges, in *2021 Twelfth International Conference on Ubiquitous and Future Networks (ICUFN).* (IEEE, 2021), pp. 413–416
47. Y. Ma, Y. Du, L. Du, J. Lin, Z. Wang, In-memory computing: the next-generation ai computing paradigm, in *Proceedings of the 2020 on Great Lakes Symposium on VLSI* (2020), pp. 265–270
48. Z. Wang, T. Luo, R.S.M. Goh, W. Zhang, W.F. Wong, Optimizing for in-memory deep learning with emerging memory technology. IEEE Trans. Neural Netw. Learn. Syst. (2023)
49. L. Alzubaidi, J. Zhang, A.J. Humaidi, A. Al-Dujaili, Y. Duan, O. Al-Shamma, J. Santamaría, M.A. Fadhel, M. Al-Amidie, L. Farhan, Review of deep learning: concepts, CNN architectures, challenges, applications, future directions. J. Big Data **8**, 1–74 (2021)
50. N. Shlezinger, Y.C. Eldar, S.P. Boyd, Model-based deep learning: on the intersection of deep learning and optimization. IEEE Access **10**, 115384–115398 (2022)
51. A. Boroumand, S. Ghose, B. Akin, R. Narayanaswami, G.F. Oliveira, X. Ma, E. Shiu, O. Mutlu, Google neural network models for edge devices: analyzing and mitigating machine learning inference bottlenecks, in *2021 30th International Conference on Parallel Architectures and Compilation Techniques (PACT)* (IEEE, 2021), pp. 159–172
52. J. Lee, N. Chirkov, E. Ignasheva, Y. Pisarchyk, M. Shieh, F. Riccardi, R. Sarokin, A. Kulik, M. Grundmann, On-device neural net inference with mobile gpus (2019). arXiv preprint arXiv: 1907.01989.
53. Q. Guo, S. Chen, X. Xie, L. Ma, Q. Hu, H. Liu, Y. Liu, J. Zhao, X. Li, An empirical study towards characterizing deep learning development and deployment across different frameworks and platforms, in *2019 34th IEEE/ACM International Conference on Automated Software Engineering (ASE)* (IEEE, 2019), pp. 810–822
54. P. Pochelu, *Deep Learning Inference Frameworks Benchmark* (2022). arXiv preprint arXiv: 2210.04323
55. H. Tabani, R. Pujol, J. Abella, F.J. Cazorla, A cross-layer review of deep learning frameworks to ease their optimization and reuse, in *2020 IEEE 23rd International Symposium on Real-Time Distributed Computing (ISORC)* (IEEE, 2020), pp. 144–145

56. S. Tavarageri, G. Goyal, S. Avancha, B. Kaul, R. Upadrasta, *AI Powered Compiler Techniques for DL Code Optimization* (2021). arXiv preprint arXiv:2104.05573
57. S. Chetlur, C. Woolley, P. Vandermersch, J. Cohen, J. Tran, B. Catanzaro, E. Shelhamer, cudnn: Efficient primitives for deep learning (2014). arXiv preprint arXiv:1410.0759
58. A. Reuther, P. Michaleas, M. Jones, V. Gadepally, S. Samsi, J. Kepner, Survey of machine learning accelerators, in *2020 IEEE High Performance Extreme Computing Conference (HPEC)* (IEEE, 2020), pp. 1–12
59. J. Li, Z. Qin, Y. Mei, J. Cui, Y. Song, C. Chen, Y. Zhang, L. Du, X. Cheng, B. Jin, D. Lavery, oneDNN graph compiler: a hybrid approach for high-performance deep learning compilation, in *2024 IEEE/ACM International Symposium on Code Generation and Optimization (CGO)* (IEEE, 2024), pp. 460–470
60. L. Tang, Y. Wang, T.L. Willke, K. Li, Scheduling computation graphs of deep learning models on manycore cpus (2018). arXiv preprint arXiv:1807.09667
61. A.V. Kumar, M. Sivathanu, Quiver: an informed storage cache for deep learning, in *18th USENIX Conference on File and Storage Technologies (FAST 20)* (2020), pp. 283–296
62. *XLA architecture.* (n.d.). OpenXLA Project. https://openxla.org/xla/architecture
63. T. Chen, T. Moreau, Z. Jiang, L. Zheng, E. Yan, H. Shen, M. Cowan, L. Wang, Y. Hu, L. Ceze, A. Krishnamurthy, {TVM}: an automated {End-to-End} optimizing compiler for deep learning, in *13th USENIX Symposium on Operating Systems Design and Implementation (OSDI 18)* (2018), pp. 578–594
64. N.A. Andriyanov, Analysis of the acceleration of neural networks inference on Intel processors based on openvino toolkit, in *2020 Systems of Signal Synchronization, Generating and Processing in Telecommunications (SYNCHROINFO)* (IEEE, 2020), pp. 1–5
65. Y. Liu, Y. Wang, R. Yu, M. Li, V. Sharma, Y. Wang, Optimizing {CNN} model inference on {CPUs}, in *2019 USENIX Annual Technical Conference (USENIX ATC 19)* (2019), pp. 1025–1040
66. M. Kirisame, S. Lyubomirsky, A. Haan, J. Brennan, M. He, J. Roesch, T. Chen, Z. Tatlock, Dynamic tensor rematerialization (2020). arXiv preprint arXiv:2006.09616
67. A. Ashari, S. Tatikonda, M. Boehm, B. Reinwald, K. Campbell, J. Keenleyside, P. Sadayappan, On optimizing machine learning workloads via kernel fusion. ACM SIGPLAN Notices **50**(8), 173–182 (2015)
68. J. Liu, D. Li, G. Kestor, J. Vetter, Runtime concurrency control and operation scheduling for high performance neural network training, in *2019 IEEE International Parallel and Distributed Processing Symposium (IPDPS)* (IEEE, 2019), pp. 188–199
69. M. Levental, Memory planning for deep neural networks (2022). arXiv preprint arXiv:2203.00448
70. J. Mohan, A. Phanishayee, A. Raniwala, V. Chidambaram, Analyzing and mitigating data stalls in DNN training (2020). arXiv preprint arXiv:2007.06775
71. Y. Pisarchyk, J. Lee, Efficient memory management for deep neural net inference (2020). arXiv preprint arXiv:2001.03288
72. N. Ström, Scalable distributed DNN training using commodity GPU cloud computing (2015)
73. A. Krizhevsky, One weird trick for parallelizing convolutional neural networks (2014). arXiv preprint arXiv:1404.5997
74. J. Dean, G. Corrado, R. Monga, K. Chen, M. Devin, M. Mao, M.A. Ranzato, A. Senior, P. Tucker, K. Yang, A. Ng, Large scale distributed deep networks. Adv. Neural Inf. Process. Syst. **25** (2012)
75. J. Hagemann, S. Weinbach, K. Dobler, M. Schall, G. de Melo, *Efficient Parallelization Layouts for Large-Scale Distributed Model Training* (2023). arXiv preprint arXiv:2311.05610
76. K. Nagrecha, Systems for parallel and distributed large-model deep learning training (2023). arXiv preprint arXiv:2301.02691.
77. W. Xu, Y. Zhang, X. Tang, Parallelizing DNN training on GPUs: Challenges and opportunities, in *Companion Proceedings of the Web Conference 2021* (2021), pp. 174–178
78. Z. Tang, S. Shi, W. Wang, B. Li, X. Chu, Communication-efficient distributed deep learning: a comprehensive survey (2020). arXiv preprint arXiv:2003.06307

79. J. Wang, G. Joshi, Adaptive communication strategies to achieve the best error-runtime trade-off in local-update SGD. Proc. Mach. Learn. Syst. **1**, 212–229 (2019)
80. J. Chen, X. Pan, R. Monga, S. Bengio, R. Jozefowicz, Revisiting distributed synchronous SGD (2016). arXiv preprint arXiv:1604.00981
81. J. Konečný, H.B. McMahan, D. Ramage, P. Richtárik, Federated optimization: Distributed machine learning for on-device intelligence (2016). arXiv preprint arXiv:1610.02527
82. J. Konečný, *Federated Learning: Strategies for Improving Communication Efficiency* (2016). arXiv preprint arXiv:1610.05492
83. B. McMahan, E. Moore, D. Ramage, S. Hampson, B.A. Arcas, Communication-efficient learning of deep networks from decentralized data, in *Artificial Intelligence and Statistics* (PMLR, 2017), pp. 1273–1282
84. C. Shorten, T.M. Khoshgoftaar, A survey on image data augmentation for deep learning. J. Big Data **6**(1), 1–48 (2019)
85. S. Yun, D. Han, S.J. Oh, S. Chun, J. Choe, Y. Yoo, Cutmix: Regularization strategy to train strong classifiers with localizable features, in *Proceedings of the IEEE/CVF International Conference on Computer Vision* (2019), pp. 6023–6032
86. H. Zhang, mixup: Beyond empirical risk minimization (2017). arXiv preprint arXiv:1710.09412
87. C. Khosla, B.S. Saini, Enhancing performance of deep learning models with different data augmentation techniques: a survey, in *2020 International Conference on Intelligent Engineering and Management (ICIEM)* (IEEE, 2020), pp. 79–85
88. M. Bayer, M.A. Kaufhold, C. Reuter, A survey on data augmentation for text classification. ACM Comput. Surv. **55**(7), 1–39 (2022)
89. S. Kobayashi, Contextual augmentation: data augmentation by words with paradigmatic relations (2018). arXiv preprint arXiv:1805.06201
90. C. Shorten, T.M. Khoshgoftaar, B. Furht, Text data augmentation for deep learning. J. Big Data **8**(1), 101 (2021)
91. Z. Tang, Y. Gao, L. Karlinsky, P. Sattigeri, R. Feris, D. Metaxas, OnlineAugment: Online data augmentation with less domain knowledge, in *Computer Vision–ECCV 2020: 16th European Conference, Glasgow, UK, August 23–28, 2020, Proceedings, Part VII 16* (Springer International Publishing, 2020), pp. 313–329
92. M. Gupta, P. Agrawal, Compression of deep learning models for text: a survey. ACM Trans. Knowl. Discov. Data (TKDD) **16**(4), 1–55 (2022)
93. T. Jenrungrot, M. Chinen, W.B. Kleijn, J. Skoglund, Z. Borsos, N. Zeghidour, M. Tagliasacchi, Lmcodec: a low bitrate speech codec with causal transformer models, in *ICASSP 2023–2023 IEEE International Conference on Acoustics, Speech and Signal Processing (ICASSP)* (IEEE, 2023), pp. 1–5
94. Y. Gong, L. Liu, M. Yang, L. Bourdev, Compressing deep convolutional networks using vector quantization (2014). arXiv preprint arXiv:1412.6115
95. R. Shu, H. Nakayama, Compressing word embeddings via deep compositional code learning (2017). arXiv preprint arXiv:1711.01068
96. J. Ballé, Efficient nonlinear transforms for lossy image compression, in *2018 Picture Coding Symposium (PCS)* (IEEE, 2018), pp. 248–252
97. M. Goyal, K. Tatwawadi, S. Chandak, I. Ochoa, Deepzip: Lossless data compression using recurrent neural networks (2018). arXiv preprint arXiv:1811.08162
98. M. Yu, L. Zhang, K. Ma, Revisiting data augmentation in model compression: an empirical and comprehensive study, in *2023 International Joint Conference on Neural Networks (IJCNN)* (IEEE, 2023), pp. 1–10
99. L. Fan, F. Zhang, H. Fan, C. Zhang, Brief review of image denoising techniques. Vis. Comput. Ind., Biomed., Art **2**(1), 7 (2019)
100. K. Killamsetty, D. Sivasubramanian, G. Ramakrishnan, R. Iyer, Glister: Generalization based data subset selection for efficient and robust learning, in *Proceedings of the AAAI Conference on Artificial Intelligence*, vol. 35, no. 9 (2021), pp. 8110–8118

101. B. Sorscher, R. Geirhos, S. Shekhar, S. Ganguli, A. Morcos, Beyond neural scaling laws: beating power law scaling via data pruning. Adv. Neural. Inf. Process. Syst. **35**, 19523–19536 (2022)
102. D. Karimi, H. Dou, S.K. Warfield, A. Gholipour, Deep learning with noisy labels: exploring techniques and remedies in medical image analysis. Med. Image Anal. **65**, 101759 (2020)
103. K. Lee, H. Lee, K. Lee, J. Shin, Training confidence-calibrated classifiers for detecting out-of-distribution samples (2017). arXiv preprint arXiv:1711.09325
104. W. Ma, M. Papadakis, A. Tsakmalis, M. Cordy, Y.L. Traon, Test selection for deep learning systems. ACM Trans. Softw. Eng. Methodol. (TOSEM) **30**(2), 1–22 (2021)
105. M. Sachan, E. Xing, Easy questions first? A case study on curriculum learning for question answering, in *Proceedings of the 54th Annual Meeting of the Association for Computational Linguistics*, vol. 1, ed. by K. Erk, N.A. Smith Long Papers (2016), pp. 453–463. https://doi.org/10.18653/v1/P16-1043
106. M. Kumar, B. Packer, D. Koller, Self-paced learning for latent variable models. Adv. Neural Inf. Process. Syst. **23** (2010)
107. Q. Jia, Y. Liu, H. Tang, K.Q. Zhu, In-sample curriculum learning by sequence completion for natural language generation (2022). arXiv preprint arXiv:2211.11297
108. C. Coleman, C. Yeh, S. Mussmann, B. Mirzasoleiman, P. Bailis, P. Liang, J. Leskovec, M. Zaharia, Selection via proxy: Efficient data selection for deep learning (2019). arXiv preprint arXiv:1906.11829
109. Y. Yang, H. Kang, B. Mirzasoleiman, Towards sustainable learning: coresets for data-efficient deep learning, in *International Conference on Machine Learning* (PMLR, 2023), pp. 39314–39330
110. I.H. Sarker, Deep learning: a comprehensive overview on techniques, taxonomy, applications and research directions. SN Comput. Sci. **2**(6), 420 (2021)
111. Y. Cheng, D. Wang, P. Zhou, T. Zhang, A survey of model compression and acceleration for deep neural networks (2017). arXiv preprint arXiv:1710.09282
112. Q. Qin, J. Ren, J. Yu, H. Wang, L. Gao, J. Zheng, Y. Feng, J. Fang, Z. Wang, To compress, or not to compress: characterizing deep learning model compression for embedded inference, in *2018 IEEE Intl Conf on Parallel & Distributed Processing with Applications, Ubiquitous Computing & Communications, Big Data & Cloud Computing, Social Computing & Networking, Sustainable Computing & Communications (ISPA/IUCC/BDCloud/SocialCom/SustainCom)* (IEEE, 2018), pp. 729–736
113. A. Gholami, S. Kim, Z. Dong, Z. Yao, M.W. Mahoney, K. Keutzer, A survey of quantization methods for efficient neural network inference, in *Low-Power Computer Vision* (Chapman and Hall/CRC, 2022), pp. 291–326
114. T. Liang, J. Glossner, L. Wang, S. Shi, X. Zhang, Pruning and quantization for deep neural network acceleration: a survey. Neurocomputing **461**, 370–403 (2021)
115. M. Nagel, M. Fournarakis, R.A. Amjad, Y. Bondarenko, M. Van Baalen, T. Blankevoort, A white paper on neural network quantization (2021). arXiv preprint arXiv:2106.08295.
116. K. Wang, Z. Liu, Y. Lin, J. Lin, S. Han, Haq: hardware-aware automated quantization with mixed precision, in *Proceedings of the IEEE/CVF Conference on Computer Vision and Pattern Recognition* (2019), pp. 8612–8620
117. J. Liu, S. Tripathi, U. Kurup, M. Shah, Pruning algorithms to accelerate convolutional neural networks for edge applications: a survey (2020). arXiv preprint arXiv:2005.04275
118. H. Cheng, M. Zhang, J.Q. Shi, A survey on deep neural network pruning: taxonomy, comparison, analysis, and recommendations. IEEE Trans. Pattern Anal. Mach. Intell. (2024)
119. M. Zhu, S. Gupta, To prune, or not to prune: exploring the efficacy of pruning for model compression (2017). arXiv preprint arXiv:1710.01878
120. G. Hinton, *Distilling the Knowledge in a Neural Network* (2015). arXiv preprint arXiv:1503.02531
121. J. Gou, B. Yu, S.J. Maybank, D. Tao, Knowledge distillation: a survey. Int. J. Comput. Vision **129**(6), 1789–1819 (2021)

122. J.H. Cho, B. Hariharan, On the efficacy of knowledge distillation, in *Proceedings of the IEEE/CVF International Conference on Computer Vision* (2019), pp. 4794–4802
123. T. Elsken, J.H. Metzen, F. Hutter, Neural architecture search: a survey. J. Mach. Learn. Res. **20**(55), 1–21 (2019)
124. C. White, M. Safari, R. Sukthanker, B. Ru, T. Elsken, A. Zela, D. Dey, F. Hutter, Neural architecture search: Insights from 1000 papers (2023). arXiv preprint arXiv:2301.08727
125. M. Wistuba, A. Rawat, T. Pedapati, A survey on neural architecture search (2019). arXiv preprint arXiv:1905.01392
126. X. He, K. Zhao, X. Chu, AutoML: a survey of the state-of-the-art. Knowl.-Based Syst. **212**, 106622 (2021)
127. X. Dong, D.J. Kedziora, K. Musial, B. Gabrys, Automated deep learning: neural architecture search is not the end. Found. Trends® Mach. Learn. **17**(5), 767–920 (2024)
128. M. Torkamani, P. Wallis, S. Shankar, A. Rooshenas, Learning compact neural networks using ordinary differential equations as activation functions (2019). arXiv preprint arXiv:1905.07685
129. C. Surianarayanan, J.J. Lawrence, P.R. Chelliah, E. Prakash, C. Hewage, A survey on optimization techniques for edge artificial intelligence (AI). Sensors **23**(3), 1279 (2023)
130. T. Li, A.K. Sahu, A. Talwalkar, V. Smith, Federated learning: challenges, methods, and future directions. IEEE Signal Process. Mag. **37**(3), 50–60 (2020)
131. H.B. McMahan, E. Moore, D. Ramage, B.A. Arcas, Federated learning of deep networks using model averaging. **2**(2) (2016). arXiv preprint arXiv:1602.05629
132. A.D.A. Garcez, L.C. Lamb, Neurosymbolic AI: the 3rd wave. Artif. Intell. Rev. **56**(11), 12387–12406 (2023)
133. M. Arunachalam, V. Sanghavi, Y.A. Yao, Y.A. Zhou, L.A. Z. Wang, N. Wen, Ammbashankar, N.W. Wang, F. Mohammad, Strategies for optimizing end-to-end artificial intelligence pipelines on Intel Xeon processors (2022). arXiv preprint arXiv:2211.00286
134. D. Xin, H. Miao, A. Parameswaran, N. Polyzotis, Production machine learning pipelines: Empirical analysis and optimization opportunities, in *Proceedings of the 2021 International Conference on Management of Data* (2021), pp. 2639–2652
135. Y. Zhou, X. Dong, B. Akin, M. Tan, D. Peng, T. Meng, A. Yazdanbakhsh, D. Huang, R. Narayanaswami, J. Laudon, Rethinking co-design of neural architectures and hardware accelerators (2021). arXiv preprint arXiv:2102.08619
136. J.R. Hu, J. Chen, B.K. Liew, Y. Wang, L. Shen, L. Cong, Systematic co-optimization from chip design, process technology to systems for GPU AI chip, in *2018 International Symposium on VLSI Design, Automation and Test (VLSI-DAT)* (IEEE, 2018), pp. 1–2
137. F. Faghri, H. Pouransari, S. Mehta, M. Farajtabar, A. Farhadi, M. Rastegari, O. Tuzel, Reinforce data, multiply impact: improved model accuracy and robustness with dataset reinforcement, in *Proceedings of the IEEE/CVF International Conference on Computer Vision* (2023) (pp. 17032–17043)
138. Z. Jia, J. Thomas, T. Warszawski, M. Gao, M. Zaharia, A. Aiken, Optimizing DNN computation with relaxed graph substitutions. Proc. Mach. Learn. Syst. **1**, 27–39 (2019)
139. W. Niu, J. Guan, Y. Wang, G. Agrawal, B. Ren, Dnnfusion: accelerating deep neural networks execution with advanced operator fusion, in *Proceedings of the 42nd ACM SIGPLAN International Conference on Programming Language Design and Implementation* (2021), pp. 883–898
140. D.J. Kedziora, K. Musial, B. Gabrys, Autonoml: towards an integrated framework for autonomous machine learning (2020). arXiv preprint arXiv:2012.12600
141. Z. Shi, C. Sakhuja, M. Hashemi, K. Swersky, C. Lin, Learned hardware/software co-design of neural accelerators (2020). arXiv preprint arXiv:2010.02075
142. Y. Yu, M. Abadi, P. Barham, E. Brevdo, M. Burrows, A. Davis, J. Dean, S. Ghemawat, T. Harley, P. Hawkins, X. Zheng, Dynamic control flow in large-scale machine learning, in *Proceedings of the Thirteenth EuroSys Conference* (2018), pp. 1–15
143. N. Jouppi, G. Kurian, S. Li, P. Ma, R. Nagarajan, L. Nai, N. Patil, S. Subramanian, A. Swing, B. Towles, D.A. Patterson, Tpu v4: An optically reconfigurable supercomputer for

machine learning with hardware support for embeddings, in *Proceedings of the 50th Annual International Symposium on Computer Architecture* (2023), pp. 1–14
144. A. Getmanskaya, DL-inferencing for 3D Cephalometric Landmarks Regression task using OpenVINO (2020)
145. H. Tabani, A. Balasubramaniam, E. Arani, B. Zonooz, Challenges and obstacles towards deploying deep learning models on mobile devices (2021). arXiv preprint arXiv:2105.02613.
146. C. Hao, J. Dotzel, J. Xiong, L. Benini, Z. Zhang, D. Chen, Enabling design methodologies and future trends for edge AI: specialization and codesign. IEEE Design & Test **38**(4), 7–26 (2021)
147. V. Dunjko, J.M. Taylor, H.J. Briegel, Quantum-enhanced machine learning. Phys. Rev. Lett. **117**(13), 130501 (2016)
148. L. Buffoni, F. Caruso, New trends in quantum machine learning (a). Europhys. Lett. **132**(6), 60004 (2021)
149. M. Cerezo, G. Verdon, H.Y. Huang, L. Cincio, P.J. Coles, Challenges and opportunities in quantum machine learning. Nat. Comput. Sci. **2**(9), 567–576 (2022)
150. M. Davies, A. Wild, G. Orchard, Y. Sandamirskaya, G.A.F. Guerra, P. Joshi, P. Plank, S.R. Risbud, Advancing neuromorphic computing with loihi: a survey of results and outlook. Proc. IEEE **109**(5), 911–934 (2021)
151. A. Mehonic, A.J. Kenyon, Brain-inspired computing needs a master plan. Nature **604**(7905), 255–260 (2022)
152. A. Zador, S. Escola, B. Richards, B. Ölveczky, Y. Bengio, K. Boahen, M. Botvinick, D. Chklovskii, A. Churchland, C. Clopath, D. Tsao, Toward next-generation artificial intelligence: Catalyzing the neuroai revolution (2022). arXiv preprint arXiv:2210.08340
153. S.S. Gill, M. Xu, C. Ottaviani, P. Patros, R. Bahsoon, A. Shaghaghi, M. Golec, V. Stankovski, H. Wu, A. Abraham, S. Uhlig, AI for next generation computing: emerging trends and future directions. IoT **19**, 100514 (2022)
154. N.J. Yadwadkar, F. Romero, Q. Li, C. Kozyrakis, A case for managed and model-less inference serving, in *Proceedings of the Workshop on Hot Topics in Operating Systems* (2019), pp. 184–191
155. R.Y. Aminabadi, S. Rajbhandari, A.A. Awan, C. Li, D. Li, E. Zheng, J. Rasley, Y. He, Deepspeed-inference: enabling efficient inference of transformer models at unprecedented scale, in *SC22: International Conference for High Performance Computing, Networking, Storage and Analysis* (IEEE, 2022), pp. 1–15
156. L. Lundberg, D. Häggander, W. Diestelkamp, Conflicts and trade-offs between software performance and maintainability, in *International Workshop on Software and Performance* (Springer Berlin Heidelberg, Berlin, Heidelberg, 2000), pp. 56–67
157. Z. Durdik, B. Klatt, H. Koziolek, K. Krogmann, J. Stammel, R. Weiss, Sustainability guidelines for long-living software systems, in *2012 28th IEEE International Conference on Software Maintenance (ICSM)* (IEEE, 2012) pp. 517–526
158. S. Krishnan, A.J. Elmore, M. Franklin, J. Paparrizos, Z. Shang, A. Dziedzic, R. Liu, Artificial intelligence in resource-constrained and shared environments. ACM SIGOPS Oper. Syst. Rev. **53**(1), 1–6 (2019)

Chapter 13
Opportunity Discovery for Effective Innovation Using Artificial Intelligence

Krunal Padwekar, Kanchan Awasthi, and Subhas Chandra Misra

Abstract To confront and alleviate challenges, innovation holds great importance as it confers a competitive advantage upon entities such as businesses and nations. Due to the vast array of problems, it may be difficult to tackle all of them. Some of these challenges have already been addressed by a multitude of individuals, while others have received less attention. Currently, there have been numerous advancements in the domain of text analysis, bolstered by Large Language Models. These models facilitate the retrieval and examination of huge amounts of textual data, yielding in the creation of new insights. Consequently, these models aid in the identification of existing problems and shed light on overlooked issues. As a result, innovators can explore novel avenues in both business and technological innovation. This chapter delves into the present progress and state-of-the-art technologies pertaining to the discovery of opportunities in innovation. The existing literature on opportunities encompasses two perspectives: opportunity creation and opportunity discovery. Opportunity creation is subjective, relying on the perception of researchers or entrepreneurs. On the other hand, opportunity discovery involves the analysis of the limitations of the current state-of-the-art. While qualitative analysis techniques have traditionally been employed in these studies, the availability of new text analytics methods and tools accelerates the process of uncovering new opportunities in both business and technology. This can lead to effective innovations that may address previously overlooked challenges.

Keywords Opportunity discovery · Opportunity creation · Artificial intelligence · Text analytics

K. Padwekar (✉) · K. Awasthi · S. C. Misra
Department of Management Sciences, Indian Institute of Technology Kanpur, Kanpur, India
e-mail: krunalp@iitk.ac.in

T. R. Benala et al. (eds.), *Boosting Software Development Using Machine Learning*, Artificial Intelligence-Enhanced Software and Systems Engineering 7,
https://doi.org/10.1007/978-3-031-88188-6_13

13.1 Introduction

Businesses are identifying opportunities in products, services, technologies, and operations to outperform competitors. They analyze existing products and activities to identify gaps and fill them with unique products or technologies. Opportunity discovery is the process of identifying latent potential and converting it into tangible, profitable ventures, which is often overlooked by other [1]. While opportunity creation involves finding or discovering untapped possibilities, looking beyond existing constraints, and transforming them into beneficial solutions [1].

Opportunity discovery is a crucial process for business growth and success [2], identifying trends and potential areas for innovation [3]. Originating in entrepreneurship, it involves identifying new markets and niches for expansion and profitability [4]. This process is essential for strategic planning and gaining a competitive edge in fields like business, technology, and environmental management [5, 6].

Traditional methods are often limited by restrictive criteria and lack of empirical evidence. AI has significantly improved opportunity identification processes by automating and expediting the analysis of large, complex data sets [7]. Techniques like natural language processing and machine learning are used to systematically evaluate options and identify promising opportunities [8, 9]. AI algorithms are also used in competitor analysis to extract and prioritize competitive factors, thereby expanding the scope of opportunity discovery [10, 11].

This study has covered the existing studies on AI and opportunity discovery to reveal the potential of opportunity discovery in the segment of technology opportunity, product and service opportunity and business opportunity. The authors can consider this study as a supportive tool for carrying out research in the field of opportunity discovery and make exemplar contribution to the field.

The article is formatted as follows: first we described the concept of opportunity and opportunity discovery in Sect. 13.2. Section 13.3 of this paper provides a detailed explanation of the methodology used and presents a complete description of the studies conducted in the areas of "technological opportunity", "product and service opportunity", and "business opportunity". Finally, limitations and future research is presented in Sect. 13.4 and conclusion in Sect. 13.5.

13.2 Background

13.2.1 Opportunity as a Concept

Opportunity is a favorable circumstance that allows actions, not generated by humans but by factors like expertise, technology, customers, and socio-political and economic conditions [12]. The literature on opportunities focuses on two perspectives: opportunity creation and discovery, highlighting how potential is realized in individuals' minds through cognition [1]. Opportunity creation is a subjective process involving experimentation and prototype development, while opportunity discovery analyzes limitations in technology or business. These approaches complement each other, enabling companies to identify gaps, develop innovative solutions, and stay ahead of competitors.

13.2.2 Opportunity Discovery

Opportunity discovery is a crucial concept in entrepreneurship, strategic management, and innovation, recognizing hidden potential in a specific environment, based on the belief that opportunities are objective realities [4]. There are two perspectives that exist in the literature related to opportunity discovery *Schumpeterian perspective (Schumpeterian (1934), as cited in De Jong and Marsili, 2010) and Kirznerian view (Kirznerian (1973), as cited in De Jong and Marsili, 2010).* In the *Kirznerian* perspective, innovation and novel combinations are not deemed essential prerequisites. Neither do such prospects necessitate macroeconomic transformations linked to fresh technologies or societal shifts. Instead, the entrepreneur is characterized as an individual who gains from information imbalances within established markets. *Kirzner's (1973)* pivotal attribute is the assertion that entrepreneurs possess the ability to identify prospects for entrepreneurial gains. In the *Schumpeterian* perspective, the entrepreneur is portrayed as the catalyst of change through inventive practices and actively establishing novel prospects. *Schumpeter's (1934)* fundamental concept revolves around innovation denoted as 'new combinations'. Hence, the entrepreneur is characterized as an individual who formulates a fresh combination and endeavours to implement it in the market. *Schumpeter* contended that shifts in technology, political dynamics, governmental regulations, macroeconomic elements, and societal patterns yield new data that entrepreneurs can leverage to devise strategies for reorganizing resources into more advantageous configurations.

Opportunity discovery is a systematic approach to understanding the market environment, identifying market voids, unfulfilled consumer needs, and inefficiencies. It involves recognizing hidden potential, converting these into profitable ventures, stimulating innovation, and nurturing entrepreneurial achievements.

13.3 Methodology

This study relies on a qualitative approach to understand the existing studies on opportunity discovery. For this objective, the current literature on opportunity and innovation is explored. Two dimensions of opportunity are identified in the extant studies: "opportunity discovery" and "opportunity creation" [13]. Since, the focus of the study is to comprehend opportunity discovery, the articles related to opportunity discovery and innovations were separated and analysed to generate insights. The visual representation of the followed approach is shown in Fig. 13.1.

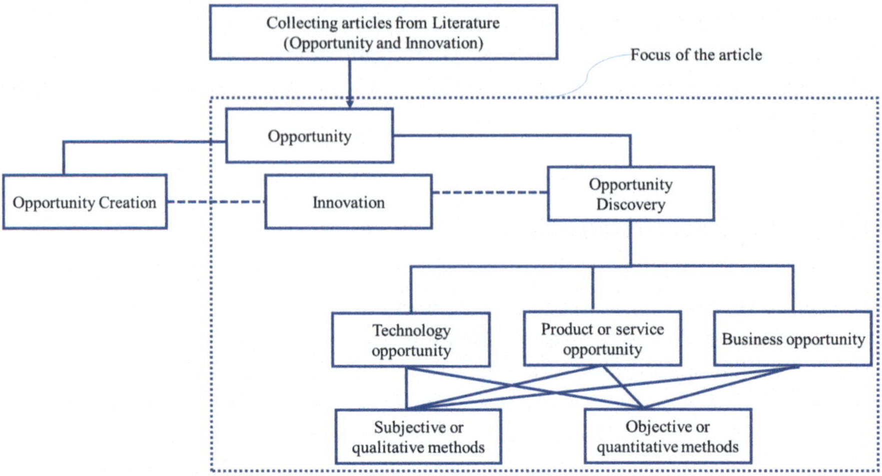

Fig. 13.1 Research approach

13.3.1 Analysing Articles on Opportunity Discovery

The existing articles on opportunity discovery mainly focuses on three dimensions: (1) technology opportunity (2) product and service opportunity (3) business opportunity. The approach followed to recognize these three opportunities are either subjective or objective. Subjective approach is mainly intuition driven or cognitive framework based whereas objective approach mainly relies on using novel artificial intelligence and machine learning approach such as text mining, natural language processing, etc. [7].

13.3.1.1 Technology Opportunity

Technology opportunity is a crucial factor in business growth and innovation in the rapidly evolving digital environment, indicating the likelihood of technological advancement in a specific field [14]. Technology Intelligence and Technology Opportunity Discovery (TOD) are methods used to track technological advancements and predict their path in various contexts, enabling businesses to identify new opportunities and pathways for sustainable business expansion [14]. The use of Artificial Intelligence (AI) in TOD has significantly improved the identification and seizing of technological opportunities, enabling enterprises to forecast future trends and consumer preferences with precision. Some of the studies in the intersection of TOD and AI are presented in Table 13.1.

Table 13.1 Existing studies on TOD and AI algorithms

Authors	Data source	Method	Findings
[15]	Patents data from several technical domain	The research introduced a function-oriented framework for four TOD trajectories, based on firms' existing technologies and products, which was applied through practical TOD instances	This approach identifies new opportunities by presenting TOD paths, including existing technology-to-modified, existing technology-to-adaptable, existing product-to-modified, and existing product-to-adaptable technology
[16]	U.S. patents related to digital information and security	A new method for identifying patent outliers is proposed, focusing on minimal clustering, limited centrality within the citation network, and reduced similarity to surrounding patents	Developed a patent clustering algorithm, and a methodology for evaluating and ranking patent outliers
[17]	U.S. patents related to information and security	A selected group of patent pairs are identified as potential convergence candidates based on their centrality within the Patent Citation Network and their disparity in conceptual linkages	This article presents a novel approach to detecting outliers at the edges in the context of identifying technology opportunities arising from the convergence of existing patents
[14]	USPTO online patent database	The study employs a future-oriented approach to predict new Technological Knowledge Flows (TKFs) across various fields, utilizing link forecasting within a directed network for convergence	The process of identifying converging technological opportunities involves identifying potential enhancements in knowledge flow links and the emergence of new connections
[6]	User reviews and patents data	'Opinion trigger' concept is used to get user needs for TOD. It analyses user needs into object and function, removing the term gap between users and technicians	The study identifies two technology opportunities: a gap between technology ability and user sentiment, and a technology with low ability to meet user needs

(continued)

Table 13.1 (continued)

Authors	Data source	Method	Findings
[18]	Derwent Patent Database	Patents are utilized to create Domain Knowledge Networks (DKNs) and analyze technological terms, conceptualizing opportunities as interconnected subnetworks within DKNs, using regression analysis and ant colony optimization	This study provides a thorough analysis of key methodological issues in TOD, focusing on the creation of knowledge frameworks and the evaluation of technological advancements
[8]	Technology text data and new technology-based firms (NTBF) technology categorization data	A novel framework for TOD that leverages deep learning and knowledge graph utilizing three primary data sources: technology, (NTBF), and investor data	The research introduced a functional technology intelligence system to provide TOD indices, establishing a foundation for TOD in presenting information on NTBF and investment
[19]	USPTO data	It constructs a technology ecology, uses a link prediction model, and ranks candidates using 13 quantitative indexes	The study investigates the impact of internal capability and external context on the adoption of technology

13.3.1.2 Product or Service Opportunity

Businesses must identify emerging trends, consumer preferences, and technological advancements to enhance market presence, boost profitability, and maintain relevance in the evolving marketplace, fostering innovation and competitiveness. Researchers have explored product or service opportunity exploration, emphasizing the importance of knowledge, cognitive mechanisms, and social connections. They used AI technology to identify new business prospects and make data-informed decisions for business prosperity. Some of the studies in the intersection of product and service opportunity and AI are presented in Table 13.2.

13.3.1.3 Business Opportunity

A business opportunity is a potential avenue for organizations to introduce new products or enhance existing ones, aiming to meet consumer needs, address market gaps, and stimulate business growth [1]. Researchers utilized AI technology to identify new business opportunities and develop data-driven strategies, enhancing competitiveness by analyzing complex datasets and making real-time decisions. Some of the studies in the intersection of product and service opportunity and AI are presented in Table 13.3.

Table 13.2 Existing studies on product and service opportunity and AI algorithms

Authors	Data source	Method	Findings
[20]	Business method patents in USPTO database	The use of Artificial Intelligence (AI) in TOD has significantly improved the identification and seizing of technological opportunities, enabling enterprises to forecast future trends and consumer preferences with precision	This study presents a practical approach to creating innovative mobile services by utilizing extensive patent data and assisting service designers in the creation process
[21]	Expert group rated service concepts (NSC) based on new service development (NSD) criteria	Morphology analysis is a method used to analyze service concepts by examining morphology matrix combinations, while genetic algorithms serve as an optimization engine and screening tool	The tool, developed using relational database and Java, aids service designers in selecting suitable service concepts, addressing regulatory violations, market competition, and resource unavailability
[22]	Business method patents data	It uses network analysis and cross-impact analysis to create a portfolio map, focusing on IT and mobile services	The study proposes a technology-centered approach for managing opportunities in analyzing patent classification, knowledge flow, cross-impact, and creating a portfolio map to explore these opportunities
[23]	USPTO database	It used text mining to extract product information from a patent database, generate product connection rules, and evaluate potential value	This paper presents a systematic method for identifying potential product opportunities within a firm, aiming to facilitate product-oriented R&D
[24]	USPTO database	Text mining, LDA, and collaborative filtering is applied to generate assignee-product portfolio vectors and visually map recommended products	The approach provides a visual recommendation map, classifying recommended products based on product heterogeneity and promise
[25]	Business method patents in USPTO database	Patent co-classification, ANP, data envelopment analysis- window analysis, and a portfolio map to manage opportunities is applied	This study offers managerial insights for establishing technology-based service strategies or policies for firms or nations

13.4 Limitations and Scope for Further Research

This study explores AI algorithms for opportunity discovery but has limitations due to methodological limitations and lack of non-AI-based articles. Improvements include presenting results from all methods, using appropriate protocols like PRISMA and SPAR-4-SLR, and using AI techniques like Latent Dirichlet Allocation, Latent Semantic Allocation, etc. [29].

Table 13.3 Existing studies on business opportunity and AI algorithms

Authors	Data source	Method	Findings
[11]	Patent data and trademark data	It employed collaborative filtering, portfolio analyses, and association mining techniques to analyse data	This study discovers new business opportunities using competitor intelligence
[26]	Patents and merger and acquisition (M&A) cases	The method uses IPC-ISIC concordance and is verified using a successful M&A case	The paper proposes a method for identifying business diversification opportunities through M&A cases, suggesting that similar industrial sectors acquired from similar companies can serve as diversification directions
[27]	KIPO trademark data	The study employs deep link prediction and competitive intelligence analysis to identify potential opportunities	The model uses co-occurrences to discover and establish diversification strategies
[1]	USPTO patents database	The study uses a local outlier factor to identify emerging areas and provides quantitative outcomes, focusing on apparatus trademarks	This study defines a business landscape as a vector space model, identifying opportunities through novelty assessment and map development
[28]	Experts' data using convenience sampling	Quantitative analysis using SPSS	The testing of technologies like Digital Twins, AI, and Blockchain-tokens reveals potential in customer relations, international transactions, home loans, and sustainable financing

13.5 Conclusion

Opportunity discovery is crucial for effective innovation, and AI can help uncover hidden growth potentials. By analyzing market trends, consumer behavior, and competitive environments, AI can identify unexplored markets and anticipate future trends with precision. This accuracy leads to sustainable business expansion, increased profitability, and a competitive advantage. Integrating AI technology in corporate activities can enhance decision-making processes and efficiency.

References

1. J. Choi, B. Jeong, J. Yoon, Identification of emerging business areas for business opportunity analysis: an approach based on language model and local outlier factor. Comput. Ind. **140**, 103677 (2022)
2. L. Martin, N. Wilson, Opportunity, discovery and creativity: a critical realist perspective. Int. Small Bus. J. **34**, 261 (2016)
3. T.H. Chiles, D.M. Vultee, V.K. Gupta, D.W. Greening, C.S. Tuggle, The philosophical foundations of a radical Austrian approach to entrepreneurship. J. Manag. Inq. **19**, 138 (2010)
4. P.G. Klein, Opportunity discovery, entrepreneurial action, and economic organization. Strateg. Entrep. J. **2**, 175 (2008)
5. E. Jeon, N. Yoon, S.Y. Sohn, Exploring new digital therapeutics technologies for psychiatric disorders using BERTopic and PatentSBERTa. Technol. Forecast. Soc. Chang. **186**, 122130 (2023)
6. T. Roh, Y. Jeong, H. Jang, B. Yoon, Technology opportunity discovery by structuring user needs based on natural language processing and machine learning. PLoS ONE **14**, e0223404 (2019)
7. R.J. Jones, A. Barnir, Properties of opportunity creation and discovery: comparing variation in contexts of innovativeness. Technovation **79**, 1 (2019)
8. M. Lee, S. Kim, H. Kim, J. Lee, Technology opportunity discovery using deep learning-based text mining and a knowledge graph. Technol. Forecast. Soc. Chang. **180**, 121718 (2022)
9. K. Song, K.S. Kim, S. Lee, Discovering new technology opportunities based on patents: text-mining and F-term analysis. Technovation **60–61**, 1 (2017)
10. N. Ko, B. Jeong, J. Yoon, C. Son, Patent-trademark linking framework for business competition analysis. Comput. Ind. **122**, 103242 (2020)
11. M. Lee, S. Lee, Identifying new business opportunities from competitor intelligence: an integrated use of patent and trademark databases. Technol. Forecast. Soc. Chang. **119**, 170 (2017)
12. R.A. Baron, M.D. Ensley, Opportunity recognition as the detection of meaningful patterns: evidence from comparisons of novice and experienced entrepreneurs. Manage. Sci. **52**, 1331 (2006)
13. J. Hong, S.A. Gill, H. Javaid, Q. Ali, M. Murad, M. Shafique, Hunting the best opportunity through the arrow of general decision-making styles: unfolding the role of social capital and entrepreneurial intention. Front. Psychol. **13**, 814424 (2022)
14. I. Park, B. Yoon, Technological opportunity discovery for technological convergence based on the prediction of technology knowledge flow in a citation network. J. Informet. **12**, 1199 (2018)
15. J. Yoon, H. Park, W. Seo, J.-M. Lee, B. Coh, J. Kim, Technology opportunity discovery (TOD) from existing technologies and products: a function-based TOD framework. Technol. Forecast. Soc. Chang. **100**, 153 (2015)
16. A. Rodriguez, A. Tosyali, B. Kim, J. Choi, J.-M. Lee, B.-Y. Coh, M.K. Jeong, Patent clustering and outlier ranking methodologies for attributed patent citation networks for technology opportunity discovery. IEEE Trans. Eng. Manage. **63**, 426 (2016)
17. B. Kim, G. Gazzola, J. Yang, J.-M. Lee, B.-Y. Coh, M.K. Jeong, Y.-S. Jeong, Two-phase edge outlier detection method for technology opportunity discovery. Scientometrics **113**, 1 (2017)
18. H. Ren, Y. Zhao, Technology opportunity discovery based on constructing, evaluating, and searching knowledge networks. Technovation **101**, 102196 (2021)
19. J. Choi, C. Lee, J. Yoon, Exploring a technology ecology for technology opportunity discovery: a link prediction approach using heterogeneous knowledge graphs. Technol. Forecast. Soc. Chang. **186**, 122161 (2023)
20. C. Kim, S. Choe, C. Choi, Y. Park, A systematic approach to new mobile service creation. Expert Syst. Appl. **35**, 762 (2008)
21. C. Lee, B. Song, Y. Park, Generation of new service concepts: a morphology analysis and genetic algorithm approach. Expert Syst. Appl. **36**, 12454 (2009)

22. C. Kim, J. Jeon, M.S. Kim, Identification and Management of opportunities for technology-based services: a patent-based portfolio approach. Innov. Manag. Policy Pract. **17**, 232 (2015)
23. W. Seo, J. Yoon, H. Park, B. Coh, J.-M. Lee, O.-J. Kwon, Product opportunity identification based on internal capabilities using text mining and association rule mining. Technol. Forecast. Soc. Chang. **105**, 94 (2016)
24. J. Yoon, W. Seo, B.-Y. Coh, I. Song, J.-M. Lee, Identifying product opportunities using collaborative filtering-based patent analysis. Comput. Ind. Eng. **107**, 376 (2017)
25. C. Kim, H. Lee, A patent-based approach for the identification of technology-based service opportunities. Comput. Ind. Eng. **144**, 106464 (2020)
26. C. Mun, Y. Kim, D. Yoo, S. Yoon, H. Hyun, N. Raghavan, H. Park, Discovering business diversification opportunities using patent information and open innovation cases. Technol. Forecast. Soc. Chang. **139**, 144 (2019)
27. B. Jeong, N. Ko, C. Son, J. Yoon, Trademark-based framework to uncover business diversification opportunities: application of deep link prediction and competitive intelligence analysis. Comput. Ind. **124**, 103356 (2021)
28. A. Mohamed, R. Faisal, Exploring metaverse-enabled innovation in banking: leveraging NFTS, blockchain, and smart contracts for transformative business opportunities. **8**, 35 (2024). https://doi.org/10.5267/j.Ijdns
29. E. Atagün, B. Hartoka, A. Albayrak, Topic modeling using LDA and BERT techniques: teknofest example, in *2021 6th International Conference on Computer Science and Engineering (UBMK)* (2021), pp. 660–664

Chapter 14
Applications of Machine Learning Algorithms in Open Innovation

Kanchan Awasthi, Krunal Padwekar, and Subhas Chandra Misra

Abstract Open innovation has emerged as an interesting topic to research in innovation management. Open innovation involves exchanging knowledge or ideas among stakeholders to improve the internal innovation process and create possibilities for external application of innovation. In this study, we are exploring the extant literature available in the domain of open innovation and machine learning to identify its relevance in the open innovation context. To attain this objective, we collected data from two databases (i.e. Scopus and Web of Science) by searching keywords related to open innovation and machine learning. After filtering and pre-processing, topic modeling is applied to discover applications of machine learning algorithms in the open innovation field. This study provides valuable inputs to researchers in the field to investigate various applications of machine learning algorithms in an open innovation context.

Keywords Open innovation · Machine learning · Topic modeling · Latent Dirichlet allocation

14.1 Introduction

In a fast-paced environment, companies must constantly innovate and generate innovative ideas to remain valuable. However, their dedicated research and development teams may not always meet customer expectations, leading to failures. This issue can be resolved by the concept *"Open Innovation" by Chesbrough. Now, the issue is comparatively easy to tackle by involving the stakeholders (customers, startups, and entrepreneurs) in the value creation process* [1]. Open innovation (OI) is a strategy that utilizes both internal and external resources, knowledge, and pathways to promote innovation and gain a competitive edge [2]. OI promotes collaboration across organizational boundaries, integrating external ideas, technologies,

K. Awasthi (✉) · K. Padwekar · S. C. Misra
Department of Management Sciences, Indian Institute of Technology Kanpur, Kanpur, India
e-mail: kanchana20@iitk.ac.in

T. R. Benala et al. (eds.), *Boosting Software Development Using Machine Learning*,
Artificial Intelligence-Enhanced Software and Systems Engineering 7,
https://doi.org/10.1007/978-3-031-88188-6_14

and skills to enhance processes and outcomes, unlike traditional R&D models [1, 3]. It has found applications in different areas including high tech enterprises, small medium enterprises (SMEs), service sectors and others [4, 5]. Open innovation benefits high-tech enterprises, SMEs, and universities by coordinating internal and external resources, generating additional profits, and enhancing regional economic competitiveness [6]. It also boosts technology transfer, fosters close relationships with regional innovation systems, and enhances research commercialization [7].

Machine learning algorithms enhance open innovation by providing data-driven insights and promoting collaborative practices, enabling companies to drive innovation and accelerate growth in this context [8, 9]. ML algorithms can enhance innovation processes by generating new ideas, optimizing workflows, and promoting creativity within organizations. Motivated by the prospects presented by machine learning algorithms in open innovation context, the present study employed one of these algorithms to pinpoint the potential research topic within the open innovation domain. Latent Dirichlet Allocation (LDA) is applied in this study to generate potential topics of research.

LDA is a widely used generative statistical model in text data for topic modeling, identifying underlying topics in documents [10]. It's used in ML and NLP for tasks like document clustering, information retrieval, and sentiment analysis [11, 12]. The model uses Bayesian statistics and Dirichlet distributions to automatically identify topics in unstructured text data, providing a valuable tool for understanding latent structure in large document corpora, and has been modified to handle large datasets [13, 14]. Advancements in LDA have improved its applicability in various research domains, making it useful for generating potential topics in open innovation contexts.

14.2 Background

14.2.1 Open Innovation and Machine Learning

ML algorithms are computational methods that permit systems to learn from data and enhance their performance over time without requiring explicit programming [15]. These algorithms are crucial for analysing huge quantities of data to recognise patterns, predict trends, and make data-driven conclusions. ML algorithms, categorized into supervised learning, unsupervised learning, and reinforcement learning, are essential for analyzing large data sets to identify patterns, predict trends, and make data-driven conclusions, aiding tasks like image identification and predictive analytics [16]. ML algorithms significantly improve operational efficiency, reduce costs, and foster cooperation in various fields, particularly in healthcare, by enabling accurate diagnoses and treatment recommendations using extensive data sets [17, 18]. Open-source machine learning models offer reduced development costs and increased customization in healthcare, particularly in precision medicine, enabling personalized treatment plans based on genomic profiles [19, 20]. Moreover, Machine learning improves intellectual property trading efficiency, reduces energy consumption, and addresses unseen classes in computer vision and NLP [21]. It also aids in public safety by predicting and mitigating risks using open data [22].

The integration of artificial intelligence in open innovation processes enhances business model innovation, collaboration, and accuracy [23]. Machine learning aids in selecting optimal variables for innovation measurement, fostering continuous improvement and advancement across multiple sectors [24].

14.3 Methodology

This article follows a mixed approach that is collecting data or articles qualitatively from existing literature and analysing the articles quantitatively using a topic modeling approach. The four major steps involved in the study are (1) Data collection (2) Data screening and filtering (3) Data cleaning and pre-processing (4) Topic modeling. The steps are shown in Fig. 14.1.

14.3.1 Data Collection

Data is collected from Scopus and Web of Science database following multiple search strings such as "Open Innovation" AND "Machine Learning", "Inbound Innovation" AND "Machine Learning", "Outbound Innovation" AND "Machine Learning", "Coupled Innovation" AND "Machine Learning", "Inside-out Innovation" AND "Machine Learning", "Outside-in Innovation" AND "Machine Learning", "Crowdsourcing" AND "Machine Learning", "Co-creation" AND "Machine Learning".

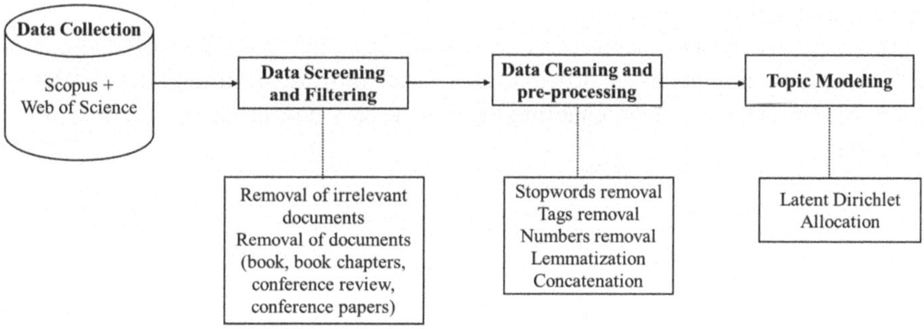

Fig. 14.1 Research approach

14.3.2 Data Screening and Filtering

This study uses two databases to extract relevant documents from various subjects, including *Computer Science, Business, Management and Accounting, Economics, Econometrics and Finance, Decision Sciences, and Social Sciences.* After filtering, 3293 documents were extracted, with 3248 in English. After removing duplicates and duplicate articles, 2033 articles were left. Only journal articles were included in the analysis, leaving 985 articles for topic modeling.

14.3.3 Data Cleaning and Pre-processing

A corpus of data is created after document screening and filtering, including title, keywords, and abstracts of articles. Preprocessing removes noise, punctuation, numbers, and common English stopwords, and converts content to lower case for improved results. This process ensures minimal noise in the results. Additionally, n-grams are converted into unigrams to ensure critical information is not omitted, and use the "_" operator to connect terms with specific meanings. Lemmatization optimizes text for topic modelling, involving three stages: converting plural nouns to singular equivalents, restoring irregular verb forms, and recovering comparative adjectives' original forms.

14.3.4 Topic Modeling

Topic modelling is a text mining technique that identifies latent topics in documents using naturally compatible words. LDA is the most popular technique in trend analysis, extracting topics without human assistance, making it useful for analyzing unstructured text information.

This article uses LDA to interpret corpus data, using the "bag-of-words" paradigm to determine document topical distribution. A probability distribution assigns a dominant topic to each document, enabling topic aggregation or grouping based on topic distribution.

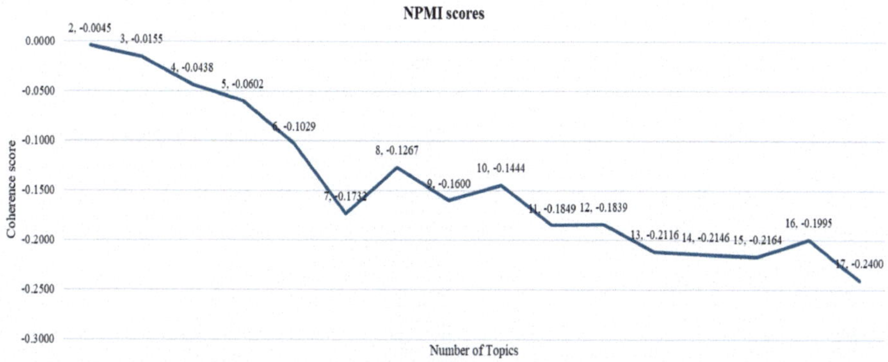

Fig. 14.2 Coherence scores

14.4 Results

14.4.1 Suitable Number of Topics

The selection of appropriate topics is influenced by various factors including human interpretability, coherence, and diversity. One of the metrics employed in elucidating topics to human beings is the coherence score. The coherence score is a probabilistic measure used to explain topics to humans by estimating the similarity of words to each other. This paper uses Normalised Pointwise Mutual Information (NPMI) for calculating topic count, an extended version of Pointwise Mutual Information, used in Word Clustering, NLP, and Information Recovery [25].

The formula for NPMI is as follows:

$$\mathrm{NPMI}(a_i, a_j) = \frac{\log \frac{p(a_i, a_j)}{p(a_i)p(a_j)}}{-\log p(a_i, a_j)} \tag{14.1}$$

where a_i and a_j are words, their probabilities are $p(a_i)$ and $p(a_j)$ and their joint probability is $p(a_i, a_j)$.

By observing NPMI score, we found that the best coherence score is -0.1732 that represents seven number of topics. The coherence score with distinct number of topics is shown in Fig. 14.2. Higher coherence score with adequate interpretation of topics is preferable.

14.4.2 Inter-Topic Distance Maps

An inter-topic distance map is generated through the utilization of the Python LDA visualization package to comprehend the association and similarity among various

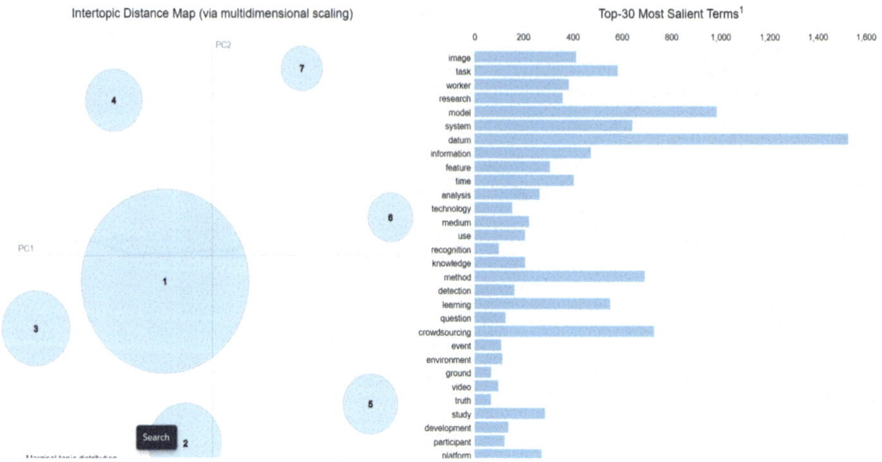

Fig. 14.3 Intertopic distance map and most salient terms

topics. The inter-topic distance map illustrates the significance of each topic within a database by displaying the distance and size of circles. Saliency and relevance are key components, with saliency referring to a word's prominence [26] and relevance assessing its alignment with information requirements [27].

The saliency and relevance are represented as:

$$saliency(term_t) = frequency(w) * \sum_t \left(p(t|w) * \log\left(\frac{p(t|w)}{p(t)}\right)\right) \tag{14.2}$$

$$relevance(termw|topict) = \lambda * \mathrm{p(w|t)} + (1 - \lambda) * \frac{\mathrm{p(w|t)}}{\mathrm{p(w)}} \tag{14.3}$$

The study used a relevance metric (λ) value of 1 to identify key terms related to open innovation and machine learning, highlighting ongoing research themes. A visualisation of intertopic distance map and salient terms is shown in Fig. 14.3 followed by visualization of most representative terms of Topic 1 at relevance metric value of 1 in Fig. 14.4.

14.4.3 Prominent Terms and Emerged Topics

The output was generated by implementing the most prevalent tokens for each topic along with the corresponding label as depicted in Table 14.1. To mitigate potential biases, the researchers independently assigned codes to the topics and subsequently arrived at the final topics following the methodology proposed by [28].

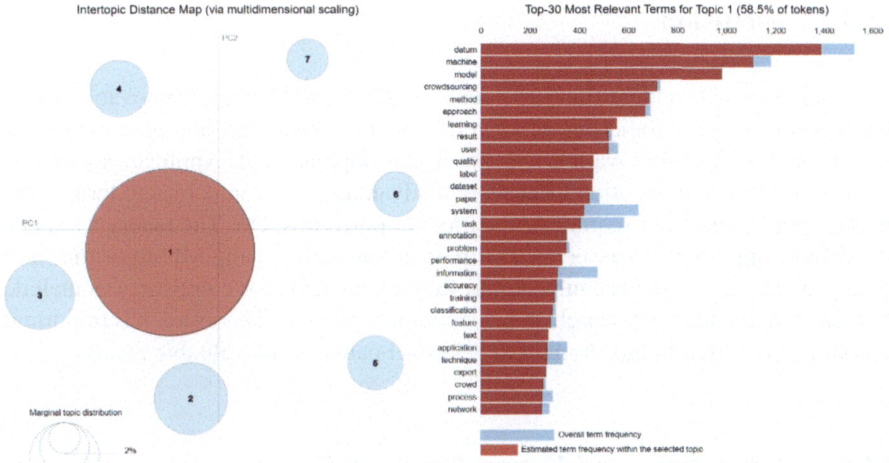

Fig. 14.4 Most representative terms of topic 1 at $\lambda = 1$

Table 14.1 Topic label with most relevant terms

Topic number	Topic label	Most relevant terms	% of tokens
1	Crowdsourced machine learning systems and user quality assessment	Machine, model, crowdsourcing, learning, user, quality, system, annotation, label	58.5
2	Technology development and analysis	Technology, medium, software, development, analysis, intelligence, energy, information, knowledge, source	11
3	Real-time event detection systems for traffic and environmental news	System, time, event, information, news, environment, road, vehicle, detection, traffic	9.8
4	Healthcare quality assessment and patient perception analysis	Perception, rule, similarity, measure, patient, benchmark, concern, survey, trust, customer	6.8
5	Enhancing dietary choices through knowledge graphs and multi-task learning	Image, worker, task, recognition, reliability, answer, object, question, food, recommendation	6.2
6	Integrating multimodal signals for enhanced emotion recognition in educational settings	Speech, app, video, teacher, emotion, learner, population, analysis, team, consumption	4.2
7	Enhancing land cover mapping accuracy using advanced metric-based strategies	Ground, truth, land, opinion, entity, citizen, game, resolution, mapping, metric	3.5

14.5 Conclusion

This study focused on identifying relevant topics in the field of open innovation using ML algorithm. These topics are of keen interest for researchers and practitioners to reflect upon as open innovation is a novel and popular field. Applications of ML algorithms in open innovation offer several advantages for companies to look at the market trends, customer requirements, product portfolios, etc. The emerging scope and diverse opportunities associated with open innovation using ML algorithms are manifold. These two fields can generate valuable insights for companies to include them in their agenda for research and development process. Therefore, it is important to continue research in this domain to get novel patterns and valuable results.

14.5.1 Limitations and Future Research Scope

Limitations associated with this study are as following: (1) challenge of interpreting the results generated by these algorithms (2) the black-box nature of some ML algorithms. While ML algorithms like LDA can uncover patterns and insights in large datasets, the complexity of the algorithms can make it difficult for non-experts to understand and interpret the results accurately. This can hinder the effective implementation of these algorithms in open innovation processes, as stakeholders may struggle to make informed decisions based on the algorithm's output. Additionally, the black-box nature of some ML algorithms can make it challenging to explain how certain conclusions or recommendations were reached, which may lead to skepticism or resistance from users in the open innovation context. Advancements in interpretability and transparency of ML algorithms can support stakeholders to make more informed decisions and effectively leverage the potential of ML for driving innovation and collaboration. Moreover, the integration of human expertise and domain knowledge with ML algorithms can provide more accuracy and relevance of the results produced.

14.5.2 Implications

First, by leveraging these algorithms, organizations can enhance their decision-making processes, drive innovation, and foster collaboration. Decision making is enhanced since ML algorithms, including LDA, can analyse large datasets to derive significant insights and trends that can aid in making informed decisions in open innovation initiatives. Second, by utilizing ML algorithms, organizations can capitalise on emerging trends, predict market demands, and uncover hidden patterns that can fuel innovation in products, services, and processes. Third, ML algorithms can facilitate collaboration among diverse stakeholders in open innovation ecosystems by providing a common platform for sharing insights, ideas, and feedback. Fourth, by applying ML algorithms, organizations can quickly learn from past experiences, successes, and failures in open innovation projects. Fifth, ML algorithms can help in optimizing resource allocation by identifying areas with the highest potential for innovation and collaboration.

References

1. H.W. Chesbrough, *Open Innovation: The New Imperative for Creating and Profiting from Technology* (Harvard Business School Press, Boston, Mass, 2003)
2. H. Chesbrough, *Open Innovation: A New Paradigm for Understanding Industrial Innovation*, in *Open Innovation*, ed. by H. Chesbrough, W. Vanhaverbeke, J. West (Oxford University PressOxford, 2006), pp. 1–12.
3. M. Elmquist, T. Fredberg, S. Ollila, Exploring the field of open innovation. Eur. J. Innov. Manag. **12**, 326 (2009)
4. H. Chesbrough, A.K. Crowther, Beyond high tech: early adopters of open innovation in other industries. R&D Management **36**, 229 (2006)
5. J. West, S. Gallagher, Challenges of open innovation: the paradox of firm investment in open-source software. R&D Management **36**, 319 (2006)
6. M. Dodgson, D. Gann, A. Salter, The role of technology in the shift towards open innovation: the case of Procter & Gamble. R&D Management **36**, 333 (2006)
7. J.B.P. Bejarano, Open innovation: A technology transfer alternative from universities. A systematic literature review (2023)
8. A. Adikari, Value co-creation for open innovation: an evidence-based study of the data driven paradigm of social media using machine learning (2021)
9. Q. Lu, H. Chesbrough, Measuring open innovation practices through topic modelling: revisiting their impact on firm financial performance. Technovation **114**, 102434 (2022)
10. D.M. Blei, A.Y. Ng, M.I. Jordan, Latent dirichlet allocation. J. Mach. Learn. Res. **3**, 993 (2003)
11. R. Agarwal, *Phrases Based Document Classification from Semi Supervised Hierarchical LDA*, in *2021 2nd International Conference on Computation, Automation and Knowledge Management (ICCAKM)* (2021), pp. 332–337
12. H. Gadekar, N. Bugalia, Automatic classification of construction safety reports using semi-supervised YAKE-Guided LDA approach. Adv. Eng. Inform. **56**, 101929 (2023)
13. F. Culasso, B. Gavurova, E. Crocco, E. Giacosa, Empirical identification of the chief digital officer role: a latent Dirichlet allocation approach. J. Bus. Res. **154**, 113301 (2023)
14. M. Janmaijaya, A.K. Shukla, P.K. Muhuri, A. Abraham, Industry 4.0: latent Dirichlet allocation and clustering based theme identification of bibliography. Eng. Appl. Artif. Intell. **103**, 104280 (2021)
15. J. Garcia, G. Villavicencio, F. Altimiras, B. Crawford, R. Soto, V. Minatogawa, M. Franco, D. Martínez-Muñoz, V. Yepes, Machine learning techniques applied to construction: a hybrid bibliometric analysis of advances and future directions. Autom. Constr. **142**, 104532 (2022)
16. B. Mahesh, Machine learning algorithms—a review, **9**, (2018)
17. N. J. Ahuja, *Adoption of Machine Learning and Open Source: Healthcare 4.0 Use Cases*, in *Application of Deep Learning Methods in Healthcare and Medical Science* (Apple Academic Press, 2022)
18. T. Li, L. Ma, Z. Liu, K. Liang, Economic granularity interval in decision tree algorithm standardization from an open innovation perspective: towards a platform for sustainable matching. J. Open Innov. Technol. Mark. Complex. **6**, 4 (2020)
19. C. Huang, R. Mezencev, J.F. McDonald, F. Vannberg, Open source machine-learning algorithms for the prediction of optimal cancer drug therapies. PLoS ONE **12**, e0186906 (2017)

20. S. Talwar, A. Dhir, N. Islam, P. Kaur, A. Almusharraf, Resistance of multiple stakeholders to e-health innovations: integration of fundamental insights and guiding research paths. J. Bus. Res. **166**, 114135 (2023)
21. J. Parmar, S. Chouhan, V. Raychoudhury, S. Rathore, Open-world machine learning: applications, challenges, and opportunities. ACM Comput. Surv. **55**, 205:1 (2023)
22. G.B. Rocca, M. Castillo-Cara, R.A. Levano, J.V. Herrera, L. Orozco-Barbosa, Citizen security using machine learning algorithms through open data, in *2016 8th IEEE Latin-American Conference on Communications (LATINCOM)* (2016), pp. 1–6
23. A. Kuzior, M. Sira, P. Brożek, Use of artificial intelligence in terms of open innovation process and management. Sustainability **15**, 9 (2023)
24. S. M. Parvin Hosseini, A. Azizi, Machine learning approach to identify predictors in an econometric model of innovation, in *Big Data Approach to Firm Level Innovation in Manufacturing: Industrial Economics*, ed. by S.M. Parvin, Hosseini, A. Azizi (Springer, Singapore, 2020), pp. 41–52
25. S.M. Watford, R.G. Grashow, V.Y. De La Rosa, R.A. Rudel, K.P. Friedman, M.T. Martin, Novel application of normalized pointwise mutual information (NPMI) to mine biomedical literature for gene sets associated with disease: use case in breast carcinogenesis. Comput. Toxicol. **7**, 46 (2018)
26. J. Chuang, C.D. Manning, J. Heer, Termite: visualization techniques for assessing textual topic models, in *Proceedings of the International Working Conference on Advanced Visual Interfaces* (ACM, Capri Island Italy, 2012), pp. 74–77
27. C. Sievert, K. Shirley, LDAvis: a method for visualizing and interpreting topics, in *Proceedings of the Workshop on Interactive Language Learning, Visualization, and Interfaces* (Association for Computational Linguistics, Baltimore, Maryland, USA, 2014), pp. 63–70
28. M.B. Miles, A.M. Huberman, J. Saldaña, *Qualitative Data Analysis: A Methods Sourcebook*, 3rd edn. (SAGE Publications Inc., Thousand Oaks, Califorinia, 2014)

The manufacturer's authorised representative in the EU is Springer Nature Customer Service Centre GmbH, Europaplatz 3, 69115 Heidelberg, Germany. If you have any concerns regarding our products, please contact ProductSafety@springernature.com

Printed and bound by CPI Group (UK) Ltd, Croydon, CR0 4YY
30/09/2026
02225569-0014